Okra: Status, Challenges and Opportunities

Surinder K. Tikoo • Sharan Angadi
Aparna Tiwari • Ramesh Kumar Yadav
B. S. Tomar • Aderibigbe Helen Adeniyi
Editors

Okra: Status, Challenges and Opportunities

Springer

Editors
Surinder K. Tikoo
Plant Breeding
ATPBR- Foundation for Advanced Training in Plant Breeding
Aurangabad, Maharashtra, India

Sharan Angadi
Genetics and Plant Breeding
ATPBR- Foundation for Advanced Training in Plant Breeding
Bangalore, Karnataka, India

Aparna Tiwari
ATPBR- Foundation for Advanced Training in Plant Breeding
Aurangabad, Maharashtra, India

Ramesh Kumar Yadav
Division of Vegetable Science
Indian Agriculture Research Institute
New Delhi, Delhi, India

B. S. Tomar
Division of Vegetable Science
Indian Agriculture Research Institute
New Delhi, Delhi, India

Aderibigbe Helen Adeniyi
Farming Systems Research Programme
National Horticultural Research Institute
Ibadan, Nigeria

ISBN 978-981-97-9962-6 ISBN 978-981-97-9963-3 (eBook)
https://doi.org/10.1007/978-981-97-9963-3

This Springer imprint is published by the registered company Springer Nature Singapore Pte Ltd.
The registered company address is: 152 Beach Road, #21-01/04 Gateway East, Singapore 189721, Singapore

Preface

Okra (*Abelmoschus esculentus L.* Moench) is a key vegetable in the Malvaceae family, thriving in tropical and subtropical regions. India leads global production, contributing over 72% (six million tonnes) from 0.5 million hectares. Okra's yield in India has risen from 4.6 t/ha in 2011 to 13.5 t/ha in 2021 due to increasing adaptation of F1 hybrids by the farmers during the last decade. The same was made possible by the seed sector breeders in India developing high-yielding hybrids that were highly resistant to an important viral disease, Yellow Vein Mosaic Virus (YVMV) and also good tolerance Okra Leaf Curl Virus (OLCV). The virus resistance sources are traced to the release of open-pollinated varieties released by public sector around 1987, viz: varieties Arka Anamika and Arka Abha from ICAR-IIHR, Bangalore and variety Parbhani Kranti by Vasantrao Naik Marathwada Krishi Vidyapeeth Parbhani, both from India. In the last 5 years, the release of the short internoded hybrid Radhika, by Advanta Seeds, India (Now Advanta Enterprises), has further revolutionised the market as farmers got ease of harvest and another 20% increase in yield as well as excellent pod quality. These events over the last few years have led to seed revenue growing from $12 m in 2011 to $112 m in 2021. The result is that okra overtook tomato and chillies for value and became the number 1 crop in India.

Besides India, Nigeria is the next big country for okra cultivation taking 15% of the global market. Okra cultivation is now growing in several countries making it a global crop. Research on okra's genetics and agronomy is increasing globally. The global seed requirement is projected to reach 6000 mt by 2030, making okra a $300 m crop.

ATPBR in collaboration with ICAR-IARI proposed to hold "Global Okra Roundtable—GORT1" to unite scientists and other stakeholders involved in okra improvement, focusing on breeding, technology, pathology, seed production and marketing. The major aim was to know the latest advances on the crop's improvement in most of the important countries of okra and deliberate on the way forward for (a) developing new and stronger sources of resistance to the viruses and other diseases, (b) new innovations possible and (c) find ways to widen the scope of the crop globally. A panel discussion session at the end of the conference was held to sum up action points for coming years and work towards developing public-private partnership projects.

The round table was held to get the latest information on all aspects of the crop, and it was decided to publish all the papers presented during GORT 1 in the form of a book, *Okra: Status, Challenges, and Opportunities*. It comprehensively covers all aspects of okra crop production, including crop improvement, pathology, agronomy, processing, sales, and marketing, providing a broad understanding of this important crop. To offer a global perspective, the authors hail from various countries, including India, Taiwan, Nigeria, South Africa, and Egypt, representing both public and private sectors.

This book will be an invaluable resource of information for scientists, researchers, breeders, pathologists, agricultural leaders, students, and decision-makers interested in okra. It provides a holistic and global prospects for the crop, addressing current challenges and exploring future opportunities.

Aurangabad, Maharashtra, India — Surinder K. Tikoo
Bangalore, Karnataka, India — Sharan Angadi
Aurangabad, Maharashtra, India — Aparna Tiwari
New Delhi, Delhi, India — Ramesh Kumar Yadav
New Delhi, Delhi, India — B. S. Tomar
Ibadan, Nigeria — Aderibigbe Helen Adeniyi

Acknowledgements

This book is a comprehensive compilation of lectures delivered by experts during the Global Okra Round Table Meeting (GORT 1), held in Delhi from October 10–12, 2022. The conference witnessed the participation of experts from around the world, both physically and virtually, who shared their extensive knowledge, wisdom, and experiences across various facets of the okra value chain. These included breeding, genomics, genetic engineering, agronomy, physiology, public-private partnership, and other crucial aspects instrumental in promoting okra as a global crop.

This book was prepared under the aegis of "Foundation for Advanced Training in Plant Breeding (ATPBR)". ATPBR extends gratitude to all contributors who played a crucial role in developing this unique repository of information on okra. Notable figures such as **Prashant Belgamwar**, Regional Lead, Asia and Africa and International Vegetables, Advanta Seeds; **Vineeta Sudhanshu**, Dy General Manager, APEDA, New Delhi; **Dr. Aderibigbe Helen Adeniyi**, Chief Research Officer, Farming Systems and Ext. Res., Ibadan, Nigeria; **Dr. Roland Schafleitner**, Head, Mol. Genetics and Flagship Leader, WorldVeg, Taiwan; **Dr. Dorcas O Ibitoye,** Asst. Director, NIHORT, Ibadan, Nigeria; **Dr. Joseph John,** Retd Principal Scientist, ICAR-NBPGR, Thrissur, India; **Dr. K.V. Bhat,** Retd. Principal Scientist, ICAR-NBPGR, New Delhi; **Dr. R.K. Yadav**, Principal Scientist, ICAR-IARI, New Delhi; **Dr. Abe S. Gerrano**, Veg and Ornamental Crops, Ag. ResmCouncil, Pretoria, South Africa; **Dr. M. Pitchaimuthu**, Principal Scientist, ICAR-IIHR, Bengaluru; **Professor Ehab Awad-Allah Ibrahim**, Chief of Research, Hortic. Research, Institute, Giza, Egypt; **Dr. Sanjeet Kumar**, Plant Breeder, Ex- ICAR and World Veg, India; **Dr. K.V. Ravishankar**, Principal Scientist, Biotechnology, ICAR-IIHR, Bengaluru; **Dr. Achuit Singh**, Scientist, ICAR-IIVR, Varanasi; **Dr. M.B. Arun Kumar**, Principal Scientist, Division of Seed Science and Technology, ICAR-IARI, New Delhi; **Dr. Suman Lata**, Scientist, ICAR-IARI, New Delhi; **Dr. S.K. Singh**, Principal Scientist, Agronomy, ICAR-IIVR, Varanasi; **Dr. B.S. Tomar,** Joint Director (Extension) and Head, Division of Vegetable Science, ICAR-IARI, New Delhi; **Dr. V.K. Pandita,** Ex-Head and Principal Scientist, ICAR-IARI Regional Station, Karnal, **Vijay Kori,** Technology Development Lead—International, Vegetables, UPL Advanta, Seeds, India; **Dr. Neena Chawla**, Punjab Agriculture University (PAU), Ludhiana, India; **Dr. Pradip Karmakar**, Scientist, ICAR-IIVR, Varanasi; **Dr. Samir Sawant,** Chief Scientist and Prof., CSIR-NBRI, Lucknow, India; **Dr. O.P. Dutta**, Ex-Head, Vegetable Crops, ICAR-IIHR, Bangalore, India;

Dr. Maneesh Bhatnagar, Growth Unit Head, Mahyco Seeds, Jalna, India; **Dr. Dushyant Baghel**, CEO, Nucleome, Hyderabad, India; and **Dr. Surinder K. Tikoo**, Independent Director—ATPBR have played instrumental roles as authors. The book owes its existence to the dedication and hard work of these contributors, without whom its formation would have been impossible.

ATPBR extends heartfelt thanks to Springer for embracing the concept of this book and patiently waiting nearly 2 years for its compilation.

Last, but certainly not the least, ATPBR expresses gratitude to the book's editors Dr. Surinder K. Tikoo along with Dr. Sharan Angadi, Dr. Aparna Tiwari, Dr. R.K. Yadav, Dr. B.S. Tomar, and Dr. Aderibigbe Helen Adeniyi. Their meticulous editing and valuable contributions have resulted in a holistic book on okra that encompasses all aspects of the okra value chain.

Contents

Editors and Contributors

About the Editors

Surinder K. Tikoo is the cofounder of Tierra Agrotech Ltd, Pune, and currently Director of the Foundation for Advanced Training in Plant Breeding Research (ATPBR). He has over five decades of experience as a successful vegetable breeder, teacher, and research and development manager. He has held global leadership positions over long periods in three MNCs in the seed sector, as well as spent 20 years with the Indian Council of Agricultural Research (ICAR) as a vegetable breeder, teacher, and research guide. He has developed several nationally and internationally released varieties and hybrids of tomato and okra, some of which are market leaders in their respective segments. He has served on several advisory committees of the Government of India and various professional bodies. Dr. Tikoo, a PhD from Bangalore University, has published over 60 research and review papers. He has been conferred the Lifetime Achievement Award for Excellence in Translational Research, jointly by the Society of Plant Biochemistry and Biotechnology and Springer among other awards. He is an elected Fellow of the Society for Promotion of Horticulture, IIHR, Bangalore.

Sharan Angadi is a plant geneticist and breeder who worked for 17 years in India, first with UAS, Bangalore, and then with the Indian Council for Agricultural Research (ICAR), contributing significantly to the genetic improvement of oilseed crops and horticultural crops. He then devoted 22 years to Pro Agro, India, and Nunhems Seeds, Netherlands, where he served as Head of Breeding for Asia Pacific and Global Coordinator for peppers, spanning several countries in Asia, Europe, and the Americas. He has experience in applying emerging technologies to viable commercial plant breeding. Dr. Angadi obtained a BSc (Agri.) from UAS, Bangalore; an MSc (Agri.) in Genetics and Plant Breeding from Tamil Nadu Agricultural University, Coimbatore; and a PhD from IARI, New Delhi. After superannuating from active service with the private sector, he continues to provide expert R&D advice to seed companies in India and abroad, besides serving on various expert bodies. The Asia and Pacific Seed Association (APSA) has honored him with the "Most Influential Plant Breeding Researcher for Horticultural Crops Award" for the year 2021.

Aparna Tiwari is a cofounder and Chief Executive Officer at the Foundation for Advanced Training in Plant Breeding (ATPBR), where she leads professional training, technology liaising, awareness sessions, and other capacity-building activities. She also serves as a conference secretary and coordinates 5F Farming national conferences annually. She has worked previously in various positions with the FAO (Food and Agriculture Organization), Monsanto India Pvt. Ltd., and ICRISAT (International Crops Research Institute for the Semi-Arid Tropics). She obtained her MSc and PhD from Wageningen University, The Netherlands, and her Bachelor's (BSc) from Banaras Hindu University. She has authored a book and published several research papers in peer-reviewed international journals and conferences.

Ramesh Kumar Yadav is working as a Professor and Principal Scientist at the Division of Vegetable Science, ICAR-Indian Agricultural Research Institute (IARI), New Delhi. He completed his graduation and post-graduation from GB Pant University of Agriculture and Technology, Pantnagar. He has been working on okra, tomato, and leafy vegetable improvement and has developed 15 vegetable varieties/hybrids, namely tomato varieties Maniliema Pusa Shakti (heat tolerant) and tomato hybrid Pusa Rakshit (for Protected condition), YVMV-resistant okra variety Pusa Bhindi-5 and Pusa okra Hybrid-1, nutritionally important bathua variety Pusa Green, spinach variety Pusa Vilayati Palak, and okra variety Pusa Lal Bhindi-1. As chairman, he has guided four PhD and seven MSc students at IARI, New Delhi. He has published more than 95 research papers in national and international journals. He received the Young Scientist Award in the year 2009 by Govt. of Uttar Pradesh. He is a Fellow of the Indian Society of Vegetable Science and the Indian Academy of Horticultural Sciences and an Honorary Fellow of The Society for Horticultural Research and Development.

B. S. Tomar is the Head of Division of Vegetable Science, ICAR-IARI, New Delhi. He also worked as Joint Director (Extn.) as an additional charge from 2021 to 2023. He has developed 35 vegetable crop varieties/hybrids and guided 25 MSc and PhD students in Seed Science Technology and Vegetable Science. He has published 125 research papers comprising of 100 research papers with more than 6 NAAS rating in 12 reputed international journals. He has also published 12 bulletins, 2 practical manuals, and 250 popular articles and edited 5 books. He is a recipient of the Jawaharlal Nehru Award for outstanding contribution to PhD Research-1995, IARI Best Teacher Award 2014, Dr. Gautam Kalloo Award for Excellence in Horticulture Research-2017, Lifetime Achievement Award in R&D in Horticulture-2023 of SHRD, and 7th NHRDF Award-2021 for contribution in onion research and seed production. He has also served as a TCDC consultant in FAO in Uzbekistan in 2009.

Aderibigbe Helen Adeniyi is a research agronomist and was an Assistant Director of Research, and Head of Farming Systems Programme at the National Horticultural Research Institute, Idi-Ishin, Ibadan, Nigeria. Due to her resilient research activities and academic prowess, Adeniyi Helen was appointed as the Honorable Commissioner (junior Minister) for Innovation, Science and Technology, Kogi State, Nigeria

(March 2024). She hails from Ekinrin-Adde, Kogi State in the North-Central part of Nigeria, from where she had her early education (Primary and Secondary). She graduated with BAgric (Crop Production) from the University of Ilorin, MSc (Agronomy) from the University of Ibadan, and PhD (Agronomy) also from the University of Ilorin, all in Nigeria. She has participated and attended many workshops and conferences both locally and internationally including 1st Global Okra Round Table (GORT I) October 10–12, 2022, ICAR—Indian Agricultural Research Institute, Pusa, New Delhi, with many publications in both local and international journals and proceedings. She facilitated the hosting of GORT 2 (October 2–4, 2024) in Nigeria at the National Horticultural Research Institute, Idi-Ishin, Ibadan. Her major research areas are in Horticulture, Farming Systems, Organic Agriculture, Innovation, and Technology.

Contributors

Aderibigbe Helen Adeniyi Farming Systems Research Programme, National Horticultural Research Institute, Ibadan, Nigeria

Helen Niegiria Aderibigbe Farming System, National Horticultural Research Institute, Ibadan, Nigeria

Hamdino M. I. Ahmed Vegetable Crops Research Department, Horticulture Research Institute, Agricultural Research Center, Giza, Egypt

Sharan Angadi Genetics and Plant Breeding, ATPBR- Foundation for Advanced Training in Plant Breeding, Bangalore, India

Mahesh Badiger Division of Vegetable Science, ICAR-Indian Agricultural Research Institute, New Delhi, India

Dushant Singh Baghel Nucleome Informatics Pvt Ltd, Hyderabad, India

T. K. Behera ICAR-Indian Institute of Vegetable Research, Varanasi, Uttar Pradesh, India

Division of Crop Improvement, Indian Institute of Vegetable Research, Varanasi, India

Prashant Belgamwar Asia & Africa and International Vegetables, Advanta Seeds UPL Ltd., Hyderabad, Telangana, India

K. V. Bhat ICAR-National Bureau of Plant Genetic Resources, Pusa Campus, New Delhi, India

Manish Bhatnagar Growth Unit Head, Mahyco Private Limited, Mumbai, India

P. B. Chaukhande Division of Vegetable Science, ICAR-Indian Agricultural Research Institute, New Delhi, India

Paresh Chaukhande ICAR-Indian Agricultural Research Institute, New Delhi, India

Neena Chawla Department of Vegetable Science, Punjab Agricultural University, Ludhiana, Punjab, India

Harshwardhan Choudhary Division of Vegetable Science, ICAR-Indian Agricultural Research Institute, New Delhi, India

O. P. Dutta Division of Vegetable Crops, ICAR-Indian Institute of Horticultural Research, Bengaluru, Karnataka, India

M. Gayathri Division of Basic Sciences, ICAR-Indian Institute of Horticultural Research, Bangalore, India

Jagadish Hosamani UAS, Dharwad, Dharwad, India

Department of Seed Science & Technology, College of Agriculture, Vijayapur, India

Issaka Housseini Horticulture Training Dov Center, Niamey, Niger

Dorcas Olubunmi Ibitoye Genetic Resources Unit, National Horticultural Research Institute (NIHORT), Ibadan, Nigeria

Ehab A. Ibrahim Vegetable Crops Research Department, Horticulture Research Institute, Agricultural Research Center, Giza, Egypt

Suresh Iyer Nucleome Informatics Pvt Ltd, Hyderabad, India

K. Joseph John ICAR-National Bureau of Plant Genetic Resources, Regional Station, Thrissur, Kerala, India

Suhas G. Karkute Division of Crop Improvement, ICAR-Indian Institute of Vegetable Research, Varanasi, Uttar Pradesh, India

Pradip Karmakar Division of Crop Improvement, Indian Institute of Vegetable Research, Varanasi, India

Simarjot Kaur Department of Biochemistry, Punjab Agricultural University, Ludhiana, Punjab, India

S. B. Kiran Division of Vegetable Science, ICAR-Indian Agricultural Research Institute, New Delhi, India

Vijay Kumar Kori Advanta Seeds UPL Ltd., Hyderabad, Telangana, India

Ram Krishna Division of Crop Improvement, ICAR-Indian Institute of Vegetable Research, Varanasi, Uttar Pradesh, India

Ashish Kumar Nucleome Informatics Pvt Ltd, Hyderabad, India

Rajeev Kumar Division of Crop Improvement, Indian Institute of Vegetable Research, Varanasi, India

Sanjeet Kumar Genetics and Plant Breeding, Acharya Narendra Deva University of Agriculture & Technology, Kumarganj, Ayodhya, India

C. J. Lakshmi ICAR-National Bureau of Plant Genetic Resources, Regional Station, Thrissur, Kerala, India

A. Lalitha Hari Dharini Nucleome Informatics Pvt Ltd, Hyderabad, India

Suman Lata Division of Vegetable Science, ICAR-Indian Agricultural Research Institute, New Delhi, India

M. Latha ICAR-National Bureau of Plant Genetic Resources, Regional Station, Thrissur, Kerala, India

Ya-Ping Lin World Vegetable Center, Shanhua, Tainan, Taiwan

Lalit Mahatma Navsari Agricultural University, Navsari, Gujarat, India

B. Mahesha Division of Crop Protection, ICAR, Indian Institute of Horticultural Research, Bengaluru, India

Sujan Majumder Division of Crop Improvement, Indian Institute of Vegetable Research, Varanasi, India

Brijesh Kumar Maurya Division of Crop Improvement, Indian Institute of Vegetable Research, Varanasi, India

Shraddha Mishra Nucleome Informatics Pvt Ltd, Hyderabad, India

C. N. Murali ICAR-National Bureau of Plant Genetic Resources, Pusa Campus, New Delhi, India

V. A. Muhammed Nissar ICAR-Indian Institute of Spices Research, Kozhikode, Kerala, India

Arun Pandey Advanta Seeds UPL Ltd., Hyderabad, Telangana, India

Chitra Devi Pandey ICAR-National Bureau of Plant Genetic Resources, Pusa Campus, New Delhi, India

Vinod K. Pandita ICAR-Indian Agricultural Research Institute, Regional Station, Karnal, India

Hukum Singh Panwar Division of Crop Improvement, ICAR-Indian Institute of Vegetable Research, Varanasi, Uttar Pradesh, India

Sathish Kumar Peddamma Nucleome Informatics Pvt Ltd, Hyderabad, India

M. Pitchaimuthu Division of Vegetable Crops, ICAR-Indian Institute of Horticultural Research, Bengaluru, Karnataka, India

P. V. Puneeth Division of Vegetable Science, ICAR-Indian Agricultural Research Institute, New Delhi, India

N. Rai Division of Crop Improvement, Indian Institute of Vegetable Research, Varanasi, India

Baskaran Ramachandran Nucleome Informatics Pvt Ltd, Hyderabad, India

K. V. Ravishankar Division of Basic Sciences, ICAR-Indian Institute of Horticultural Research, Bangalore, India

M. Krishna Reddy Promolecular Sciences Pvt. Ltd., Bengaluru, Karnataka, India

Venu Gopal Reddy Advanta Seeds UPL Ltd., Hyderabad, Telangana, India

Manjsh Saakre ICAR-National Institute of Plant Biotechnology, New Delhi, India

Vidya Sagar Division of Crop Improvement, Indian Institute of Vegetable Research, Varanasi, India

Aichatou Salifou Ainoma Seed Farm, Niamey, Niger

K. Subramanya Sastry ICAR-Indian Institute of Millets Research, Rajendranagar, Hyderabad, India

Roland Schafleitner World Vegetable Center, Shanhua, Tainan, Taiwan

Sandeep Shakya Department of Biochemistry, Punjab Agricultural University, Ludhiana, Punjab, India

Mahipal Sharma Advanta Seeds UPL Ltd., Hyderabad, Telangana, India

Achuit K. Singh Division of Crop Improvement, ICAR-Indian Institute of Vegetable Research, Varanasi, Uttar Pradesh, India

Himanshu Singh Division of Crop Improvement, Indian Institute of Vegetable Research, Varanasi, India

P. M. Singh Division of Crop Improvement, Indian Institute of Vegetable Research, Varanasi, India

S. K. Singh ICAR-Indian Institute of Vegetable Research, Varanasi, Uttar Pradesh, India

Vidhya Singh Nucleome Informatics Pvt Ltd, Hyderabad, India

Vinita Sudhanshu Agricultural and Processed Food Products Export Development Authority (APEDA), New Delhi, India

A. Suma ICAR-National Bureau of Plant Genetic Resources, Regional Station, Thrissur, Kerala, India

Surinder K. Tikoo Plant Breeding, ATPBR- Foundation for Advanced Training in Plant Breeding, Aurangabad, India

Aparna Tiwari ATPBR- Foundation for Advanced Training in Plant Breeding, Aurangabad, India

B. S. Tomar Division of Vegetable Science, ICAR-Indian Agricultural Research Institute, New Delhi, India

V. Venkataravanappa Division of Crop Protection, ICAR-Indian Institute of Horticultural Research, Bengaluru, India

N. D. Vinay Division of Vegetable Science, ICAR-Indian Agricultural Research Institute, New Delhi, India

R. K. Yadav Division of Vegetable Science, ICAR-Indian Agricultural Research Institute, New Delhi, India

Sweta Yadav Nucleome Informatics Pvt Ltd, Hyderabad, India

V. R. Yalmalle ICAR-Indian Agricultural Research Institute, New Delhi, India

Introduction

1

Surinder K. Tikoo

Okra (*Abelmoschus esculentus* L. Moench) is an important vegetable crop belonging to the family Malvaceae. Okra is grown throughout the tropical and subtropical regions of the world and in the warmer parts of the temperate zone. India is the largest producer of okra globally, contributing more than 72% (six million tonnes) from an area of 0.5 million hectares (NHB 2020). Okra has vast potential for earning foreign exchange, as it has a significant share in fresh vegetable exports (APEDA 2020). Tender pods of okra are used as a delicious vegetable. Its leaves, buds, and flowers are also consumed, especially in West Africa. It is used in canned, dehydrated, and frozen forms to a limited extent. Okra mucilage is used in food, non-food, and medicinal products. Okra dry seeds are a rich source of oil (18–20%) and protein (20–23%). During the sugar production process, okra roots and stems are used to clean the cane juice. Okra development in India had taken pace with the release of Pusa Sawani, the first YVMV (Yellow Vein Mosaic Virus) resistant variety by Dr. Harbhajan Singh at IARI, New Delhi, in the late 50s. Pusa Sawani remained a ruling variety for several decades worldwide. Many private seed companies in India, like Mahyco, started their seed business with this variety. In the last few years okra has now become the number one crop in India as per seed revenue which is over 700 crores plus.

Later on, with the development of Parbhani Kranti by Drs. Jambhale and Nerker and Arka Anamika by Dr. O.P. Dutta in the 80s, okra development took a further boost. "With the expansion of private seed companies in India after 80s, okra research and development resulted into advent of hybrids both by private and public sector on a large scale, e.g Vijay and Varha (IAHS), Sun 40 (Pioneer), Mahyco-10 (Mahyco), 152 (Syngenta) Shakti (Nunhems), CO-OH 4 (TNAU), Arka Nikita-GMS Based (IIHR) Kashi Bhairav (IIVR), Pusa Okra Hyb-1 (IARI) etc

S. K. Tikoo (✉)
Plant Breeding, ATPBR - Foundation for Advanced Training in Plant Breeding, Aurangabad, India

S. K. Tikoo et al. (eds.), *Okra: Status, Challenges and Opportunities*,
https://doi.org/10.1007/978-981-97-9963-3_1

were key products." In the last 5 years, okra has become the most valuable vegetable crop in India, both in terms of annual revenue and increased popularity. Hybrids that possess strong resistance to diseases and pests and have better yielding capacity have become more popular among farmers as they ensure a high return on their investments.

Several hybrids have been developed by both the public and private sectors, namely Mahyco-10, Shakti, CO-4, Arka Nikita (GMS-based), Kashi Bhairav, Pusa Okra Hybrid-1, etc. Some of these hybrids have contributed significantly to tackling the problem of YVMV, a devastating virus disease, face by farmers. The introduction of the short internode, strong virus-resistant hybrid Radhika (15–20% more yield) by Advanta (2018) enhanced the popularity of hybrid okra across India. Adaptation of hybrids has been easier as no enhanced agro-technique and additional costs were required to grow the hybrid vs open-pollinated (OP) varieties. The history of okra improvement in India through variety and hybrid development is presented in Fig. 1.1.

In the last 5 years, okra (*Abelmoschus esculentus*) has become the most valuable vegetable crop in India, both in terms of increased popularity and high revenues for not only the seed sector but also the small farmer. The yield per unit area have seen huge increases in the last decade, rising from 4.6 t/ha in 2011 to 6.8 t/ha in 2018 and 13.5 t/ha in 2021, as per FAO Stats. Hence, the production increase has been largely due to genetic gain for yield and not just increased acreage (Table 1.1). From a seed revenue of $12 m in 2011, the same has reached $112 m in 2021. This huge change has been possible due to the breeding of hybrids resistant to whitefly-transmitted viruses, such as Yellow Vein Mosaic Virus (YVMV), over the last three decades by the seed sector in India. The game changer has been the release of the short-internoded, strongly virus-resistant hybrids, like Radhika, where good tolerance to the other

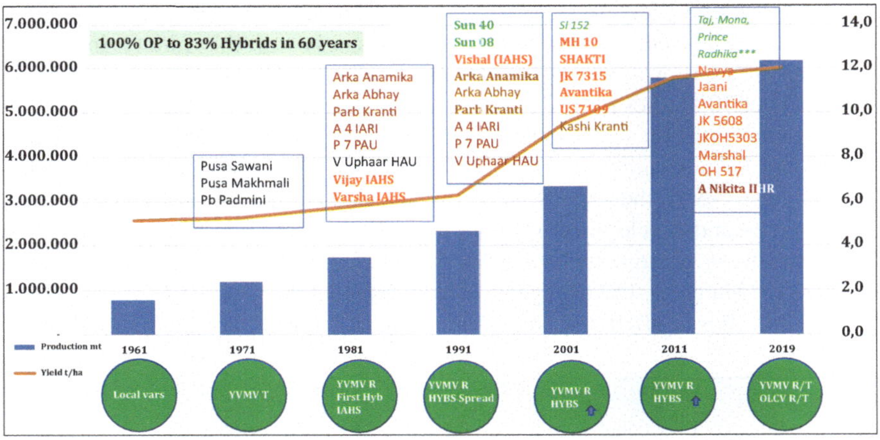

Fig. 1.1 Okra history of product and technology development in India 1961–2019. (Graphic S.K.Tikoo, Yield & area data FAO Stats)

Table 1.1 Okra area, yield, and seed quantity per year from 2011 to 2021 in India

	Area 000 ha	Yield. t/ha	Seed Qty mt
2010–11	431	4.6	2000
2018–19	513	6.8	3500
2020–21	526	13.5	4800
% over 2010–11	122%	291%	240%

important menace, Okra Leaf Curl Virus (OLCV), has been combined with YVMV by Advanta, which has helped increase yields by about 25%. The spread to the smallest farmers, who grow 80% of our food, has been relatively easy, as no changes in their agro-techniques were needed for them to adopt the new hybrids.

In more than 30 West and Central African countries, okra is cultivated as both rain-fed and irrigated crops and is the most valued and popular vegetable, consumed in fresh and dried forms. The last 5 years have seen okra gaining ground as a global crop as its nutritional values are being recognized by consumers. This is especially after the COVID pandemic that a healthy and balanced diet is becoming embedded in the global mindset. Okra seeds are being exported to over 20 countries from India. Recently, an increased number of research papers on its genetics, breeding, genomics, and agronomy from Asia and African countries have appeared. This is a further indication of it becoming a global crop. Over 80% of the market is covered by hybrid seeds in India. Global seed requirements are expected to be 6000 mt by 2030, making it a $300 m crop soon.

As per FAO estimates, around 75% of the okra market is in India, and 12% is in Nigeria. The last 5 years have seen okra gaining ground as a global crop as its nutritional value is being recognized by consumers, with the pandemic during the last 2 years having increased its importance manifold. The last few years have also seen increased research paper publications on its genetics, breeding, genomics, and agronomy from several countries, especially from African countries, Egypt, Turkey, and India. While 4800 mt are required annually in India, of which around 82% is hybrid, global seed requirements are expected to be 6000 mt by 2030, making it a $300 m-plus crop soon. The round table will help understand the requirements of other countries too.

References

APEDA Annual Report 2020–21. https://apeda.gov.in/apedawebsite/Annual_Reports/Apeda_Annual_Report_English_2020-21.pdf

FAOSTAT (2020) Statistical database. https://www.scirp.org/reference/referencespapers

National Horticulture Board (2020) Area and production statistics. https://www.nhb.gov.in

Part I

Okra a Global Crop

Okra Emerging Global Crop: Perspective and Challenges

2

Prashant Belgamwar

Okra (*Abelmoschus esculentus* L.), belonging to the cotton family Malvaceae, originated in the Abyssinian centre of origin of cultivated plants, an area that includes present-day Ethiopia, the mountainous or plateau portion of Eritrea, and the eastern, higher part of the Anglo-Egyptian Sudan. The crop was probably taken into Egypt by Moslems from the East who conquered Egypt in the seventh century. It is believed that by 700 AD, okra cultivation had spread throughout North Africa, the Middle East and India (Zeven and Zhukovsky 1975).

About 50 species have been described by taxonomists in the genus *Abelmoschus*, but as per the classification by IBPGR 1991, eight species—*moschatus, manihot, esculentus, tuberculatus, ficulneus, crinitus, angulosus* and *caillei*—are reported, out of which the first three are wild and cultivated, and the remaining are all wild forms (IBPGR 1991). There are significant variations in the chromosome numbers and ploidy levels of different species in the genus *Abelmoschus*. The lowest number reported is $2n = 56$ for *A. angulosus* (Ford 1938), whereas the highest chromosome numbers reported are close to 200 for *A. manihot var. caillei* (Singh and Bhatnagar 1975; Siemonsmo 1982a, b).

If we look back into the history of okra cultivation in India, in the 1950s it was grown like a field crop under high-density planting with seed drills from home-grown seed. In 1955, Pusa Makhmali was the first variety developed through selection, and in 1958, the first resistant variety, Pusa Sawani, was released by ICAR-IARI, following pedigree selection from a cross between IC1542 and Pusa Makhmali by Dr H. B. Singh. Between 1970 and 1980, many improved varieties were released by public sector research institutions: Punjab Padmini, P7, Punjab 8 (PAU), Harbhajan, Selection-2 (IARI), Parbhani Kranti (MAU), Co1, MDU1 (TNAU), Gujarat Bhendi

P. Belgamwar (✉)
Asia & Africa and International Vegetables, Advanta Seeds UPL Ltd.,
Hyderabad, Telangana, India
e-mail: prashant.belgamwar@advantaseeds.com

S. K. Tikoo et al. (eds.), *Okra: Status, Challenges and Opportunities*,
https://doi.org/10.1007/978-981-97-9963-3_2

1 (AAU) and Arka Anamika (ICAR-IIHR). In the mid-1980s, the first private sector okra hybrids named 'Vijay' and 'Varsha' were released by Indo-American Hybrid Seeds. The decade of 1990s marked the globalization of the seed sector in India and opened global doors for the import and export of vegetable seeds and products, including okra. During the decade of 2000–2010, okra cultivation marked for adoption of hybrids having high resistance to Yellow Vein Mosaic Virus (YVMV) and good fruit quality. After 2010, hybrids under cultivation had high yield potential due to YVMV resistance, but at the same time, an increasing occurrence was observed for another *Begomovirus* called Okra Enation Leaf Curl (OELCV), which caused significant economic yield losses under severe infestation (Sanwal et al. 2014; Venkataravanappa et al. 2015).

The year 2014 was historic for okra cultivation when Advanta Seeds (UPL Ltd.) introduced the first combined resistant hybrid for both the viruses YVMV and OELCV along with good fruit quality. The year 2018 was remarkable, when Advanta Seeds introduced new generation hybrids having high Combined Virus Resistance coupled with excellent fruit quality and an innovative plant type (dwarf with fruiting at shorter internodes), which enabled okra farmers to adopt precision production technologies like the usage of mulch, drip and water-soluble fertilizers, immensely benefitted growers with high returns on investment. With improved genetics of Advanta, the seed rates per acre have gone down. By adopting advanced agronomic practices, farmers are getting a higher number of pickings (up to 60), resulting in the highest return on investment (Advanta Seeds, Unpublished Report).

2.1 Global Okra Market (Figs. 2.1 and 2.2)

The key drivers for high-potential okra markets are: huge domestic consumption driven by population growth, high-performing genetics (strong resistance to viruses), adoption of improved agronomic practices and increasing export demand for South Asian communities globally.

From a market perspective, okra is segmented based on pod shape, colour, local preferences with food habits and its usage. The major types are Indian type, Clemson type, tubular and baby okra, which are further subdivided into green, light green and red colours.

2.2 Advanta's Combo Virus Resistance Story

High pressure of both the viruses (YVMV and OELCV) was a big challenge for okra-growing farmers, resulting in a yield loss of 50–90% depending on severity, and the hybrids available in the market had fair resistance to YVMV but not to OELCV, which was prevalent in almost all okra-growing areas of the country. Therefore, combined high virus resistance was the need of okra growers. Advanta Seeds addressed the same by delivering the first combo virus resistance hybrids named Taz, Mona and Venus Plus in year 2014, followed by Prince and Lavanya in

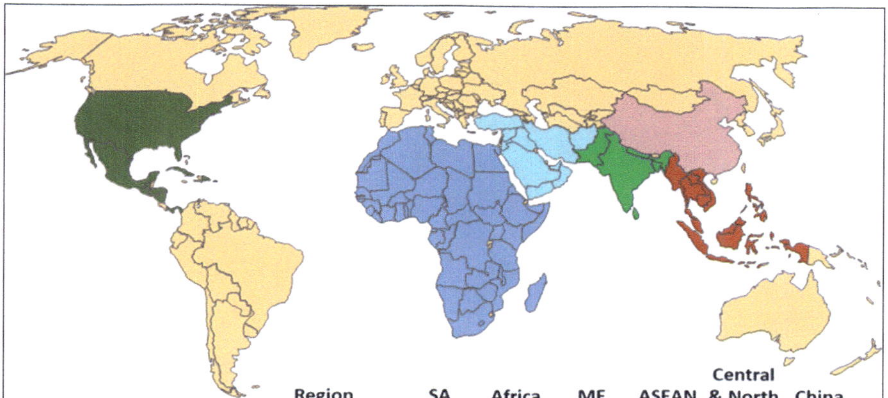

Region	SA	Africa	ME	ASEAN	Central & North America	China
Volume (MT)	5700	6850	300	115	75	55
Hybridisation (%)	80	38	75	66	0	100

Fig. 2.1 Global okra market. (Advanta Seeds, Unpublished Report)

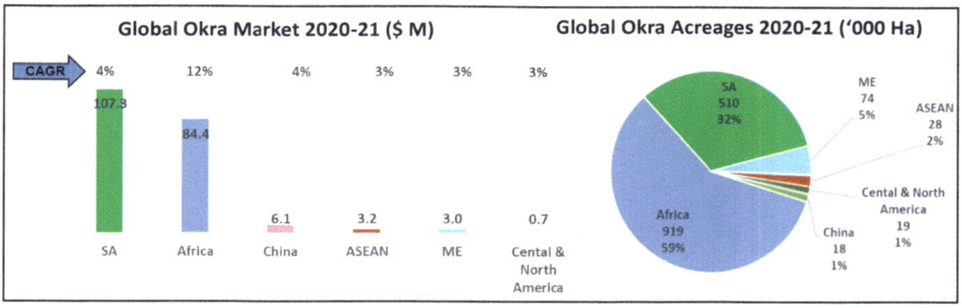

Fig. 2.2 Global okra market and acreages. (Advanta Seeds, Unpublished Report)

2016. In year 2018, Advanta delivered its benchmark hybrids named Radhika and Navya Jaani, which have an innovative plant type stature—short internodes, very high combined virus resistance (for YVMV+OELCV), coupled with excellent fruit quality (Pictures 2.1 and 2.2).

These innovative hybrids of Advanta resulted in an extended crop life cycle of up to 120 days, enabling farmers to go for a very high number of pickings, a smaller number of chemical sprays, saving the cost of cultivation by approximately 20%, and an overall increase in yield by 20–50%, subject to disease severity. Considering the high return on investment, the majority of the vegetable growers adopted okra as a cash crop, and Advanta's hybrids became the industry benchmark (Figs. 2.3 and 2.4).

Picture 2.1 YVMV resistance vs susceptible

Picture 2.2 OELCV resistance vs susceptible

2.3 Road to Global Crop

Okra is a common man's vegetable crop grown in all seasons, easily available and consumer-friendly. Okra consumption is increasing at a CAGR of 5%, with price realization to the farmers increasing at a 6% CAGR. Okra is easy to cook and is used for various local preparations, from daily diets to party dishes.

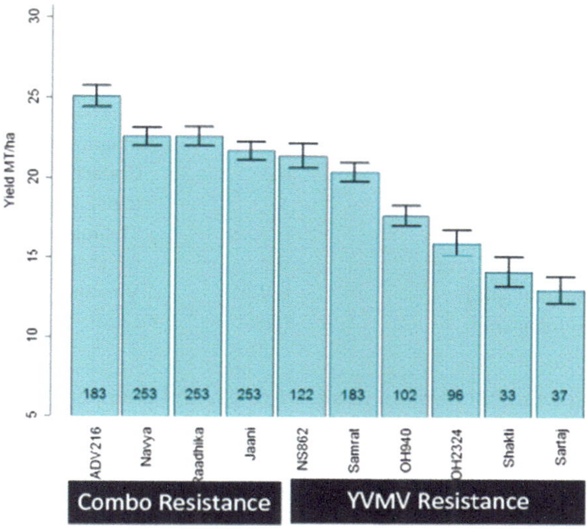

Fig. 2.3 Yield difference between combo virus resistance vs single virus (YVMV) resistance. (Advanta Seeds Internal Trials Data, Unpublished Report)

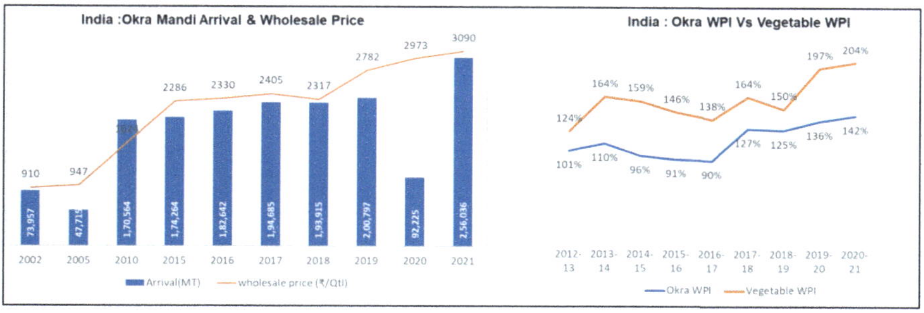

Fig. 2.4 Trend of okra arrival to market yard and wholesale price. (Source web-eaindustry.nic.in; ministry of commerce & industry, Govt of India)

Okra is known as a nutritional powerhouse for its high health properties like antioxidants, antibacterial, anti-diabetic, gastro-protective, lipid-lowering agents, immunomodulatory and neuropharmacological agents, which are gaining attention globally so as to include in daily dietary patterns. With its large number of minerals in optimum quantity, okra is known as a 'Mineral Bank' (Petropoulos et al. 2018) (Table 2.1).

Red okra hybrid 'Kumkum' recently launched by Advanta Seeds is gaining popularity among consumers and farmers as a value-added product for high return on investment. The red okra has more polyunsaturated fatty acids (helps to reduce bad cholesterol), more sodium (controls blood pressure), more iron (for proper function of haemoglobin) and more protein (vital in metabolism) than green okra.

Nowadays, many value-added products in terms of ready-to-eat okra are gaining popularity among consumers. Okra chips, spiced okra, crispy bhendi, okra in

Table 2.1 Nutritional value: composition per 100 g of edible okra portion

Calories	35.0	Calcium (mg)	66.0
Moisture (g)	89.6	Iron (mg)	0.35
Carbohydrates (g)	6.4	Potassium (mg)	103.0
Protein (g)	1.9	Magnesium (mg)	53.0
Fat (g)	0.2	Copper (mg)	0.19
Fibre (g)	1.2	Riboflavin (mg)	0.01
Minerals (g)	0.7	Thiamine (mg)	0.07
Phosphorus (mg)	56.0	Nictonic acid (mg)	0.06
Sodium (mg)	6.9	Vitamin C (mg)	13.10
Sulphur (mg)	30.0	Oxalic acid (mg)	8.0

vinegar, cut okra, okra sliced rings, and chopped and frozen okra are readily available in markets. Okra is also being preferred in the menu of the majority of restaurants, star hotels, cafés, etc., around the globe.

If we look into the global fresh okra supply chain, produce from India is being exported to the UK, European Union, USA, Canada, Middle East and Malaysia. Other fresh okra exporting countries are Pakistan, Thailand, Egypt, Jordan, Kenya and Ethiopia, which mainly export to the UK and European Union.

2.4 Future Trends and Challenges

Considering the increasing popularity of red okra in relation to its health benefits, future biofortified okra research has great scope for researchers/breeders. Due to frequent changes in existing viruses and the emergence of new viruses, it will be required to breed new multiple virus-resistant varieties using new sources of resistance. Further increases in productivity should also be a target for okra breeders, besides bringing value-added products that are highly suitable for ready-to-eat okra and for the processors. Considering climate change, breeding for abiotic stresses will also be the need of the future; combining resistance to biotic stress and tolerance to abiotic stresses, coupled with fruit quality, will be the key challenge for future breeding. There is a need to work on customized mechanization to be used from cultivation to harvesting and post-harvest handling as well. Currently, okra is a crop that has very limited support from molecular breeding technologies; hence, the molecular breeders and biotechnologists should focus more on the research and development of okra.

References

Ford CE (1938) A contribution to a cytogenetical survey of the Malvaceae. Genetica 20:431–452

International Board for Plant Genetic Resources IBPGR (1991) International crop network series, report of an international workshop on okra genetic resources. International Board for Plant Genetic Resources, Rome, p 5

Petropoulos S, Fernandes Â, Barros L, Ferreira ICFR (2018) Chemical composition, nutritional value and antioxidant properties of Mediterranean okra genotypes in relation to harvest stage. Food Chem 242:466–474

Sanwal SK, Singh M, Singh B, Naik PS (2014) Resistance to yellow vein mosaic virus and okra enation leaf curl virus: challenges and future strategies. Curr Sci 106:470–471

Siemonsmo JS (1982a) La culture du gombo (*Abelmoschus* spp) legume fruit. Thesis Univ Wageningen, The Netherlands

Siemonsmo JS (1982b) West African okra. Morphological and cytological indications for the existence of a natural amphiploid of *Abelmoschus esculentus* (L.) Moench and *A. Manihot* (L.) Medikus. Euphytica 31(1):241–252

Singh HB, Bhatnagar A (1975) Chromosome number in an okra from Ghana. Indian J Genet Plant Breed 36:26–27

Venkataravanappa V, Reddy CNL, Jalali S, Briddon RW, Reddy MK (2015) Molecular identification and biological characterisation of a begomovirus associated with okra enation leaf curl disease in India. Eur J Plant Pathol 141:217–235

Zeven A, Zhukovsky P (1975) Dictionary of cultivated plants and their centres of diversity. Excluding ornamentals, forest trees and lower plants. PUDOC, Wageningen

Export Potential of Fresh Okra from India

3

Vinita Sudhanshu

Okra (*Abelmoschus esculentus* L. (Moench), HS code 07099990) is known in many English-speaking countries as *lady's finger, bhindi, ochro, or gumbo*. It is an economically important vegetable crop grown in tropical and subtropical parts of the world and is most popular in India, Nigeria, Iraq, Pakistan, Cameroon, Iraq, and Ghana (Swamy 2023). Though it is virtually not grown in Europe and North America, yet, lot of people in these countries have started to like this vegetable because of its good amount of vitamin A and folic acid, besides carbohydrates, phosphorus, magnesium, and potassium (Sivakumar et al. 2013). India is the *largest producer* of okra in the world (Table 3.1), with a total production of *6466 thousand MTs* during the year 2020–21 (NHB, Final estimate 2020–21).

3.1 Production Scenario: India

The global production of okra during the year 2020 stood at 10.54 million metric tons. India ranks first in the world with 6.46 million metric tons (**64.41%** of the total world production) (Source: FAOSTAT 2020) of okra production. State-wise production of okra for the year 2021–22 is mentioned in Table 3.2.

V. Sudhanshu (✉)
Agricultural and Processed Food Products Export Development Authority (APEDA), New Delhi, India
e-mail: vinitas@apeda.gov.in

S. K. Tikoo et al. (eds.), *Okra: Status, Challenges and Opportunities*,
https://doi.org/10.1007/978-981-97-9963-3_3

Table 3.1 A glimpse of Okra area and economic importance, 2021–22 (1st Adv. Estimate)

India's Production (Source: DAC&FW)	Qty: 6.46 Mill. MT
World ranking in Production	**1st**
India's share in World Production	**64.41%** (FAO 2020)
India's Export (Source: DGCIS)	Qty: 3925.88 Million MT Value: Rs. 23.78 Crore USD 3.17 Million
World ranking in Exports	**8th** (FAO 2020)
Major Export Destination (Source: DGCIS)	United Kingdom, UAE, Qatar, Germany, Kuwait, Canada, and Singapore
India's Opportunity	United States, France, Hong Kong, Netherlands, Belgium, and Switzerland

Table 3.2 Indian production of okra/lady's finger in the year 2021–22 (State-wise)

2021–22			
Sr No.	State	Production (000 Tons)	Share (%)
1	Gujarat	1019.42	15.89
2	West Bengal	893.96	13.93
3	Bihar	794.10	12.38
4	Madhya Pradesh	754.09	11.75
5	Orissa	662.64	10.33
6	Chhattisgarh	357.07	5.57
7	Uttar Pradesh	335.86	5.23
8	Andhra Pradesh	219.26	3.42
9	Tamil Nadu	208.25	3.25
10	Assam	200.99	3.13
11	Haryana	183.45	2.86
12	Jammu & Kashmir	143.68	2.24
13	Maharashtra	139.28	2.17
14	Jharkhand	110.43	1.72
15	Telangana	83.44	1.30
16	Karnataka	67.85	1.06
17	Himachal Pradesh	60.95	0.95
18	Punjab	54.88	0.86
19	Uttarakhand	27.73	0.43
20	Mizoram	25.02	0.39
21	Rajasthan	19.73	0.31
22	Tripura	19.09	0.30
23	Meghalaya	4.00	0.06
24	Manipur	2.75	0.04
25	Kerala	2.73	0.04
26	Nagaland	1.52	0.02
27	Arunachal Pradesh	0.39	0.01
28	Sikkim	0.33	0.01
29	Others	23.42	0.37
	Page Total	**6416.31**	

Source: National Horticulture Board (NHB)

3.2 Varieties Suitable for Export

For export purposes, mostly hybrid varieties are used. A few important hybrid varieties are given below.

Variety
Kashi-Mohini
Kashi-Mangali
Kashi-Vibhuti
Kashi-Pragati
Kashi-Lila
Hybrids
Shitala Uphar
Shitla Jyoti
Kashi-Bhairav
Kashi-Mahima

3.3 Sowing Season: Kharif, Rabi, and Summer Season

Okra is available throughout the year in West Bengal, Orissa, and Gujarat. In the rest of the states, availability of okra is maximum in the period of April to June.

3.4 Harvesting

Harvesting normally continues 45–60 days after planting. Okra must be harvested every 1–2 days to maintain the pod size in a marketable range.

3.5 Yield per Acre

Normally, an average of 4–5.6 tons are obtained per acre in India.

3.6 Market Influencing Factor

- The okra produced in West Bengal, Orissa, Bihar, and Northern States was consumed in this region itself due to high demand.
- Major competition in okra export is from Nigeria, Ghana, Pakistan, and Egypt. These countries are also near to Europe and the Middle East than India.

3.7 Strengths for Exporting Okra

- India is the largest producer of okra in the world.
- In India, several superior cultivars and even hybrids are available for cultivation, with productivity ranging between 15 and 20 tons/ha.
- Okra is available in India throughout the year, and production can be tailored according to demand.
- There is excellent research support for okra because IIHR, Bangalore; IARI, New Delhi; IIVR, Varanasi; and SAUs are in different regions of the country to provide solutions to various problems in its cultivation.
- The acceptance of India's agricultural produce is increasing worldwide due to concurrent developments in the areas of state-of-the-art cold chain infrastructure and quality assurance measures. To enhance exports from India, the Agri Export policy has been introduced by the Government of India, under which various production clusters have been formed. Under AEP, the following initiatives have been taken:
  - Export from landlocked and difficult areas
  - Venturing into facilitating direct export from Farmer Producer Company
  - Encouraging technology interventions and start-ups through new distribution e-traders.

3.8 Global Trade Scenario

The value of global imports of vegetables, n.e.s., fresh/chilled (HS code 070990) stood at USD **2897.25** million (Table 3.3). Germany is the largest importer of okra, followed by the United States, France, Hong Kong, the United Kingdom, the Netherlands, Canada, UAE, Belgium, Switzerland, Singapore, Austria, Kuwait, the Czech Republic, and Sweden. France, the United Kingdom, Germany, and the Netherlands have a share of 82.85% (*Source: UNCOMTRADE*).

3.9 Indian Trade Scenario

India, despite being the number one country in okra production, occupying **64.41%** share in global production, has a very insignificant share in global export at 1.74% in terms of quantity and 3.2% in terms of money {(US million dollars) (*Source: FAOSTAT- and UNCOMTRADE*)}. The United Kingdom is the leading importer of okra (HS code-07099990—Other Vegetables Fresh or Chilled) from India, occupying a 23.7% share in terms of USD million during 2021–22 (*Source: DGCIS*), followed by the UAE with an 18.3% share in terms of USD million during 2021–22. The export data for the years 2020–21 and 2021–22 is given in Table 3.4.

Table 3.3 Major importing countries of okra in the World

Major importing countries of okra in the World						
HS Code: "070999 Other leguminous vegetables, shelled or unshelled"						
Unit: Qty In MT Value in US$ Mill						
Importing country	**2018**		**2019**		**2020**	
	Qty	**Value**	**Qty**	**Value**	**Qty**	**Value**
Germany	127546.02	316.82	138338.3	341.37	140569.42	359.76
USA	354305.29	284.9	385237.75	315.15	404107.51	349.02
France	79497.95	212.17	143996.22	256.51	123185.33	270.08
Hong Kong	91379.55	65.8	125709.52	131.68	176851.68	264.1
UK	117368.49	246.46	101998.24	211.26	99368.72	221.66
Netherland	73418.36	150.15	95744.91	177.67	78240.72	180.02
Canada	24879.99	158.06	117103.38	157.52	111737.83	169.23
United Arab Emirates	205178.95	100.86	258236.85	98.92	354775.27	93.66
Belgium	53854.35	87.17	39933.83	85.62	45528.16	89.83
Switzerland	27858.14	66.27	29776.05	68.99	32213.22	75.94
Singapore	66281.17	66.94	65943.62	67.2	25718.51	74.17
Austria	34266.95	96.06	38426.26	91.12	29282.28	71.72
Kuwait	28873.26	50.43	0	0	33007.82	70.69
Czech Republic	26868.71	35.17	22855.99	33.41	36094.41	61.35
Sweden	23163.92	48.73	23296.22	46.74	21954.29	49.14
Total of top 15 countries	**1334741.1**	**1985.99**	**1586597.14**	**2083.16**	**1712635.17**	**2400.37**
Other countries	**485568.23**	**602.47**	**413393.24**	**506.05**	**391123.97**	**496.88**
Total	**1820309.33**	**2588.46**	**1999990.38**	**2589.21**	**2103759.14**	**2897.25**
Share of top 15 (%)	**73.32**	**76.72**	**79.33**	**80.46**	**81.41**	**82.85**

Source: UN Comtrade, as reported by the Importing countries

Table 3.4 Export data for the years 2020–21 and 2021–22

	2020–21			2021–22		
Country	Qty	Rs. Crore	US$ Mill	Qty	Rs. Crore	US$ Mill
UK	664.11	5.88	0.8	471.1	5.64	0.75
United Arab Emirates (UAE)	686.18	4.87	0.66	742.44	4.35	0.58
Qatar	325	2.2	0.3	514.32	2.95	0.4
Germany	429.92	3.39	0.46	337.95	2.6	0.35
Kuwait	200.32	1.13	0.15	395.64	2.41	0.32
Canada	295.87	2.57	0.35	191.56	1.73	0.23
Singapore	898.32	5.79	0.78	221.84	1.09	0.15
Nepal	1434.74	2.05	0.28	680.77	0.75	0.1
Bhutan	46.23	0.28	0.04	196.78	0.72	0.1
Sweden	12.63	0.13	0.02	33.12	0.45	0.06
Others	215.1	1.47	0.18	140.36	1.09	0.13
Total	5208.42	29.76	4.02	3925.88	23.78	3.17

Source: DGCIS

3.10 Process Flow of Export of Okra

Chain of events (Packhouse up to shipment)

Harvesting (in morning)

↓

Handling and Transportation

↓

Unloading

↓

Grading and Sorting

↓

Packing

↓

Pre-cooling

↓

Cold Storage

↓

Palletizing

↓

Refrigerated Truck

↓

Shipment

3.11 APEDA's Activities and Initiatives for Promotion of Horticulture Produce

(i) **Collaboration with Missions/Embassies**: APEDA has been regularly interacting with the Indian Missions abroad, which enabled it to tap the opportunities in the importing countries' requirements arising on account of the Covid pandemic.

(a) **Virtual Trade Fairs (VTFs):** To leverage the opportunity due to the pandemic situation and to sustain the existing markets, APEDA took the pioneer step and introduced its own platform through an in-house team for organizing Virtual Trade Fairs to establish contact between Indian exporters and importers.

A Virtual Trade Fair for Indian Fruits, Vegetables, and Floriculture Show was held from 26th to 28th May 2021.

(b) **Virtual Buyer Seller Meets:** In a bid to boost the export potential of its mandated agricultural products, APEDA has organized a series of Virtual Buyer Seller Meets (V-BSM) in association with Indian Missions abroad. The meet brought together key stakeholders from the respective Governments and trade on a common platform for strengthening strategic cooperation between India and Bhutan in the Agri-Food Sector.

Since April 2020 till date, 49 V-BSMs have been organized by APEDA in association with Indian Missions with UAE, Kuwait, Indonesia, Switzerland, Belgium, Iran, Canada (Organic products), the UAE and the United States (for GI products), Germany, South Africa, Australia, Thailand, Oman, Nepal, Bhutan, Azerbaijan, Uzbekistan, Vietnam, Netherlands, Brunei, Cambodia, South Korea, Sudan, Japan, GCC Countries, Bahrain, Angola, Egypt, Algeria, Togo, Ghana, Venezuela, etc.

(ii) **Promotion of GI-tagged, indigenous, ethnic agricultural products:** In tandem with Hon'ble PM's call for "vocal for local" and "Atmanirbhar Bharat," APEDA has been focusing on the promotion of exports of locally sourced GI (Geographical Indications) tagged as well as indigenous, ethnic agricultural products. New products and new export destinations have been identified, and accordingly, the trial shipments have been facilitated. The list has been placed at Annexure A.

(iii) **Market access:** Through sustained discussions and bilateral talks, efforts have been made in opening new markets for agricultural products by pursuing with the potential countries for gaining market access for various products.

(iv) **Sensitization/Outreach/Capacity Building programs:** A series of Capacity-Building/training programs have been carried out in the Agri Clusters and States in association with the State departments, SAUs, and KVKs for the FPOs/FPCs/SHGs/exporters in order to provide linkage of the Farmer groups to exports and to promote the entrepreneurs for becoming prospective exporters.

With the support of NCDC and VAMNICOM, several programs have been organized for Cooperatives to enable the development of Cooperatives as agri exporters.

(v) **Partnering with Line Ministries for cluster development:** Efforts have been made for having a synergy with the identified Clusters of MoA and FW, MoFPI and DGFT. The cluster development activities are being aligned with the Clusters of MIDH (NHB) identified under the Horticulture Cluster development program and ODOP (PMFME) scheme of MoFPI for seeking convergence with their financial assistance schemes.

Towards integration with the DEH initiative of DGFT, APEDA, in association with Invest India, coordinated with the respective Indian Missions to organize a series of events.

(vi) **Special focus on Himalayan States and NER:** APEDA managed to push exports from Jammu & Kashmir, as the first consignment of Mishri variety cherries was shipped from Srinagar to Dubai. Saffron and dry fruits from the UT of Jammu and Kashmir to Saudi Arabia. FMCG Lulu Group International has been channelized along with sending samples. Further, the export of walnuts and apples has been facilitated from J&K alongside apricots from Ladakh.

The export of Himalayan millets, Himalayan goat meat, and fresh vegetables has been facilitated from Uttarakhand, along with apples from Himachal Pradesh (in association with HPMC). Fresh jackfruit (from Tripura), Burmese grapes, red rice/tender jackfruit (Assam), and GI-tagged King chilli (Nagaland) have been sourced from the farmers in NER and exported to multiple destinations.

(vii) **Infrastructure upgradation:** APEDA has been assisting exporters under its financial assistance scheme under the components of Infrastructure Development, market promotion, and quality development. The scheme has been beneficial for setting up modern integrated packhouses having facilities as per the requirements of importing countries. In addition, the scheme covers assistance for the purchase of refrigerated vehicles to support the supply chain in horticulture. The above two components have largely helped exporters of fresh horticulture produce like mangoes, pomegranates, grapes, bananas, fresh vegetables, and the floriculture sector. The efforts made by APEDA stand out in the establishment of around 250 export-oriented EU-compliant packhouses in the private sector. Financial assistance provided to State Governments under Infrastructure Development has also enhanced the State Governments' capacities to export.

(viii) **Quality initiatives:** APEDA pioneered its first online traceability system for the export of grapes to EU countries, i.e. GrapeNet. After the success of the first traceability system in the grape sector, APEDA has replicated similar systems for peanuts (PeanutNet), organic products (Tracenet), and meat products (MeatNet). Residue monitoring through export procedures for fresh grapes, pomegranates, okra, green chillies, bitter gourd, drumstick, brinjal, curry leaves, peanuts, and peanut products was introduced.

Considering the importance of laboratory testing, APEDA has expanded its laboratory network to 230 for carrying out sampling and analysis of APEDA scheduled products for export certification.

(ix) **Agri export policy implementation and Cluster development:** State Governments have been engaged to finalize the State-specific agri export action plan as per the template provided by APEDA for the implementation of the Agri export policy. Twenty-two States and two UTs (Ladakh, A&N Islands), viz. Maharashtra, U.P., Kerala, Nagaland, Tamil Nadu, Assam, Punjab, Karnataka, Gujarat, Rajasthan, Andhra Pradesh, Telangana, Manipur, Sikkim, Uttarakhand, M.P., Mizoram, Meghalaya, Tripura, Arunachal Pradesh, and Himachal Pradesh, have finalized the State-specific action plan, while other States are in the process of finalization.

Subsequent to cluster visits undertaken in the initial phase, several rounds of stakeholder VC meetings and other activities at the cluster level have been organized to activate all the clusters. As an outcome of the cluster development programs and Buyer Seller meet in the cluster districts, exports of horticulture produce took place by air/sea mode from the Clusters of Varanasi

(Fresh fruits/vegetables), Lucknow (Mango), Nagpur (Oranges), and Anantpur and Theni (Banana) to the Middle East and London.

(x) **Resource convergence with Ministry of Railways (MoR):** Cooperation has been sought towards strengthening of export infrastructure by rail.

3.12 Intervention Required for Enhancement of Export

The measures that need to be taken for enhancing competitiveness are as follows:

- Productivity needs to be increased.
- Quality needs to be enhanced so that India can make quality products available at a competitive price.
- Freshness and eye appeal must be enhanced in the product in order to be competitive.
- Packaging needs to be improved.
- Backward linkage with a focus on extension needs to be given. Farmers need to be trained on the quality requirements of importing countries.
- The product needs to be branded.
- New markets need to be explored.

References

Food and Agriculture Organization of the United Nations (FAO) (2020) World food and agriculture: Statistical yearbook 2020. ISBN: 978-92-5-133394-5. http://www.fao.org/3/cb1329en/CB1329EN.pdf

Sivakumar K et al (2013) Co-inoculation effect of AM fungi and phosphobacteria on the growth and yield of rhizosphere of Bhendi (*Abelmoschus esculentus L.*) as influenced by chemicals and biopesticides. Int J Pharm Biol Arch 4(2):375–378

Swamy KRM (2023) Origin, distribution, taxonomy, botanical description, cytogenetics, genetic diversity and breeding of okra (*Abelmoschus esculentus (L.) Moench.*). Int J Dev Res 13(03):62026–62046

4 Okra Area, Productivity and Key Improvements in Nigeria

Aderibigbe Helen Adeniyi

4.1 Introduction

Okra, *Abelmoschus esculentus*, is a flowering annual that bears edible pods and grows in all parts of Nigeria. Okra (*Abelmoschus esculentus*) is a multipurpose vegetable crop and a good source of many nutrients (Hegazi and Hamideldin 2010), grown throughout the tropics and warmer temperate regions of the World (Patil et al. 2015). It constitutes a major economic crop in the West African subregion owing to its importance as a component of various recipes in many local delicacies and vegetable dishes (Kumar et al. 2010).

Okra is a popular and nutritious crop whose cultivation and consumption cut across tribes, cultures and religions in Nigeria. Okra is consumed majorly as draw soup or as a stew thickener; this is because most of our foods are swallowed in the form of small balls or boluses. Examples of foods usually eaten with okra soup are iyan, amala, fufu, eba, agidi, tuwo, etc., which are products of processed yam, cassava and maize, respectively. The flower and leaf are also used in making soup in some parts of the country, but the fruit is the one that is generally acceptable.

Due to the fact that the storage of fresh vegetables is grossly inadequate in the country, okra is usually processed into dry form through dehydration, either as whole or sliced, and thereafter milled into powdered form, which can be stored for a while and is also used in making draw (mucilaginous) soup.

However, okra is regarded as a minor crop, and farmers hardly see its cultivation as a big business; as such, adequate and abundant production to meet local consumption has been a Herculean task.

A. H. Adeniyi (✉)
Farming Systems Research Programme, National Horticultural Research Institute, Idi-Ishin, Jericho, Ibadan, Nigeria

S. K. Tikoo et al. (eds.), *Okra: Status, Challenges and Opportunities*,
https://doi.org/10.1007/978-981-97-9963-3_4

4.2 Okra Production

The global production of okra in the World is estimated at 9.96 million tonnes, with India and Nigeria being the most predominant producers of 6.18 million tonnes and 1.88 million tonnes, respectively (FAOSTAT 2020). Of the total production of okra in the World, India and Nigeria are known to contribute more than 70% and 15%, respectively. Okra is found in almost every market in Nigeria (Akoroda et al. 1985). Nigeria recorded the highest okra fruit yield in 2010, with an average yield of 27, 275 kg/ha, but this has been on the decline, with the sharpest drop recorded in 2011, with an average yield of 8, 735 kg/ha (Ibitoye and Kolawole 2022). Lack of improved varieties and complex interactions of biotic and abiotic stresses might be part of the reasons for the decrease in fruit yield over the years. Its total cultivated area is about 1.5 million ha (Adejonwo et al. 1989). It has a considerable area under cultivation in Nigeria and it is traditionally and majorly cultivated as a rainy season crop in farms, even as a backyard or home/roadside garden crop (Adeniji et al. 2007).

It is called different names by Nigerians, depending on the ecological zone and tribe; for instance, it is called 'ila' in Yoruba, kubewa in Hausa and okuru Ibo. Its consumption cuts across gender, religion, age brackets, ethnicity, tribe, low- and high-income earners, as well as rural and urban communities. All parts, except the stem, are edible in Nigeria. The stem is put to other uses, such as rope and sack making.

Okra pods are a powerhouse of valuable nutrients (Adetuyi et al. 2011) and an affordable source of protein, carbohydrates, minerals, vitamins and dietary fibre. Nutritionally, the tender green pods of okra are important sources of Vitamins A, B, C and K, folic acid, Potassium, Magnesium, Calcium and trace elements such as Copper, Manganesse, Iron, Zinc, Nickel and Iodine (Habtamu et al. 2014), which are often lacking in the diets of people in most developing countries like our Nation, Nigeria.

The composition of okra pods per 100 g edible portion:

- Water 88.6 g, energy 144.00 kJ (36 kcal)
- Protein 2.10 g
- Carbohydrate 8.20 g
- Fat 0.20 g
- Fibre 1.70 g
- Ca 84.00 mg
- P 90.00 mg
- Fe 1.20 mg
- β-carotene 185.00 μg
- Riboflavin 0.08 mg
- Rhiamin 0.04 mg
- Niacin 0.60 mg
- Ascorbic acid 47.00 mg

Succulent leaves are also used in soup making just as the pod.

The composition of okra leaves per 100 g edible portion is:

- Water 81.50 g
- Energy 235.00 kJ (56.00 kcal)
- Protein 4.40 g, fat 0.60 g
- Carbohydrate 11.30 g
- Fibre 2.10 g
- Ca 532.00 mg, P 70.00 mg, Fe 0.70 mg
- Ascorbic acid 59.00 mg, β-carotene 385.00 μg
- Thiamin 0.25 mg
- Riboflavin 2.80 mg
- Niacin 0.20 mg (Gopalan et al. 2007; Varmudy 2011)
- Carbohydrates are mainly present in the form of mucilage (Liu et al. 2005; Kumar et al. 2010)

Okra flowers are used in the same way as the pod and leaves, but in addition, they are used as herbal tea.

In whichever way okra is consumed, the most important quality been considered in this part of the world is its high mucilage and sliming content.

4.3 Factors Affecting Okra Production

Okra pod yield may be affected by physiological and agronomic characters acting singly or in interaction with each other.

Several authors, including Osalusi et al. (2019) and Ekunwe et al. (2018), have identified the various problems that affect okra production in Nigeria to include the following:

- Poor access roads and inadequate transportation facilities were considered
- most serious by some okra farmers
- Inadequate capital and difficulty in getting funds
- Problems of pests, diseases and weed control
- Difficulty in getting land
- High cost of inputs and difficulty in getting improved seeds, agrochemicals and fertilizers
- Difficulty in finding labour
- Poor storage facilities
- Weather conditions and decline in soil fertility

Even in cases where high-yield cultivars have been grown, the inherently low fertility status of the soils, coupled with inadequate application of fertilizers, remains the principal limiting factor to okra production, especially in the savannah regions of Nigeria. As a result of these constraints, yields of 2–3 tonnes/ha of green fruits have been reported (Muhammad et al. 2020). The low yield may also be attributed to poor-yielding genotypes that are used by the local farmers (Manyong et al. 2002). Seasonal weather variation also has a direct influence on the quantity and quality of agricultural production. It could also be noted that factors affecting okra production in Nigeria differ from one agro-geographical zone to another (Fig. 4.1). The six agro-ecological zones are: *Sahel, Sudan, Northern Guinea, Southern Guinea, forest and Coast*.

Varieties planted are also agro-ecology dependent, as are uses and trait preferences.

Babalola et al. (2020) observed that land acquisition, size of farmland, land preparation method and source of capital play major roles in okra farming among smallholder farmers in the Kabba-Bunu Area of Kogi State, North Central Nigeria. In Rivers State, South–South Nigeria, okra production is affected by waterlogged

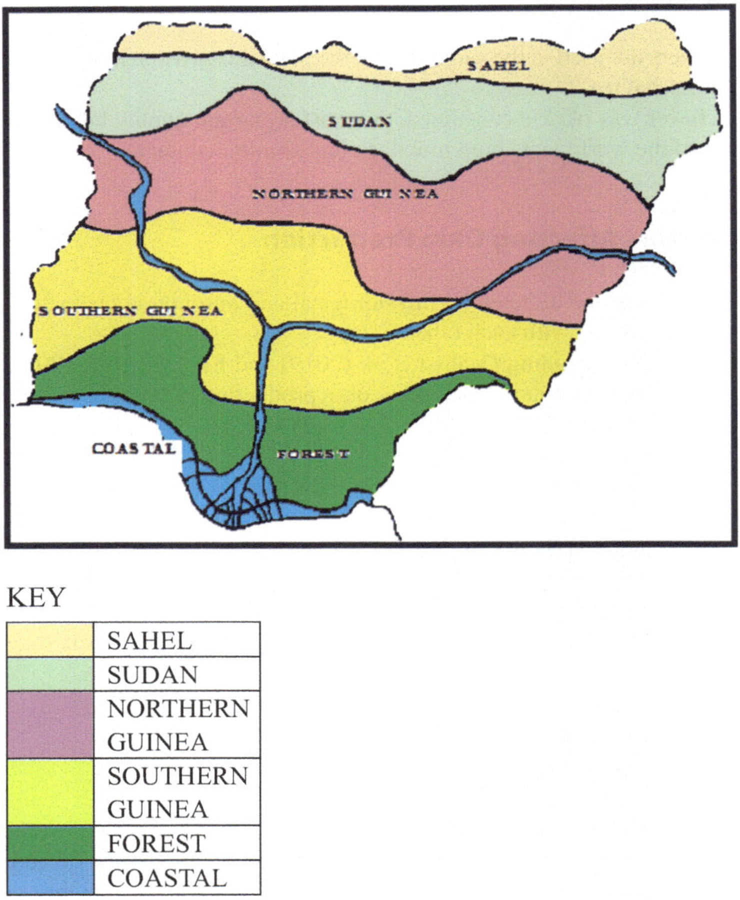

Fig. 4.1 Map of Nigeria showing the different agro-ecological zones. (Source: Damisa et al. (2010))

environments, according to Dimkpa et al. (2019). Amadi et al. (2014) revealed that okra fruit rot (fungi) was quite rampant in their surveyed location of Awka South Iga in Anambra State, South-Eastern Nigeria.

The result of the analysis of constraints encountered by okra farmers carried out in the Akinyele local government area, Oyo State, by Osalusi et al. (2019), ranked from most critical to the least, showed that pests and diseases took the lead, followed by the inadequate transportation facilities, weather conditions and the high cost of inputs as the major constraints. It was noted that these constraints, identified as most important, sum up to over half (58.2%) of the problems of vegetable farmers in the study area. It was then concluded that if these constraints are looked into, other impediments may cease to exist or be reduced to a minimum in the study area. However, Adebayo et al. (2020) found out that in the Ido area of Ibadan, Oyo State, good sources of income, labour availability, farmland and access to improved varieties had greatly improved farmers' production in that area. He also discovered that 50% of respondents adopted the improved variety NHAe 47-4 because of its high yield and rapid growth. Hence, for optimum yield, seed treatment, improved seed, appropriate planting time, seed rate, planting space, good cultural practices and the correct use of fertilizer should be encouraged.

Ibitoye and Kolawole (2022), through participatory rural appraisal (PRA), also affirmed that farmers' constraints include the lack of improved varieties, diseases, insect pests, market acceptability, drought tolerance, among others. There is, therefore, a need for concerted efforts for a market-driven breeding programme for improved cultivars, which must also be based on farmers'/consumers'/stakeholders' preferred traits such as earliness, yield, resistance to pests and diseases, among others. The scarcity of conventional fertilizer, associated with its high cost, has also created a lot of problems in arable crop production in Nigeria.

4.4 Suitable Soil and Climate for Okra Production in Nigeria

Okra is adapted to wide varieties of soil, but a deep fertile sandy loam with good drainage and a soil pH of 6.0–6.8 is optimum (Eke et al. 2008; Aboyeji et al. 2021).

The climate of Nigeria is characterized into two seasons: the Wet and Dry. The wet season (summer) is normally from April to October, while the dry season (winter) is from November to March. The wet season is characterized by longer days and warm, humid weather, while the dry season is associated with relatively shorter days, high day and low night temperatures (Katung 2007). An average monthly temperature range of 20–30 °C is considered optimum for growth, flowering and fruiting. Okra is sensitive to low temperatures and develops poorly below 15 °C. For the optimum productivity of okra in Nigeria, the wet season environment is considered more appropriate. Dry season farming is practiced where there are adequate irrigation practices. It is more economically viable at this season because fewer farmers are involved in dry season farming.

4.5 Cultural Practices

The seeds of okra are usually sown directly using the dibbling method, between 2 and 3 seeds/hole at a depth of 1.5–2.0 cm. Okra seeds have a hard seed coat that can impede proper germination. As a result, the seeds may need to be scarified and soaked prior to planting. To scarify the seeds, simply rub them with sandpaper and then soak the seeds for 12–24 hours prior to planting. Under ideal conditions, your okra should germinate within 7 days. Seeds can be sown directly, as is been practiced by the majority of farmers in Nigeria, if the soil temperature is warm enough and there are enough days in the season for plants to reach maturity. Planting can be done at a spacing of 20–30 cm. Thinning may be done at 2 weeks after sowing (WAS) to maintain an appropriate plant population and density. Plants that grow too close to each other compete for the same resources, such as sunlight, water and soil nutrients. Overcrowded plants begin to grow upward to receive light and may be thinner than expected. Fertilizer application is a necessary procedure for optimum yield, especially in areas where soil tests have indicated certain deficiency in required nutrients.

Okra is picked when immature, at about 3–4 inches in length, for peak flavour and tenderness if the intention is for fresh consumption rather than seed production.

Fresh okra fruits have been reported to be a good source of vitamins, minerals and plant proteins (Eke et al. 2008).

4.6 Crop Improvement and Seed Registration

Plant breeding aims to improve the characteristics of plants so that they become more desirable agronomically and economically. Generally, a successful variety is one with a total balance of traits that makes it more profitable for the growers and acceptable to the consumers than any other one they might choose.

According to guidelines for the registration and release of new crop varieties in Nigeria, for a crop to be registered and released as a new variety in Nigeria, it must have passed the tests for Distinctiveness, Uniformity and Stability (DUS), as well as for Value for Cultivation and Use (VCU).

1. The Distinctiveness, Uniformity and Stability (DUS) is an acronym for:
    - Distinctiveness: A new variety should be clearly distinguishable from any other existing varieties in Nigeria.
    - Uniformity: Individual plants of the new variety should be sufficiently uniform at the same growth stages in the expression of their characteristics.
    - Stability: The genetic traits of the new variety should be stable through generations.

    The DUS test is an internationally recognized procedure, standardized mainly for plant variety protection. The test is used to assess the novelty of new varieties and is also used to gather information regarding the botanical description of new varieties. In DUS test plots, candidate varieties are compared with similar existing varieties (check varieties), where agronomic characteristics (like yield, maturity date, height), morphological characteristics (like colour, shape and size), physio-genetical characteristics (like tolerance to pests and diseases) and general tolerance to biotic and abiotic stresses are evaluated following the International Union for the Protection of New Varieties of Plants (UPOV) guidelines.

2. Value for Cultivation and Use (VCU): Implies that for a crop to be registered as a new variety, it must be beneficial to farming and industrial communities. The objective of the VCU test for a candidate variety to be registered is to know:
   - The adaptation and stability across varied environments
   - Agronomic performance
   - Reaction to pests and diseases
   - Resistant/tolerant to abiotic stresses when compared with existing varieties
3. The variety must have superior/trait performance over the existing ones with the farmers or currently registered varieties in Nigeria.
4. The variety or its progeny is not detrimental to human or animal health and the safety of the environment when grown and used as intended.
5. The variety name is not likely to be confused with the name of a variety that had previously been or is currently being registered.
6. The variety name is not likely to offend the public.
7. The variety meets the standard for variety purity established by the International Seed Testing Association (ISTA) for a variety of that species.
8. The variety was monitored by representatives of the Registrar and Technical Sub-Committee on crops during the on-farm trial.
9. The variety is officially nominated to the Technical Sub-Committee (Crops) for registration and release and defended by the developing scientist.
10. No false statements or misleading information are submitted in support of the nomination for variety registration and release.

Table 4.1 shows some of the okra varieties that have successfully passed through the above stages and processes, as stated below.

Further efforts on the development, registration and release of new and improved okra varieties are ongoing. Worthy of note are those that are at various stages in the National Horticultural Institute, Ibadan, listed below:

- Okra variety LD88 was also developed in NIHORT by Dr. O. A. Denton (Retired) but has not been officially registered. Efforts towards its registration are ongoing.
- Dr. O. Adetula (NIHORT) also developed and has registered a NHOKRA1 variety recently.
- New materials from Dr. Chris Anyaoha (NHIK7 and NHIK11) are in advanced breeding lines at F5 and F8; NHIK7 is early maturing, while NHIK11 is high yielding.

Both have shown some reasonable field resistance to whitefly and *prodagrica*.

Table 4.1 Okra varieties that have successfully passed through the test for Distinctiveness, Uniformity and Stability (DUS) and for Value for Cultivation and Use (VCU), are stated below

	NHAe 47-4	V2	V35
National code	NGAE-00-03	NGAE-96-1	NGAE-96-2
Variety name	NHAR 47-4	V2	V35
Original name	NHAR 47-4	V2	V35
Outstanding trait	Early maturing, good draw property	Fruit slender, smooth, bell-shaped, high-yielding	High-yielding, bigger fruit
Origin/source	Ilorin	IAR&T, Ibadan	IAR&T, Ibadan
Developing institute	NIHORT	IAR&T, Ibadan	IAR&T, Ibadan
Breeders/ collaborator	Dr. O. A. Denton and Prem Nath	Dr. A. O. Ojomo	Dr. A. O. Ojomo
Year of release	1985	1973	1973
Year of registration	2000	1996	1996

Source: *NSPI* Nigerian Seed Portal Initiative

There are many different types of okra to choose from, though flavour is generally mild (Nan Schiller 2019), but there are other variations to be considered; numerous landraces and cultivars differing in fruit shape, fruit colour, plant architecture, mucilaginous properties and days to maturity, with varying levels of tolerance to biotic and abiotic factors affecting okra production, can be found in Nigeria due to preferential selection over the years by farmers (Anyaoha et al. 2018). Understanding the nature and extent of variation within new germplasm collections is an important prerequisite to identifying superior cultivars for hybridization. The level of vegetable consumption in Nigeria is rising annually owing to the greater appreciation of their food value (Haruna 2003).

4.7 Conclusion

The importance of okra as a vegetable cannot be overemphasized. Nigeria is a major producer of okra, majorly for its fruits and seeds. It has a considerable area under cultivation in Nigeria. Factors affecting okra production in Nigeria differ from one agro-geographical zone to another but are generally responsible for the low yield that is be recorded. Okra varieties have been evaluated for Distinctiveness, Uniformity and Stability (DUS), as well as for Value for Cultivation and Use (VCU), and some have been released and registered with the Nigeria Agricultural Seed Council.

References

Aboyeji CM, Olaniyan D, Dunsin O, Adekoya AO, Okunola FO, Ejue SW, Adebiyi OTV, Olofintoye TAJ, Aremu OC, Afolayan JO, Adeniyi HA (2021) Physiological growth, yield and quality response of okra to sole and combined application of green biomass, poultry manure and inorganic fertilizers. Plant Physiol Rep (INPP). Indian Society for Plant Physiology, INPP-D-21-00223R2. https://doi.org/10.1007/s40502-021-00628-7. Published online on November 9, 2021

Adebayo AS, Azeez FA, Abegunrin OO, Ogunwale OG, Oyewole OO, Ojo FF (2020) Introduction of improved okra (NHe 47-4) variety as a means of economic enhancement of farmers in Ido Village, Ido Local Government Area, Oyo State, Nigeria. J Appl Sci Environ Manage 24(12):2054–2060

Adejonwo KO, Ahmed MK, Lagoke STO, Karikari SK (1989) Effects of variety and period of weed interference on growth and yield of okra (*Abelmosvhus esculentus*). Niger J Weed Sci 21(2):21–28

Adeniji OT, Kehinde OB, Ajala MO, Adebisi MA (2007) Genetic studies on seed yield of West African okra [*Abelmoschus caillei* (A. Chev.) Stevels]. J Trop Agric 45:36–41

Adetuyi FO, Osagie AU, Adekunle AT (2011) Nutrient, antinutrient, minireal and zinc bioavailability of okra *Abelmoschus esculentus* (L.) Monech Variety. Am J Food Nutr 1(2):49–54

Akoroda MO, Ayin OA, Tijani IOA (1985) Edible fruit productivity and harvest duration of okra in southern Nigeria. Trop Agric (Trinidad) 63(20):110–112

Amadi JE, Nnamani C, Ozonkonko CO, Eze CS (2014) Survey of the incidence and severity of okra (*Abelmoschus esculentus* L. Moench) fruit rot in Awka South Iga, Anambra State, Nigeria. Int J Curr Microbiol App Sci 3(4):1114–1121

Anyaoha CO, Orkpeh U, Oguntolu OO, Ikoro JI, Ibitoye DB (2018) Agromorphological characterization of twelve okra (Abelmoschus Esculentus (L.) Moench) genotypes from South-Western Nigeria. Niger Agric J 49(1):247–252. Available online at: http://www.ajol.info/index.php/naj

Babalola TS, Alemoru MO, Lawal JA (2020) Evaluation of okra production among smallholder farmers in Kaba-Bunu area of Kogi State Nigeria. East Afr Sch J Agric Life Sci 3(8):260–264

Damisa MA, Sanni SA, Abdoulaye TA, Kamara Y, Ayanwale A (2010) Household typology based analysis of livelihood strategies and poverty status in the Sudan Savannah of Nigeria: baseline conditions. Learn Publics J Agric Environ Stud 2(1):146–160. ISSN-L: 2026–5654

Dimkpa SON, Tobin-West MD, Baraka RE, Ogbowu AC (2019) Field evaluation of some okra (*Abelmoschus esculentus* L. Moench) varieties in the humid tropics, Rivers State. Global J Agric Res 7(2):21–34

Eke KA, Essien BA, Ogbu JU (2008) Determination of optimum planting time of okra (Abelmoschus esculentus) Cultivars in the derived savannah. In: Proceedings of the 42nd annual conference of agricultural society of Nigeria (ASN). October 19th to 23rd in Ebonyi State University, pp 242–245

Ekunwe PA, Alufohai G, Adolue CF (2018) Economic viability of okra (Abelmoschus esculentus) production in Ika South and North East local government areas of Delta State, Nigeria. Agro-Sci J Trop Agric Food Environ Ext 17(1):57–62. ISSN 1119–7455

FAOSTAT (2020) Food and Agricultural Organization Statistics. https://www.fao.org/faostat/en/#data/QCL. Accessed Aug 2020

Gopalan C, Sastri SBV, Balasubramanian S (2007) Nutritive value of Indian foods @ National Institute of Nutrition (NIN), ICMR, India.

Habtamu FG, Ratta N, Haki GD, Woldegiorgis AZ, Beyene F (2014) Nutritional quality and health benefits of okra (*Abelmoschus esculentus*): a review. J Food Sci Qual Manage 33:87–96

Haruna U (2003) Strategic Options for Profitable Marketing of Fadama Crops. A paper presented at the MTRM, SAPD Headquarters, Bauchi. February 17–18. Pp 8

Hegazi AZ, Hamideldin N (2010) The effect of gamma irradiation on enhancement of growth and seed yield of okra [*Abelmoschus esculentus* (L.) Monech] and associated molecular changes. *Academic*. J Hortic For 2(3):038–051

Ibitoye DO, Kolawole AO (2022) Farmer's appraisal on okra [*Abelmoschus esculentus* (L)] production and phenotypic characterization: a synergetic approach for improvement. Front Plant Sci 13:787577. https://doi.org/10.3389/fpls.2022.787757

Katung MD (2007) Productivity of okra varieties as influenced by seasonal changes in Northern Nigeria. Bot Hort Agrobot Cluj 35(1): Print ISSN 0255–965X; Electronic ISSN 1842–4309

Kumar S, Dagnoko S, Haougui S, Ratnadass A, Pasternak D, Kouame C (2010) Okra (*Abelmoschus spp.*) in West Africa: potential and progress on its improvement. Afr J Agric Sci 5:3590–3598

Liu IM, Liou SS, Lan, TW, Hsu FL, Cheng JT (2005) Myricetin as the active principle of Abelmoschus moschatus to lower plasma glucosein streptozotocin-induced diabetic rats. Planta Medica 71:617–621

Manyong JM, Makinde VKO, Ogunbile AOG (2002) Agricultural transformation and the land use in the cereal-based system of the Northern in Guinea Savannah, Nigeria. IN Integrated Plant Nutrient Management in Sub-Saharan Africa (eds B. Vanlauwe, J. Diels, N. Sanginga and R. Merckx); CAB International 2002

Muhammad M, Kutawa AB, Tadda SA, Muhammad A, Adamu M (2020) Productivity of okra (Abelmoschus esculentus L. Moench) as Influence by NPK Fertilizer and poultry manure in Northern Sudan Savanna Region of Nigeria. International Journal of Forest, Animal and Fisheries Research (IJFAF) Vol-4, Issue-6, ISSN: 2456–8791

Nan Schiller (2019). 13 Okra Varieties to Love. © Ask the Experts, LLC. gardner'spath.com

Osalusi CS, Akanni-John R, Okeke EN, Ogunsola JO (2019) Analysis of the profitability of okra production among small holder okra farmers in Akinyele local government area, Oyo State, Nigeria. Int J Environ Agric Biotechnol (IJEAB) 4(5):1377–1381. https://doi.org/10.22161/ijeab.45.13. ISSN: 2456–1878. www.ijeab.com

Patil P, Sutar S, Joseph JK, Malik S, Rao S, Yadav S, et al (2015) A systematic review of the genus Abelmoschus (Malvaceae). Rheedea 1:14–30

Varmudy V (2011) Marking survey need to boost okra exports. @Department of economics, Vivekananda College, Puttur, Karnataka, India.

Part II

Okra Germplasm Diversity and Disease Resistance Breeding

Genomic Resources for Okra Breeding 5

Roland Schafleitner and Ya-Ping Lin

5.1 Background and Objectives

Molecular markers are versatile tools for many applications in plant science and crop improvement. They support germplasm management through genetic diversity and population structure analyses, and they are also very helpful in verifying species annotation and defining gene pools. Molecular markers are broadly applied in plant breeding for marker-assisted selection of associated traits, fore- and background selection in backcrossing programs, trait mapping, and increasingly also for genomic selection. Single nucleotide polymorphism (SNP) markers have become the most universally applied marker type. Development and use of SNP markers work best when a species has a well-established reference sequence. Based on a reference sequence, SNPs can be called for individuals or groups, and then their physical location can be inferred based on their position in the reference genome to facilitate marker development and functional genetic studies. It is also possible to call SNPs using assembled sequencing read catalogs in the absence of a reference sequence simply by comparing sequencing reads from different individuals. But this process is error-prone. SNP calling in the absence of a reference sequence is susceptible to homologous genomic segments, as highly similar sequences derived from different genomic segments may be assembled into the same catalog, misidentifying true monomorphisms as false positive polymorphisms. This type of error becomes more frequent when genome duplication and polyploidization are present in the genome. Although de novo SNP calling is achievable in polyploid okra, this approach may introduce many false heterozygous sites (Lin et al. 2022).

For most crop species, reference sequences have been produced but were not necessarily assembled to the chromosome level. The assembly of an okra reference

R. Schafleitner (✉) · Y.-P. Lin
World Vegetable Center, Shanhua, Tainan, Taiwan
e-mail: roland.schafleitner@worldveg.org

S. K. Tikoo et al. (eds.), *Okra: Status, Challenges and Opportunities*,
https://doi.org/10.1007/978-981-97-9963-3_5

sequence was only recently published (Nieuwenhuis et al. 2021), and the sequence itself seems not yet to be publicly available. The usefulness of a single reference sequence for marker development in okra may be limited due to the large differences in genome size observed between individual lines of this crop. Salameh (2014) reported genome sizes of okra landraces ranging between 3.897 and 6.539 Gb (average 5 Gb), while the genome size of a single landrace was estimated at as large as 17.3 Gb. Okra chromosome counts were reported to range from $2n = 62$ to $2n = 144$ (Benchasri 2012; Merita et al. 2012). Under these circumstances, presence/absence polymorphisms, which are not captured by reference genome-based SNP calling, may be more informative for genetic studies than SNPs. Here, we propose a cost-effective tool to develop and apply molecular markers for okra, serving for germplasm management and breeding of this polyploid crop.

5.2 Methods

K-mer mapping has been proposed as an alternative approach for molecular breeding without the need for reference sequences (Rahman et al. 2018; Voichek and Weigel 2020). This method is based on directly comparing short sequences (k-mers) among the individuals under investigation to simultaneously detect sequence and presence/absence polymorphisms. Although k-mer mapping was designed for whole genome re-sequencing studies, it can also be applied to selective genotyping studies, such as Restriction enzyme Associated-DNA (RAD) sequencing, significantly reducing the costs for analyses compared to whole genome re-sequencing approaches. Through creating and counting k-mers, presence/absence polymorphisms can be detected without any reference genome sequence. The present/absent k-mer table derived from sequencing reads can be used directly for genetic analyses, including association studies of desired traits. It has been shown that reference-free genome-wide association studies (GWAS) based on k-mers are as effective or superior to SNP-based GWAS (Voichek and Weigel 2020).

5.3 Results

We tested k-mer mapping in association studies using genebank accessions of various vegetable species of the World Vegetable Center genebank. It showed that k-mer-based GWAS with 75,000 k-mers had not only a higher detecting power than SNP-based GWAS with 8,000 SNPs developed from the same sequencing reads as the k-mers, but also identified new loci that were not revealed by SNPs. Mapping the k-mers back to reference sequences enabled the identification of QTL locations. Moreover, when the sequencing coverages are sufficiently high, assembling the significant k-mers could identify new genes that are not present in the reference genome (Rahman et al. 2018; Voichek and Weigel 2020). This preliminary experiment suggested that k-mer mapping could be a cost-effective approach to developing molecular markers for species with complex genomes, such as okra.

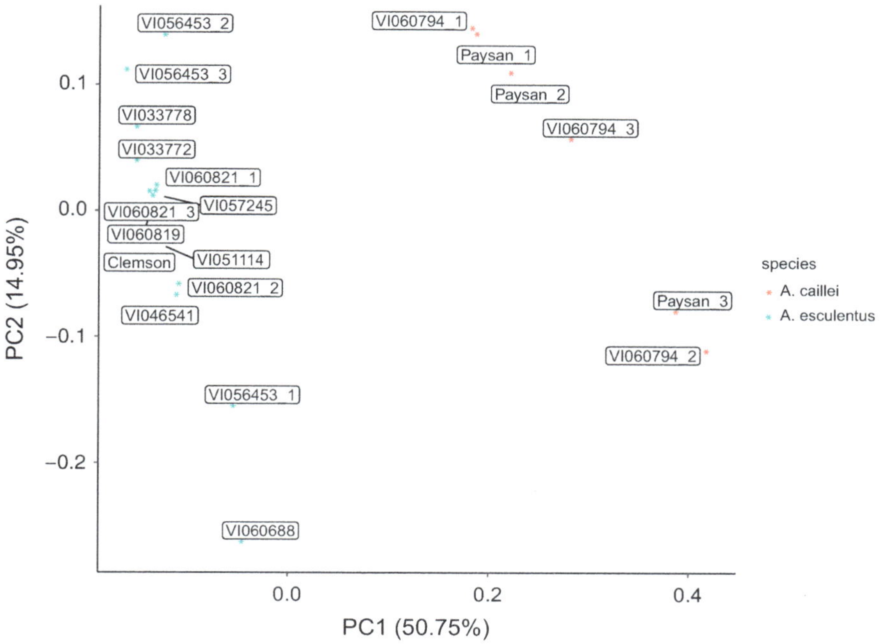

Fig. 5.1 Genetic diversity of 20 okra samples based on 1.4 million k-mers. The two genetic clusters correspond to the taxonomy classification

We performed a pilot experiment using genotyping-by-sequencing reads to capture k-mer variants in 20 okra samples. A total of 12 Gb of sequence information was obtained from an average of six million reads per sample. Based on this sequence information, about 1.4 million k-mers were created and merged as a single genotype table. The genetic relationship between the samples was created by the k-mer variants, resulting in two genetic clusters corresponding to the taxonomy classification according to the passport information (Fig. 5.1).

5.4 Conclusion

Our results imply that k-mer variants are capable of profiling genetic diversity for okra without any reference genome. With more genotypes and phenotyping data, k-mers could be used to generate k-mer–trait associations. Significant k-mers could be mapped to an okra transcriptomics pan-genome derived from several okra accessions to develop molecular markers and characterize the functional genes as well.

Acknowledgments This research was supported by the long-term strategic donors to the World Vegetable Center: Taiwan (Ministry of Agriculture and Ministry of Foreign Affairs), UK aid of the United Kingdom, United States Agency for International Development, Australian Centre for International Agricultural Research, BMZ of Germany, Thailand, the Philippines, Korea, and Japan.

References

Benchasri S (2012) Okra (*Abelmoschus esculentus* (L.) Moench) as a valuable vegetable of the world. Ratar Povrt 49:105–112

Lin YP, Wu TH, Chan YK et al (2022) *De novo* SNP calling reveals the genetic differentiation and morphological divergence in genus *Amaranthus*. Plant Genome 15:e20206

Merita K, Kattekunnel JJ, Yadav SR et al (2012) Chromosome counts in wild and cultivated species of *Abelmoschus* Medikus. From the Indian sub-continent. J Hortic Sci Biotech 87(3):593–599

Nieuwenhuis R, Hesselink T, van den Broeck HC et al (2021) Genome and transcriptome architecture of allopolyploid okra (*Abelmoschus esculentus*). bioRxiv

Rahman A, Hallgrímsdóttir I, Eisen M et al (2018) Association mapping from sequencing reads using k-mers. eLife 7:e32920

Salameh NM (2014) Flow cytometric analysis of nuclear DNA between okra landraces (*Abelmoschus esculentus* L.). Am J Agric Biol Sci 9:245–250

Voichek Y, Weigel D (2020) Identifying genetic variants underlying phenotypic variation in plants without complete genomes. Nat Genet 52:534–540

Review on Virus Diseases of Okra

6

K. Subramanya Sastry, V. Venkataravanappa, B. Mahesha, M. Krishna Reddy, and Lalit Mahatma

6.1 Introduction

Okra is an important vegetable crop belonging to the genus *Abelmoschus* of the family Malvaceae. The crop is grown worldwide, with India leading in both cultivation area and production (FAOSTAT 2014). This multipurpose crop is valued for its tender pods, while its foliage is used for biomass and its dried stems for paper pulp or fuel. However, okra is highly susceptible to over 27 begomoviruses (Singh and Dutta 1986; Swanson and Harrison 1993; Venkataravanappa et al. 2013b; Sanwal et al. 2014), primarily due to overlapping cultivation with other vegetables, weather changes, and the whitefly vector. Notably, Yellow vein mosaic disease (YVMD) and Enation leaf curl Disease (ELCD) are major threats, significantly impacting yield and fruit quality. YVMD, first reported in India (Kulkarni 1924), is caused by the Yellow Vein Mosaic Virus (YVMV) (Uppal et al. 1940). The YVMD-infected plants exhibit symptoms of vein chlorosis, leaf yellowing, reduced fruit size, and stunted growth (Sastry and Singh 1974; Venkataravanappa et al. 2012a).

The Geminiviridae family includes 15 genera with over 525 species (Fiallo-Olivé et al. 2021), among which begomoviruses are the primary plant pathogens causing significant economic losses in many crops worldwide. These viruses have a circular single-stranded DNA genome (2.7–3 kb) within twinned

K. S. Sastry
ICAR-Indian Institute of Millets Research, Rajendranagar, Hyderabad, India

V. Venkataravanappa · B. Mahesha
Division of Crop Protection, ICAR, Indian Institute of Horticultural Research, Bengaluru, India

M. K. Reddy (✉)
Promolecular Sciences Pvt. Ltd., Bengaluru, Karnataka, India

L. Mahatma
Navsari Agricultural University, Navsari, Gujarat, India

S. K. Tikoo et al. (eds.), *Okra: Status, Challenges and Opportunities*,
https://doi.org/10.1007/978-981-97-9963-3_6

para-icosahedral particles (22 × 38 nm). In India, Yellow vein mosaic disease (YVMD) and Okra enation leaf curl disease (OELCuD) are the major begomoviruses affecting okra cultivation. YVMD is linked to mono- and bipartite begomoviruses with DNA satellites, while OELCuD is caused by monopartite begomoviruses, both leading to substantial crop losses (Venkataravanappa et al. 2011, 2012a). These diseases have rapidly spread with the expansion of okra farming in the region (Singh 1990; Biswas et al. 2018). Various begomoviruses infecting okra have been documented globally (Fauquet et al. 2008; Kon et al. 2009; Hernandez-Zepeda et al. 2010). In India, distinctive monopartite begomoviruses (Fauquet et al. 2008; Venkataravanappa et al. 2012b, 2013a, b; Brown et al. 2012) and a novel bipartite species linked to Yellow vein mosaic disease (YVMD) have been identified (Venkataravanappa et al. 2012a). Yield losses due to these viruses can reach up to 50–94%, depending on infection timing (Sastry and Singh 1974; Pun and Doraiswamy 1999; Singh et al. 2013). YVMD is marked by vein yellowing, twisting, stunting, and reduced fruit size (Venkataravanappa et al. 2012a). Okra enation leaf curl disease (OELCuD) presents with pinhead-like enations, leaf curling, stem twisting, vein thickening, and deformed fruits (Fajinmi and Fajinmi 2010; Venkataravanappa et al. 2013b, 2014b) (Table 6.1).

6.2 Oka Yellow Vein Virus and Okra Enation Leaf Curl Virus Diversity

Between 1990 and 2000, several okra varieties were developed to resist OYVMD, but none maintained complete resistance long-term (Usha 2008). Varieties such as Arka Anamika, Parbhani Kranti, Arka Abhay, and P-7 showed resistance breakdown, attributed to new viral strains, mono-cropping practices, and emerging cryptic whitefly species (Sanwal et al. 2014; Venkataravanappa et al. 2011, 2012a, b, 2013a, b, 2014b, 2015a; Jamir et al. 2020). Among okra viruses, Bhendi yellow vein mosaic virus (BYVMV) is the most severe in India and has been reported in several countries, including Sri Lanka, Pakistan, Chad, Côte d'Ivoire, Nigeria, China, Sudan, and Thailand (Kulkarni 1924; Uppal et al. 1940; Capoor and Varma 1950; Zhou et al. 1998; Pun and Doraiswamy 1999; Mansoor et al. 2001; Jose and Usha 2003; Rishi 2009; Xie et al. 2010; Tsai et al. 2013; Sayed et al. 2014; Kumar et al. 2017; Tharmila et al. 2017; Naresh et al. 2019). BYVMV virions are twinned, incomplete icosahedra measuring 22 × 38 nm ($T = 1$) and have a monopartite circular ssDNA genome (DNA-A) of about 2751 nucleotides (FJ515747) (Venkataravanappa et al. 2012a; Brown et al. 2015; Zerbini et al. 2017). These monopartite begomoviruses are usually associated with a betasatellite, which is crucial for okra infection (Jose and Usha 2003). Various BYVMV isolates are listed in Table 6.2.

Table 6.1 Different virus diseases reported at global level infecting okra

Virus species	Taxonomy	Transmission	Symptoms observed on plants	Distribution and reference
Chino del tomate virus (CdTV) (Syn.) Okra yellow mottle Iguala virus (OYMoIgV)	Family: Geminiviridae Genus: *Begomovirus*	Whitefly	Yellow mottle symptoms	Mexico (Diaz-Rivera et al. 2006)
Chickpea chlorotic dwarf virus (CpCDV)	Family: Geminiviridae Genus: *Mastrevirus*	Leafhopper	Leaf curling, yellowing, and vein thickening	Pakistan (Zia-Ur-Rehman et al. 2017)
Cotton leaf curl Gezira virus (CLCuGeV)	Family: Geminiviridae Genus: *Begomovirus*	Whitefly	Leaf veins are dark green and swollen, and severely affected plants become stunted and unproductive	In Burkina Faso, Pakistan, Cameroon, Egypt, Oman, Mali, Sudan, Cote d'Ivoire, and Niger (Kon et al. 2009; Shih et al. 2009; Tiendrebeogo et al. 2010; Leke et al. 2013; Sattar 2012; Akhtar et al. 2014; Idris et al. 2014)
Okra enation leaf curl virus (OELCuV)	Family: Geminiviridae Genus: *Begomovirus*	Whitefly	Enation leaf curl symptoms	India and Pakistan (Singh and Dutta 1986; Chakraborty et al. 1997; Chandran et al. 2013; Venkataravanappa et al. 2012a; Serfraz et al. 2014; Shruti et al. 2016)
Okra mosaic virus (OKMV)	Family: Tymoviridae Genus: *Tymovirus*	Beetle vector, whitefly	Vein chlorosis, vein-banding, mosaic, and stunting	Cote d'Ivoire (Givord et al. 1972; Givord and Hirth 1973); West Africa and Brazil (Igwegbe 1983; Atiri 1984; Fauquet and Thouvenel 1987; Martelli et al. 2002; Stephan et al. 2008; Fajinmi and Fajinmi 2010)

(continued)

Table 6.1 (continued)

Virus species	Taxonomy	Transmission	Symptoms observed on plants	Distribution and reference
Hibiscus chlorotic ring spot virus (HCRSV)	Genus: *Betacarmovirus*		Leaf chlorosis symptoms	Fiji, Papua New Guinea, South Pacific Islands, Solomon Islands, Vanuatu, and the USA (Jones et al. 1998; Brunt and Spence 2000)
Turnip mosaic virus (TuMV)	Family: Potyviridae Genus: *Potyvirus*	Aphid	Chlorosis, vein-clearing, necrosis, and growth reduction	Israel (Gera et al. 2001)
Okra mottle virus (OMoV)	Family: Geminiviridae Genus: *Begomovirus*	Whitefly	Leaf mottling and chlorotic spot symptoms	Brazil (Albuquerque et al. 2013)
Okra leaf curl Oman virus (OLCOV)	Family: Geminiviridae Genus: *Begomovirus*	Whitefly		Oman (Akhtar et al. 2014)
Okra yellow crinkle virus (OYCrV)	Family: Geminiviridae Genus: *Begomovirus*	Whitefly	Plants exhibit foliar yellowing, crinkling, and cupping symptoms	Mali (Shih et al. 2007) and thereafter reported from Cameroon and Cote d'Ivoire (Leke et al. 2013; Seka et al. 2016).
Okra yellow mosaic Mexico virus (OYMV)	Family: Geminiviridae Genus: *Begomovirus*	Whitefly	Irregular yellow patches on leaves and distinctive yellow borders on leaf edges and chlorosis of subsequently developing leaves.	USA and Mexico (De La Torre-Almaraz et al. 2006; Hernandez-Zepeda et al. 2010)
Radish leaf curl virus (RaLCuV)	Family: Geminiviridae Genus: *Begomovirus*	Whitefly	Leaf curling and overall stunting of plants that bear no fruits	India (Kumar et al. 2012)
Sida golden mosaic virus	Family: Geminiviridae Genus: *Begomovirus*	–	Yellow streak, distortion of fruits, and foliar yellow mottle	Africa, Mexico (De La Torre-Almaráz et al. 2003)
Sida micrantha mosaic virus (SiMMV)	Family: Geminiviridae Genus: *Begomovirus*	Whitefly	Leaf mosaic and chlorotic spot symptoms	Brazil (Aranha et al. 2011; Inoue-Nagata et al. 2016)

(continued)

Table 6.1 (continued)

Virus species	Taxonomy	Transmission	Symptoms observed on plants	Distribution and reference
Tobacco streak virus (TSV)	Family: Bromoviridae Genus: *Illarvirus*	Thrips and pollens	Fruits distortion, chlorotic spots, chlorotic leaf blotches, distortion of leaves, chlorotic streaking	India (Krishnareddy et al. 2003; Rajamanickam and Karthikeyan 2014)

Table 6.2 Different virus species of genus *Begomovirus* infecting the okra crop

Name of virus	Place and author	Genome details
Bhendi yellow vein Delhivirus (BYVDV)	Delhi; Venkataravanappa et al. (2012a, 2018)	Bipartite
Bhendi yellow vein Bhubhaneswar virus (BYVBhV)	Bhubhaneswar (India); Venkataravanappa et al. (2013a)	Monoparite
Bhendi yellow vein Haryana virus (BYVHaV)	Haryana, India, Venkataravanappa (2008)	Monoparite
Cotton leaf curl Bangalore virus (CLCuBaV)	India Venkataravanappa et al. (2013b)	Monoparite
Cotton leaf curl Alabad virus (CLCuAV)	North India Venkataravanappa et al. (2012b)	Monoparite
Okra enation leaf curl virus (OELCuV)	Venkataravanappa et al. (2018)	Monoparite

A survey reported the average incidence of yellow vein mosaic disease (YVMD) in various states of India as follows: Karnataka (23–67%), Andhra Pradesh (45–56%), Tamil Nadu (23–75%), Kerala (42–75%), Maharashtra (23–85%), Haryana (24–65%), Uttar Pradesh (35–57%), Delhi (45%), Chandigarh (67%), and Rajasthan (45–66%) (Bhagabati and Goswami 1992; Pun and Doraiswamy 1999; Venkataravanappa 2008; Sindhumole and Manju 2013; Kumar et al. 2017; Nagendran et al. 2017).

In India, an extensive survey identified nine begomoviruses causing yellow vein mosaic disease (YVMD) in okra. Most are known species or strains from the region that affect various crops (Kumar et al. 2012; Venkataravanappa et al. 2012b, 2013b, 2014a, b, 2015a, b, 2018). These viruses often interact with four types of betasatellites across different agro-ecological zones (Venkataravanappa et al. 2011). Novel begomoviruses associated with YVMD include bhendi yellow vein Bhubaneswar virus (BYVBhV) (Venkataravanappa et al. 2013a), bhendi yellow vein Haryana virus (Venkataravanappa et al. 2014b), and a new bipartite species, bhendi yellow vein Delhi virus, alongside BYVMV and ToLCNDV DNA B component (Venkataravanappa et al. 2012a, 2015b), are predominantly associated with YVMD of okra in India (Fig. 6.1).

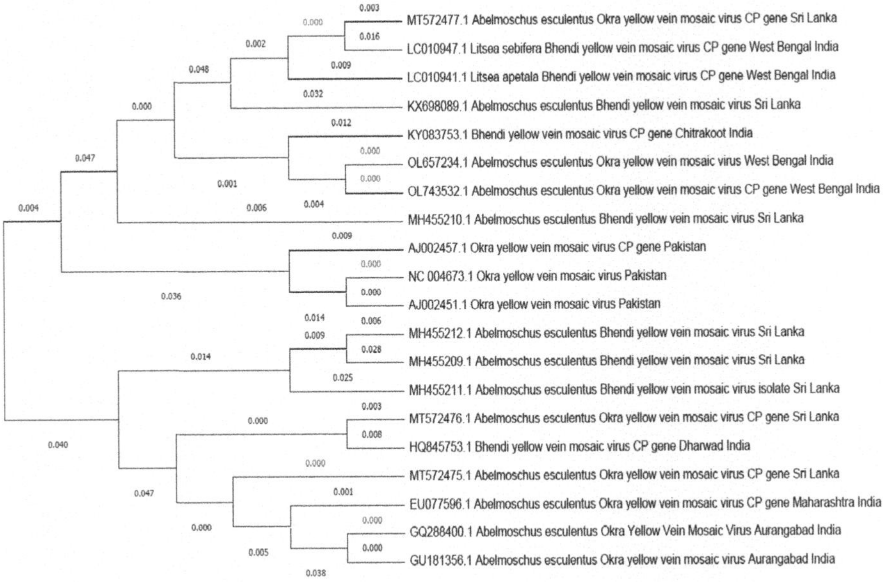

Fig. 6.1 BYVMV Phylogenetic tree

In India, diverse bipartite DNA viruses, including those causing YVMD in okra, have been reported. Koch's postulates have been fulfilled only for BYVMV associated with a betasatellite (Jose and Usha 2003). The DNA-B component of bipartite BYVMV isolates shows high sequence identity with ToLCNDV, a bipartite virus with a broad host range also reported in okra (Ghosh et al. 2008; Zhou 2013; Venkataravanappa et al. 2012a, 2014a, b; Brown et al. 2015; Zerbini et al. 2017). Recent studies have documented the association of an alphasatellite with okra enation leaf curl disease (OELCuD) and mesta yellow vein mosaic virus (MeYVMV) with YVMD in India. Increased variability in BYVMV and OELCuV sequences is linked to a broader geographical range (Naresh et al. 2019). The high-frequency association of OELCuV with OELCuD has been demonstrated (Kumar et al. 2017, 2019), and Koch's postulates were fulfilled using an infectious OELCuV DNA-A component (Emmanuel et al. 2020). Agroinoculation with both BYVMV and BYVMB has shown that BYVMB is crucial for disease manifestation in okra (Jose and Usha 2003).

6.3 Transmission

Begomoviruses are primarily transmitted by the whitefly *Bemisia tabaci*. In India, this vector has different biotypes complex, each spreading different begomoviral diseases in different crops (Venkataravanappa et al. 2017). For effective transmission of BYVMD, at least two MEAM-1 and three Asia-1 biotype whiteflies per plant are required. The minimum acquisition and inoculation access periods are 15 minutes for MEAM-1 and 20 minutes for Asia-1 (Varma 1952). Female whiteflies are more efficient transmitters than males, and MEAM-1 is more effective than Asia-1 (Nath et al. 1993; Gray and Banerjee 1999; Czosnek et al. 2001; Brown et al. 2012; Venkataravanappa et al. 2017).

While some begomoviruses have been found to be seed-transmitted in various plants (Sastry 2013; Kothandaraman et al. 2016; Sangeetha et al. 2018; Suruthi et al. 2018; Renukadevi et al. 2022), seed transmission of okra yellow vein mosaic virus has not been observed (Sisodia and Mahatma 2020; Mahatma et al. 2016, 2021). The potential for seed transmission of begomoviruses remains complex and debated, with many believing they are present but not transmitted through seeds (Mahatma and Pawar 2015; Pérez-Padilla et al. 2020; Patel and Mahatma 2022).

6.4 Virion Properties

The virions of this family are typically twinned, which are presumed to contain a single copy of circular ssDNA, ranging in size from 2.5 to 3.0 kb. For viruses with bipartite genomes, two virions containing different genomic components are required for infection, and half-size defective components and ssDNA satellites are also encapsidated, possibly in isometric virions.

6.5 Genomic Characters

Begomoviruses from the New World have exclusively bipartite genomes, consisting of DNA-A and DNA-B components. In contrast, Old World begomoviruses can be either bipartite or monopartite, the latter having only a DNA-A component (Brown et al. 2012). The BYVMV genome encodes six genes similar to monopartite and Old World bipartite begomoviruses, with two genes in the virion sense and four in the complementary sense, though their functions are not well studied. The monopartite BYVMV associates with a betasatellite (Jose and Usha 2003). Begomoviruses can have either a bipartite configuration (DNA-A and DNA-B) or a monopartite configuration (single DNA-A molecule) (Fondong 2013). Most monopartite begomoviruses are linked with betasatellites, formerly known as DNAβ (Briddon and Stanley 2006; Lozano et al. 2016). BYVMD results from infection by a begomovirus and a betasatellite (Kumar et al. 2017).

In bipartite begomoviruses, the DNA-B component has one open reading frame (ORF) on each strand: BV1 on the sense strand and BC1 on the complementary strand. The intergenic region between coding genes contains cis-acting elements crucial for gene expression, including the conserved nonanucleotide sequence TAATATTAC, essential for initiating rolling circle replication (Fondong 2013).

6.6 Diagnositic Approaches for Okra Viruses

Before developing a management program, it is crucial to confirm the viral cause of a disease. Foliar symptoms may suggest viral infection, but they are not specific enough for an accurate diagnosis. Therefore, diagnostic tests are essential for identifying the exact virus. For viruses like groundnut bud necrosis virus (GBNV) and tobacco streak virus (TSV), serological assays such as ELISA, Dot immunobinding assay (DIBA), and lateral flow immunoassay (LFIA) are effective for detection (Table 6.3). Currently, most begomovirus studies rely on molecular methods. Early detection used dot- and squash-blot hybridization with cloned geminivirus DNAs as probes (Gilbertson et al. 1991). The advent of polymerase chain reaction (PCR) technology, combined with DNA sequencing, has significantly advanced the detection and characterization of begomoviruses in okra (Table 6.3).

6.7 Management of Virus Diseases of Okra

Viral diseases significantly impact okra, a major crop in India, but no commercial viricides are available, unlike fungicides and bactericides (Thresh 2006). Weeds act as reservoirs for both viruses and vectors. The *Bemisia tabaci* biotype complex has introduced various begomoviruses into new plant species, increasing disease incidence in correlation with vector populations (N'guessant et al. 1992; Sanwal et al. 2016). These biotypes, which have a broad host range, have led to geminivirus infections in previously unaffected crops (Singh et al. 2013; Sanwal et al. 2016).

Recent advancements in okra breeding using genomic tools—such as chromosome engineering, RNA interference (RNAi), marker-assisted recurrent selection (MARS), genome-wide selection (GWS), targeted gene replacement, next-generation sequencing (NGS), and nanobiotechnology—have led to the development of viral-resistant okra varieties. However, the application of biotechnology is limited by the lack of polymorphic molecular markers, molecular genetic maps, and the complexity of the polyploid okra genome (Sastry and Zitter 2014; Schafleitner et al. 2013). This review highlights current biotechnological advancements and begomovirus management strategies and emphasizes the need for more research into transgenic varieties using coat protein (CP) genes to confer resistance to Yellow Vein Mosaic Virus (YVMV) and Okra Enation Leaf Curl Virus (OELCuV). Monopartite begomoviruses, which are more common than bipartite types, are often associated with betasatellites (DNA-B, 1350 bp) that are crucial for symptom induction, host range, replication, encapsidation, and transmission (Saunders et al. 2000; Briddon et al. 2004; Briddon and Stanley 2006). YVMD is caused by at least 16 begomoviruses and 4 betasatellite types (Singh et al. 2013).

Some begomoviruses also associate with alphasatellites, small ssDNA molecules (about 1.4 kDa) that encode a replication initiator protein but require a helper virus

Table 6.3 The major RNA and DNA viruses infecting okra and their vector and diagnostic methods

Disease	Group	Transmission	Diagnostic techniques	Reference
Bhendi yellow vein mosaic virus	*Begomovirus*	*Bemisia tabaci*	PCR/ sequencing/ NASH	Kulkarni (1924), Venkataravanappa (2008), and Venkataravanappa et al. (2013a, 2014a)
Bhendi yellow vein mosaic Delhi virus	*Begomovirus*	*Bemisia tabaci*	PCR/ sequencing	Venkataravanappa et al. (2012a)
Bhendi yellow vein Haryana virus	Biparite	*Bemisia tabaci*	PCR/ sequencing	Venkataravanappa et al. (2014a, b)
Bhendi yellow vein Bhubhaneswar virus	*Begomovirus*	*Bemisia tabaci*	PCR/ sequencing	Venkataravanappa et al. (2013a)
Tomato leaf curl New Delhi virus	*Begomovirus*	*Bemisia tabaci*	PCR/ sequencing/ NSH	Venkataravanappa et al. (2018)
Bhendi yellow vein Maharastra virus	*Begomovirus*	*Bemisia tabaci*	PCR/ sequencing	Venkataravanappa et al. (2015a)
Bhendi yellow vein Madurai virus	*Begomovirus*	*Bemisia tabaci*	PCR/ sequencing	Venkataravanappa et al. (2015a)
Okra yellow vein mosaic virus	*Begomovirus*	*Bemisia tabaci*	PCR/ sequencing	Zhou et al. (1998)
Cotton leaf curl Bangalore virus	*Begomovirus*	*Bemisia tabaci*	PCR/ sequencing	Venkataravanappa et al. (2013a)
Okra leaf curl geminivirus	*Begomovirus*	*Bemisia tabaci*	PCR/ sequencing	Swanson and Harrison (1993)
Enation leaf curl of okra	*Begomovirus*	*Bemisia tabaci*	PCR/ sequencing	Singh and Dutta (1986) and Chakraborty et al. (1997)
Mesta yellow vein mosaic virus	*Begomovirus*	*Bemisia tabaci*	PCR/ sequencing	
Radish leaf curl virus	*Begomovirus*	*Bemisia tabaci*	PCR	Kumar et al. (2012)
Okra yellow vein mosaic virus	*Begomovirus*	*Bemisia tabaci*	PCR	Zhou et al. (1998)
Bhendi yellow vein mosaic	*Begomovirus*	*Bemisia tabaci*	PCR	Zhou et al. (1998)

(continued)

Table 6.3 (continued)

Disease	Group	Transmission	Diagnostic techniques	Reference
Okra leaf curl geminivirus	*Begomovirus*	*Bemisia tabaci*	PCR	Swanson and Harrison (1993)
Okra enation leaf curl virus	*Begomovirus*	*Bemisia tabaci*	PCR/NASH	Singh and Dutta (1986), Chakraborty et al. (1997), and Venkataravanappa et al. (2014a, b)
Cotton leaf curl Gezira virus	*Begomovirus*	*Bemisia tabaci*	PCR/ sequencing	Idris and Brown (2002)
Okra yellow crinkle virus	*Begomovirus*	*Bemisia tabaci*	PCR/ sequencing	Kon et al. (2009)
Okra yellow mottle virus	*Begomovirus*	*Bemisia tabaci*	PCR/ sequencing	De La Torre-Almaraz et al. (2006)
Okra yellow mosaic Mexico virus	*Begomovirus*	*Bemisia tabaci*	PCR/ sequencing	Hernandez-Zepeda et al. (2010)
Okra yellow mosaic Igula virus	*Begomovirus*	*Bemisia tabaci*	PCR/ sequencing	Fauquet et al. (2008)
Okra yellow crinkle Mali virus	*Begomovirus*	*Bemisia tabaci*	PCR/ sequencing	Shih et al. (2007)
Sida micrantha mosaic virus	*Begomovirus*	*Bemisia tabaci*	PCR/ sequencing	Aranha et al. (2011)
Okra mottle virus	*Begomovirus*	*Bemisia tabaci*	PCR/ sequencing	Albuquerque et al. (2013)
Okra leaf curl virus	*Begomovirus*	*Bemisia tabaci*	PCR/ sequencing	Ghanem (2003)
Cotton leaf crumple virus disease	*Begomovirus*	*Bemisia tabaci*	PCR/ sequencing	Butler Jr et al. (1985)
Okra leaf curl Cameroon virus	*Begomovirus*	*Bemisia tabaci*	PCR/ sequencing	Sattar (2012)
Okra mosaic virus	*Tymovirus*	Beetle	PCR/ sequencing	Kucharek (2004)
Tobacco streak virus	*Ilarvirus*	Thrips	PCR/ sequencing	Krishnareddy et al. (2003)

for infection and transmission. The role of alphasatellites is not fully understood, though they may help overcome host defenses (Briddon and Stanley 2006; Nawaz-ul-Rehman et al. 2010). An alphasatellite of 1350 nucleotides associated with OELCuV in okra showed 79.7% similarity to a symptomless alphasatellite from hollyhock yellow vein virus (Chandran et al. 2013).

While most begomoviruses from okra are monopartite (Jose and Usha 2003), bipartite types, such as tomato leaf curl New Delhi virus (ToLCNDV) and bhendi yellow vein Delhi virus (BYVDeV), have also been found (Venkataravanappa et al. 2012a). Evidence suggests that bhendi yellow vein mosaic virus (BYVMV) may co-infect with either BYVB or a DNA-B-like sequence of ToLCNDV (Venkataravanappa et al. 2015b).

6.8 Recombination and Evolutionary Divergence of Begomoviruses Infecting Okra

The diversity of begomoviruses in mixed infections promotes recombination and pseudo-recombination, leading to new, impactful begomoviruses affecting okra (Padidam et al. 1999; Seal et al. 2006). Recombination has been crucial in the evolution of geminiviruses, including OELCuV, which derives sequences from CLCuBaV, MeYVMV, and BYVMV (Venkataravanappa et al. 2015a). OELCuD results from OELCuV infection with distinct betasatellites, BYVB and OLCuB (N'guessant 1991; Serfraz et al. 2014).

Recombination studies in okra leaf curl betasatellites and associated betasatellites identified six major and minor events (Sohrab et al. 2015). OYBHUV's recombination suggests its origin from exchanges between CYVMV, BYVMV, CLCuMuV, and MeYVMV (Venkataravanappa et al. 2013b). CLCuBaV, originally identified in cotton, is now characterized as an okra virus (Venkataravanappa et al. 2015a). Koch's postulates for YVMD in okra were confirmed by agro-inoculation with BYVMV and BYVB (Jose and Usha 2003). Sequence comparisons reveal recombinant genome fragments in various begomoviruses (Venkataravanappa et al. 2014a, b).

In southern India, okra exhibits either leaf curl or yellow vein mosaic symptoms, while northern regions show both symptoms on the same plants, possibly due to new viral strains (Sohrab et al. 2013). Screening in hotspot areas is recommended, and further analysis using infectious clones is needed to understand the contributions of individual viral components (Sanwal et al. 2014; Venkataravanappa et al. 2015a) (Fig. 6.2).

6.9 Genetics of Begomovirus Resistance in Okra

Research on the inheritance of resistance to Yellow Vein Mosaic Virus Disease (YVMVD) in okra has been ongoing since 1962, but findings have often been contradictory, likely due to the disease's complexity and environmental influences on vector establishment (Shetty et al. 2013). Among wild okra relatives, *Abelmoschus angulosus* exhibits complete YVMV resistance, controlled by two complementary genes, potentially derived from the wild subspecies *manihot* or the symptomless carrier genotype IC1542 (Dhankhar et al. 2005). Begomoviruses like cotton leaf curl burewala virus (CLCuBaV) and cotton leaf curl multan virus (CLCuMuV), which infect both cotton and okra, have diverse genetic controls in cotton, ranging from two dominant genes to more complex interactions involving up to five factors (Khan et al. 2007; Rahman et al. 2005; Ahuja et al. 2007). Given that research on okra-infecting begomoviruses has mostly focused on BYVMV, future studies should explore the genetic basis of resistance to other begomoviruses in okra, considering environmental conditions and genetic diversity, including wild species. Two *Abelmoschus* species, *A. manihot* (L.) Medik and *A. manihot* (L.) Medik ssp. *manihot*, known for their resistance to Yellow Vein Mosaic Virus (YVMV), were

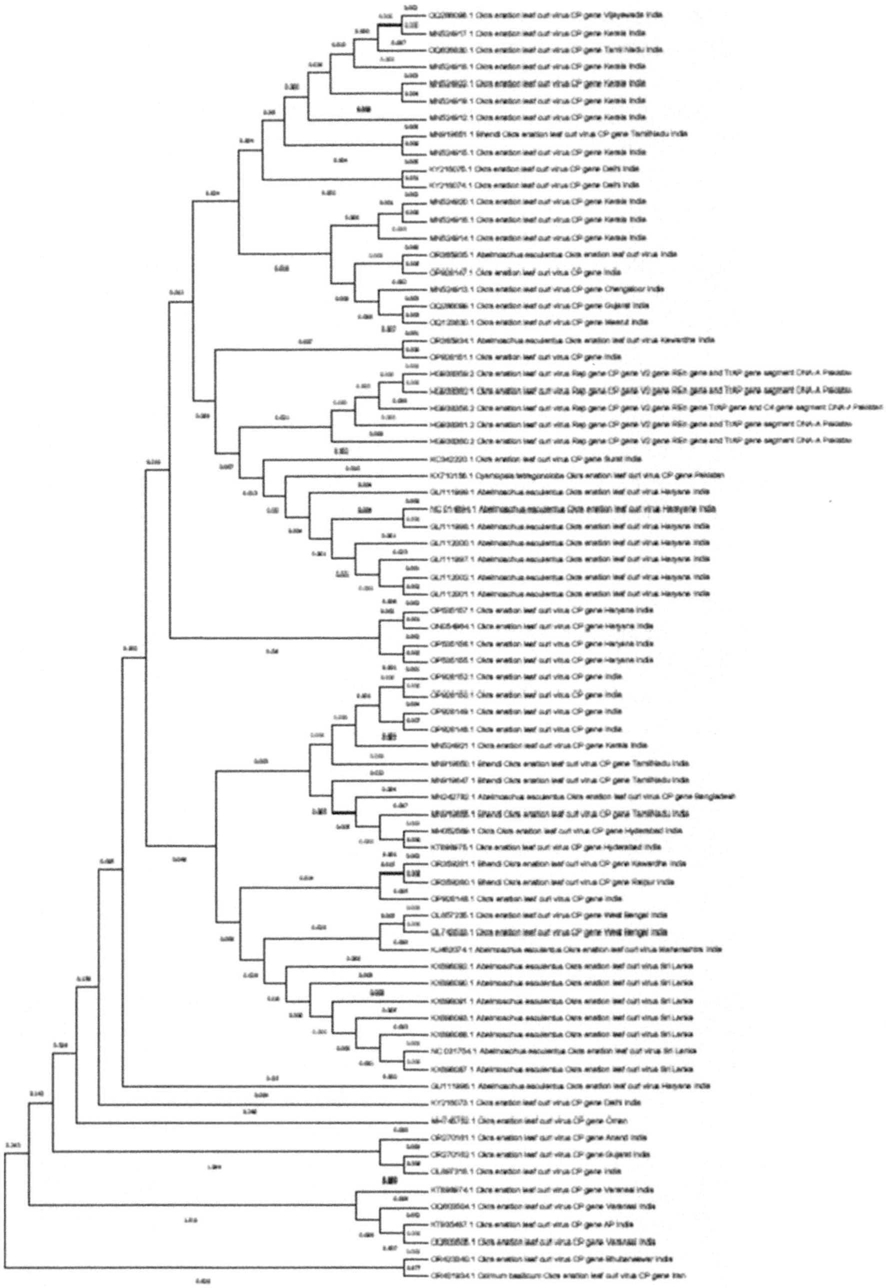

Fig. 6.2 OELCuV Phylogenetic tree

crossed with the susceptible *A. esculentus* cultivar 'Pusa Sawani'. The resulting hybrids displayed resistance to YVMV and showed partial fertility. Segregation patterns in the F2, BC1, and later generations indicated that YVMV resistance in both species is controlled by a single dominant gene (Jambhale and Nerkar 1981).

Narendran et al. (2013) from IVRI (India) developed 21 transgenic okra lines incorporating the *Bacillus thuringiensis* cry1Ac gene. The presence of the transgene was confirmed, and double-antibody sandwich ELISA analysis revealed Cry1Ac protein expression in the transgenic plants. All 21 transgenic lines were phenotypically normal and fertile, enabling the rapid development of transgenic okra varieties with new agronomic traits.

6.10 Whitefly as Vector and Factors Influencing Begomovirus Incidence in Okra

Whiteflies cause heavy damage to okra, both by direct loss of plant vitality through feeding on cell sap and also by transmitting the various viral diseases (Singh 1990). The extent of yield losses due to YVMV depends on the variety used, the stage/age of the plant, and also the virus strain (Sastry and Singh 1974; Pun and Doraiswamy 1999).

6.11 Management of Begomoviruses in Okra

Worldwide okra production faces significant challenges, including the rapid evolution of novel begomovirus strains and the increasing prevalence of viruliferous whitefly vectors (Venkataravanappa et al. 2015a). Both traditional and genomic approaches are being employed to manage these issues, as discussed in this section.

In conventional breeding, enhancing disease resistance relies on identifying appropriate sources of resistance to the various viral strains of YVMD that have emerged globally (Venkataravanappa et al. 2012a). However, no cultivated okra variety or hybrid has demonstrated complete resistance. Potential sources of resistance may be found in wild species like *A. manihot*, *A. crinitus*, *A. angulosus*, and certain landraces of *A. tetraphyllus* (Samarajeewa and Rathnayaka 2004; Singh et al. 2007). However, transferring resistance genes from these wild species to *A. esculentus* is challenging due to botanical barriers and the formation of intermediate forms through natural hybridization, which breeders have only partially exploited (Singh et al. 2007).

Some progress has been made in transferring YVMV resistance genes into cultivated okra, resulting in resistant cultivars such as P7, Parbhani Kranti, and selections 4 and 10 (Jambhale and Nerkar 1981). However, resistance to begomoviruses remains unstable, potentially due to pathogenic variability, the selection of symptomless carrier lines, or the emergence of the polyphagous "B" biotype of *B. tabaci*. Control strategies include reducing whitefly populations through insecticides, physical barriers, or removing symptomatic plants, although whiteflies are developing resistance to insecticides. Consequently, nonconventional breeding methods, including biotechnological tools, are necessary to develop pre-breeding lines resistant to viral stresses (Sanwal et al. 2014).

Das et al. (2013) recommended the hybrids ‘VNR Green’ × ‘Shagun’ and ‘Barsha Laxmi’ × ‘Parbhani Kranti’ for their high yield and low disease index values for YVMV. Vashisht et al. (2001) studied the genetics of YVMV resistance in okra, finding that additive gene effects were more significant than dominance effects. The variety Saloni F1 showed high resistance, Subz Pari was moderately resistant, and Lush Green was moderately susceptible (Irfan Ali et al. 2012). Imidacloprid proved effective in controlling whitefly populations and OYVMV. A significant correlation was also found between environmental factors and disease severity (Irfan Ali et al. 2012).

6.12 Epidemiology

Environmental factors play a key role in the development of plant diseases and the behavior of insect vectors. Lower minimum temperatures favor the development of okra yellow vein mosaic virus (OYVMV), while higher relative humidity negatively affects whitefly populations (Irfan Ali et al. 2012). Hot, dry weather promotes the spread of OYVMV and the multiplication of Bemisia tabaci, whereas cooler, humid conditions and rainfall hinder them (Singh 1990).

Studies have shown that okra sown in May has a lower disease incidence compared to June or August sowings (Nath et al. 1993; Irfan Ali et al. 2012). In North India, June sowing resulted in only 4% YVMV infection, compared to 92.3% in July sowings, with whitefly populations peaking in August (Chattopadhyay et al. 2011). High temperatures and humidity during the rainy season can support disease development, but a drop in temperature later may reduce vector populations and disease spread (Sanwal et al. 2016). Bright sunshine hours positively correlate with YVMV incidence, while lower temperatures and higher humidity negatively impact whitefly populations and disease severity (Ali et al. 2005).

6.13 Management of Vector Population by Insecticides and Oils

Numerous studies recommend using insecticides and oils to control whitefly populations, the primary vectors of okra yellow vein mosaic virus (OYVMV). Sastry and Singh (1973a, b) highlighted monocrotophos, while Irfan Ali et al. (2012) identified Imidacloprid as the most effective insecticide for managing both whitefly populations and OYVMV incidence.

Chemical treatments have shown success in controlling the virus. Gowdar et al. (2007) recommended agrochemicals like Acetamiprid and Trizophos for effective YVMV control, with Acetamiprid 20SP at 40 g a.i/ha significantly reducing disease incidence and boosting okra yield. Alam et al. (2010) found that eco-friendly methods, such as using oils with soap, marigold as a trap crop, and Imidacloprid (Admire) sprays, were effective, with Admire outperforming neem and mustard oils.

Ansar et al. (2014) reported that Imidacloprid 17.8% SL, applied as a spray and seed treatment, reduced pest populations by up to 90.2%, especially when combined with neem oil and a maize border. Azadirachtin spray also significantly reduced whitefly populations by 79.2%.

Plant growth-promoting rhizobacteria (PGPR) offer an eco-friendly alternative by inducing systemic resistance, reducing YVMV incidence by up to 86.6% (Patil et al. 2011). Biswas et al. (2008) found that Crozophera oil, followed by Palmrosa oil, reduced whitefly populations and disease incidence while increasing fruit yield. Chaudhary et al. (2016) demonstrated that neem extract at 5% was the most effective among the tested plant extracts for reducing whitefly populations and OYVMV incidence.

6.14 Management of Virus and Vector Population by Cultural Practices

Roguing, the removal of virus-infected plants early in the growing season (up to 30–45 days after transplanting), is effective in reducing virus spread, especially in environments with low disease incidence. Regular inspections are recommended until flowering, after which roguing should be stopped (Lapidot et al. 2014).

Reflective plastic mulches, like polythene, help disorient insect vectors, reducing disease severity and increasing crop yields. For example, polythene mulch resulted in significantly lower disease severity (7.41%) compared to no mulch (99.98%) (Abdulraheem and Moshood 2019).

To minimize losses, it is important to select fields away from current okra production and avoid planting during peak whitefly populations (Dahal et al. 1992). In northern India, early planted okra fields showed much lower begomovirus incidence (4%) compared to later-planted crops (92.3%) (Chattopadhyay et al. 2011). Monitoring and managing whitefly populations with insecticides throughout the season is crucial.

6.15 Use of Resistant Sources

Developing durable resistance to begomoviruses in okra is challenging due to the high diversity and recombination of these viruses. Novel biotechnological tools like marker-assisted selection and genetic engineering are essential for progress. While tomato has seen significant advances against TYLCV, okra lags behind due to limited genomic resources (Lapidot and Friedmann 2002; Sanwal et al. 2014).

Recombinant DNA technology has enabled strategies like pathogen-derived resistance (PDR) and antisense RNA technology in okra, showing promise for managing begomoviruses. For instance, genetic engineering techniques used to combat okra shoot and fruit borers could be applied to viral diseases like YVMV and ELCV (Narendran et al. 2013).

Okra genome sequencing is complex due to its large genome size and polyploidy, but sequencing its probable diploid ancestors could provide valuable insights. Collaborative efforts are needed to identify and validate markers linked to begomovirus resistance and develop a comprehensive okra linkage map (Sanwal et al. 2014).

Effective, cost-efficient tools for identifying begomovirus strains are crucial. While a nonradioactive DNA probe has been used for geminivirus diagnosis in okra (Venkataravanappa and Reddy 2013), more user-friendly DNA-based diagnostics are needed. Techniques like genotyping-by-sequencing (GBS) could also be applied to okra (Wang et al. 2015).

Biotechnological approaches like RNA interference (RNAi) and CRISPR/Cas9, already successful in other crops, could be adapted for okra to develop durable resistance against multiple begomovirus strains. Combining conventional breeding with these advanced techniques offers the potential to create resistant okra varieties.

6.16 Conclusions

The sustainable management of okra begomoviral diseases worldwide will depend on the use of an integrated pest management (IPM) approach. For a complete understanding of the epidemiology and for developing appropriate control measures, the identification of alternative hosts is an important aspect of the study, which has been somewhat neglected in the past. Presently, the begomovirus reservoirs remain largely unknown. The epidemiology of okra begomovirus diseases, climatic conditions like temperature, wind, rainfall, RH (%), light, sex of whiteflies, and plant age affect the incidence and development of viral diseases.

The integration of emerging technologies such as RNA interference (RNAi), marker-assisted recurrent selection (MARS), genome-wide selection (GWS), and CRISPR/Cas9 offers promising avenues for developing okra varieties with enhanced viral resistance. However, these advancements require increased funding and long-term public sector investment, particularly from India and major okra-producing African countries, to realize their full potential.

Continued interdisciplinary research and development are essential for overcoming the challenges of okra viral disease management. Durable resistance to begomoviruses poses a serious challenge to both breeders and pathologists as these viruses are highly diverse and constantly generate new forms via recombination. Consequently, the use of novel biotechnological tools will help in achieving resistance against begomovirus in okra. Identification of markers linked to the resistance genes and its pyramiding for combining multiple disease resistance genes in various backgrounds should be done.

References

Abdulraheem MI, Moshood AY (2019) Incidence and severity of virus diseases of Okra (Abelmoschus esculentus L. Moench) under different mulching types. Biomed J Sci Tech Res 19(1):13988–13993

Ahuja SL, Monga D, Dhayal LS (2007) Genetics of resistance to Cotton leaf curl disease in *Gossypium hirsutum* L. under field conditions. J Hered 98:79–83

Akhtar S, Khan AJ, Singh AS, Briddon RW (2014) Identification of a disease complex involving a novel monopartite begomovirus with beta and alphasatellites with okra leaf curl disease in Oman. Arch Virol 159:1199–1205

Alam MM, Hoque MZ, Khalequzzaman KM, Humayun MR, Akter R (2010) Eco-friendly management agents of okra yellow vein clearing mosaic virus of okra (Abelmoschus esculentus L. Moench). Bangladesh J Agric 35:11–16

Albuquerque LC, Aranha SA, Fernandes FR, Inoue-Nagata AK (2013) Further evidence reveals that Okra mottle virus arose from a double recombination event. Arch Virol 158:181–186

Ali S, Khan MA, Habib A, Rasheed S, Iftikhar Y (2005) Correlation of environmental conditions with okra yellow vein mosaic virus and Bemisia tabaci population density. Int J Agric Biol 7:142–144

Ansar M, Saha T, Sarkhel S, Bhagat AP (2014) Epidemiology of okra yellow vein mosaic disease and its interaction with insecticide modules. Trends Biosci 7:4157–4160

Aranha SA, Albuquerque LC, Boiteux L, Inoue-Nagata AK (2011) Detection and complete genome characterization of a Begomovirus infecting okra (Abelmoschus esculentus) in Brazil. Trop Plant Pathol 36:14–20

Atiri GI (1984) The occurrence and importance of okra mosaic virus in Nigerian weeds. Ann Appl Biol 104:261–265

Bhagabati KN, Goswami BK (1992) Incidence of yellow vein mosaic disease on okra in relation to white fly population and different sowing time. Indian J Virol 8:37–39

Biswas NK, Nath PS, Srikanta D, De BK (2008) Management of yellow vein mosaic virus disease of okra (Abelmoschus esculentusL. Moench) through different plant oils. Res Crops 9:345–347

Biswas KK, Palchoudhury S, Shukla P, Godara S, Balram N, Bhattacharyya UK, Makeshkumar T (2018) DNA-A sequences of whitefly transmitted begomovirus infecting okra (*Abelmoschus esculentus*) in India are extensively diverse. Indian Phytopathol 71:249–256

Briddon RW, Stanley J (2006) Sub-viral agents associated with plant single-stranded DNA viruses. Virology 344:198–210

Briddon RW, Bull SE, Amin I, Mansoor S, Bedford ID, Rishi N, Siwatch SS, Zafar Y, Abdel-Salam AM, Markham PG (2004) Diversity of DNA 1: a satellite-like molecule associated with monopartite begomovirus-DNA ß complexes. Virology 324(2):462–474

Brown JK, Fauquet CM, Briddon RW, Zerbini M, Moriones E, Navas-Castillo J (2012) Geminiviridae. In: King AMQ, Adams MJ, Carstens EB, Lefkowitz EJ (eds) Virus taxonomy-ninth report of the international committee on taxonomy of viruses. Elsevier Academic Press, San Diego, CA, pp 351–373

Brown JK, Zerbini FM, Navas-Castillo J, Moriones E, Ramos-Sobrinho R, Silva JC, Fiallo-Olivé E, Briddon RW, Hernández-Zepeda C, Idris A, Malathi VG, Martin DP, Rivera-Bustamante R, Ueda S, Varsani A (2015) Revision of Begomovirus taxonomy based on pairwise sequence comparisons. Arch Virol 160:1593–1619

Brunt AA, Spence NJ (2000) The natural occurrence of Hibiscus chlorotic ringspot virus (Carmovirus; Tombusviridae) in aibika or bele (*Abelmoschus manihot*) in some South Pacific Island countries. Plant Pathol 49:798

Butler GD Jr, Wilson FD, Henneberry TJ (1985) Cotton leaf crumple virus disease in okra – leaf and normal – leaf cottons. J Econ Entomol 78(6):1500–1502

Capoor SP, Varma PM (1950) Yellow vein mosaic of *H. esculentus* L. Indian J Agric Sci 20:217–230

Chakraborty S, Pandey PK, Singh B (1997) Okra enation leaf curl disease – a threat to cultivation of okra (*Abelmoschus esculents* L.) Moench. Veg Sci 24:52–54

Chandran SA, Packialakshmi RM, Subhalakshmi K, Prakash C, Poovannan K, Nixon Prabu A, Gopal P, Usha R (2013) First report of an alphasatellite associated with Okra enation leaf curl virus. Virus Genes 46:585–587

Chattopadhyay A, Dutta S, Shatterjee S (2011) Seed yield and quality of okra as influenced by sowing dates. Afr J Biotechnol 28:461–467

Chaudhary A, Khan MA, Yousaf B (2016) Management of okra yellow vein mosaic virus and its vector through plant extracts. J Plant Pathol Microbiol 8:393

Czosnek H, Ghanim M, Morin S, Rubinstein G, Fridman V, Zeidan M (2001) Whiteflies; vectors, and victims (?) of geminiviruses. Adv Virus Res 57:291–322

Dahal G, Neupane FP, Baral DR (1992) Effect of planting and insecticide and spread of yellow vein mosaic of okra in Nepal. Int J Trop Plant Dis 10:109–124

Das S, Chattopadhyay A, Dutta S, Chattopadhyay SB, Hazra P (2013) Breeding okra for higher productivity and yellow vein mosaic tolerance. Int J Veg Sci 19:58–77

De La Torre-Almaráz R, Monsalvo-Reyes AC, Rivera-Bustamante RF, Méndez-Lozano J (2003) First report of a Geminivirus inducing yellow mottle in Okra (*Abelmoschus esculentus*) in Mexico. Plant Dis 87:202

De La Torre-Almaraz R, Monsalvo-Reyes A, Romero-Rodriguez A, Arguello-Astorga GR, Ambriz-Granados S (2006) A new Begomovirus inducing yellow mottle in okra crops in Mexico is related to Sida yellow vein virus. Plant Dis 90:378

Dhankhar SK, Dhankhar BS, Yadava RK (2005) Inheritance of resistance to yellow vein mosaic virus in an interspecific cross of okra (*Abelmoschus esculentus*). Indian J Agr Sci 75:87–89

Diaz-Rivera JB, Zelaya-Molina LX, Silva-Rojas HV, Cardenas E, Nahum Marban-Mendoza N, Valadez-Moctezuma E (2006) Okra yellow mottle caused by Chino del Tomate Virus in Iguala, Mexico. Phytopathology 96(9):1046

Emmanuel CJ, Manohara S, Shaw MW (2020) Molecular characterization of begomovirus-betasatellite-alphasatellite complex associated with okra enation leaf curl disease in Northern Sri Lanka. 3 Biotech 10(12):506

Fajinmi AA, Fajinmi OB (2010) Incidence of Okra mosaic virus at different growth stages of okra plants (*Abelmoschus esculentus* (L.) Moench) under tropical condition. J Gen Mol Virol 2:28–31

FAOSTAT (2014) Available online at: http://faostat.fao.org/. Accessed 8 Feb 2017

Fauquet C, Thouvenel JC (1987) Plant viral diseases in the Ivory Coast. ORSTOM: Institut Francais de Recherche Scientifique pour le developpement en cooperation, collection, initiations-documentations techniques no 46, Paris

Fauquet CM, Briddon RW, Brown JK, Moriones E, Stanley J, Zerbini M, Zhou X (2008) Geminivirus strain demarcation and nomenclature. Arch Virol 153:783–821

Fiallo-Olivé E, Lett JM, Martin DP, Roumagnac P, Varsani A, Zerbini FM, Navas-Castillo J (2021) ICTV virus taxonomy profile: Geminiviridae 2021. J Gen Virol 102(12):001696

Fondong VN (2013) Geminivirus protein structure and function. Mol Plant Pathol 14:635–649

Gera A, Lampel M, Cohen J, Rosner A (2001) Okra (*Hibiscus esculentus*) – a new host of Turnip mosaic virus in Israel. Plant Dis 85:336

Ghanem GAM (2003) Okra leaf curl virus: a monopartite begomovirus infecting okra crop in Saudi Arabia. Arab J Biotechnol 6:139–152

Ghosh R, Paul S, Palit P, Das S, Das A, Acharya S, Mir JI, Ghosk SK, Roy A (2008) Molecular characterization and computational analysis of coat protein gene of an East Indian isolate of Bhendi yellow vein mosaic virus and associated DNA – beta satellite. Indian J Virol 19:173–182

Gilbertson RL, Hidayat SH, Martinez RT, Leong SA, Faria JC, Morales F, Maxwell DP (1991) Differentiation of bean infecting geminiviruses by nucleic acid hybridization probes and aspects of bean golden mosaic in Brazil. Plant Dis 75:336–342

Givord L, Hirth L (1973) Identification, purification and some properties of a mosaic virus of okra (*Hibiscus esculentus*). Ann Appl Biol 74:359–370

Givord L, Pfeiffer P, Hirth L (1972) Un nouveau virus du groupe de la mosaique jaune de navet: Le virus de la mosaique du gombo. CR Hebd Seances Acad Sci Ser D 275:1563–1566

Gowdar SB, Ramesh Babu HN, Aswathanarayana Reddy N (2007) Efficacy of insecticides on okra yellow vein mosaic mirus and whitefly vector, Bemisia tabaci (Guenn.). Ann Plant Prot Sci 15:116–119

Gray SM, Banerjee N (1999) Mechanisms of arthropod transmission of plant and animal viruses. Microbiol Mol Biol Rev 63:128–148

Hernandez-Zepeda C, Isakeit T, Scott A Jr, Farms R, Brown JK (2010) First report of Okra yellow mosaic Mexico virus in okra in the United States. Plant Dis 94:924

Idris AM, Brown JK (2002) Molecular analysis of Cotton leaf curl virus-Sudan reveals an evolutionary history of recombination. Virus Genes 24(3):249–256

Idris A, Al-Saleh M, Amer M, Abdalla O, Brown J (2014) Introduction of Cotton leaf curl Gezira virus into the United Arab Emirates. Plant Dis 98:1593–1593

Igwegbe ECK (1983) New strain of Okra mosaic virus in Nigeria. Plant Dis 67:320–322

Inoue-Nagata AK, Lima MF, Gilbertson RL (2016) A review of geminivirus (Begomovirus) diseases in vegetables and other crops in Brazil: current status and approaches for management. Hortic Bras 34:8–18

Irfan Ali M, Aslam Khan M, Rashid A, Ehetisham-ul-haq M, Talha Javed M, Sajid M (2012) Epidemiology of Okra yellow nein mosaic virus (OYVMV) and its management through tracer, mycotal and imidacloprid. Am J Plant Sci 3:1741–1745

Jambhale ND, Nerkar YS (1981) Inheritance of resistance to okra yellow vein mosaic disease in interspecific crosses of *Abelmoschus*. Theor Appl Genet 60:313–316

Jamir I, Mandal AK, Devi P, Bhattacharjee T, Maurya PK, Dutta S, Chattopadhyay A, Pramanik K, Banik S (2020) Screening of genotypes against viral diseases and assessment of yield loss due to yellow vein mosaic virus in okra grown in the eastern part of India. Indian Phytopathol 73:125–133

Jones P, Devonshire J, Dabek A, Howells C (1998) First report of Hibiscus chlorotic ringspot carmovirus in Tuvalu. Plant Dis 82:591

Jose J, Usha R (2003) Bhendi yellow vein mosaic disease in India is caused by association of a DNA b satellite with a begomovirus. Virology 305:310–317

Khan AI, Hussain M, Rauf S, Khan TM (2007) Inheritance of resistance to Cotton leaf curl virus in cotton (*Gossypium hirsutum* L.). Plant Prot Sci 43:5–9

Kon T, Rojas MR, Abdourhamane IK, Gilbertson RL (2009) Roles and interactions of begomoviruses and satellite DNAs associated with Okra leaf curl disease in Mali, West Africa. J Gen Virol 90:1001–1013

Kothandaraman SV, Devadason A, Ganesan MV (2016) Seed-borne nature of a begomovirus, Mung bean yellow mosaic virus in black gram. Appl Microbial Biotechnol 100:1925–1933

Krishnareddy M, Salil J, Samuel DK (2003) Fruit distortion mosaic disease of okra in India. Plant Dis 87:1395

Kucharek T (2004) Florida plant disease management guide: Okra. Plant Pathology Department document PDMG-V3-41. Florida Cooperative Extension Service, Institute of Food and Agricultural Sciences, University of Florida, Gainesville

Kulkarni CS (1924) Mosaic and other related diseases of crops in the Bombay Presidency. Poona Agric Coll Mag 16:6–12

Kumar J, Kumar A, Singh SP, Roy JK, Lalit A, Parmar D, Sharma NC, Tuli R (2012) First report of Radish leaf curl virus infecting okra in India. New Dis Rep 25:9

Kumar RV, Prasanna HC, Singh AK, Ragunathan D, Garg GK, Chakraborty S (2017) Molecular genetic analysis and evolution of begomoviruses and betasatellites causing yellow mosaic disease of bhendi. Virus Genes 53:275–285

Kumar R, Esakky R, Palicherla SR (2019) The emergence of okra enation leaf curl virus—an important begomovirus infecting okra in several states across India. Arch Phytopathol Plant Protect 52:234–238

Lapidot M, Friedmann M (2002) Breeding for resistance to whitefly-transmitted geminiviruses. Ann Appl Biol 140:109–127

Lapidot M, Legg JP, Wintermantel WM, Polston JE (2014) Management of whitefly-transmitted viruses in open-field production systems. Adv Virus Res 90:147–206

Leke WN, Sattar MN, Ngane EB, Ngeve JM, Kvarnheden A, Brown JK (2013) Molecular characterization of begomoviruses and DNA satellites associated with Okra leaf curl disease in Cameroon. Virus Res 174:116–125

Lozano G, Trenado HP, Olivé EF, Chirinos D, Pouey FG, Briddon RW, Castillo JN (2016) Characterization of non-coding DNA satellites associated with sweepoviruses (genus Begomovirus, Geminiviridae) – definition of a distinct class of begomovirus associated satellites. Front Microbiol 7:162

Mahatma L, Pawar DM (2015) Role of seed in the epidemiology of yellow mosaic virus disease in mungbean (Vigna radiata L.). J Mycol Plant Pathol 45(4):324–329

Mahatma L, Mahatma MK, Pandya JR, Solanki RK, Solanki VA (2016) Epidemiology of begomoviruses: a global perspective. In: Plant viruses: Evolution and management. Springer, Singapore, pp 171–188. https://doi.org/10.1007/978-981-10-1406-2_10

Mahatma L, Pawar D, Patel V, Patel H, Patel N, Patel R, Mahatma M (2021) Insight in seed transmission of Begomoviruses. Acad Lett:2303. https://doi.org/10.20935/AL2303

Mansoor S, Amin I, Hussain M, Zafar Y, Bull S, Briddon RW, Markham PG (2001) Association of a disease complex involving a Begomovirus, DNA 1 and a distinct DNA ß with leaf curl disease of okra in Pakistan. Plant Dis 85:922

Martelli GP, Sabanadzovic S, Abou-Ghanem Sabanadzovic N, Edwards MC, Dreher T (2002) The family Tymoviridae. Arch Virol 147:1837–1846

N'guessant KP (1991) Occurrence and spread of okra leaf curl virus (OLCD) disease in Côte d'Ivoire. Agron Afr 13:35–43

N'guessant KP, Fargettet D, Fauquet C, Thouvenel JC (1992) Aspects of the epidemiology of okra leaf curl virus in Côte d'Ivoire. Trop Pest Manag 38:122–126

Nagendran K, Mohankumar S, Aravintharai R, Balaji CG, Manoranjitham SK, Singh AK, Rai AB, Singh B, Karthikeyan G (2017) The occurrence and distribution of major viruses infecting cucurbits in Tamil Nadu state, India. Crop Prot 99:10–16

Narendran M, Deole SG, Harkude S, Shirale D, Nanote A, Bihani P, Parimi S, Char BR, Zehr UB (2013) Efficient genetic transformation of okra (*Abelmoschus esculentus* (L.) Moench) and generation of insect-resistant transgenic plants expressing the *cry1Ac* gene. Plant Cell Rep 32:1191–1198

Naresh M, Khan ZA, Kumar R, Kale SP, Patil VM, Rajput JC, Dasgupta I (2019) Occurrence and variability of begomoviruses associated with bhendi yellow vein mosaic and okra enation leaf curl diseases in south-western India. Virus Dis 30:511–525

Nath PD, Gupta MK, Bora P (1993) Influence of sowing time on the incidence of yellow vein mosaic and whitefly (*Bemisia tabaci*) population on okra. Indian J Virol 8:45–48

Nawaz-ul-Rehman MS, Nahid N, Mansoor S, Briddon RW, Fauquet CM (2010) Post-transcriptional gene silencing suppressor activity of the alpha-rep of non-pathogenic alphasatellites associated with begomoviruses. Virology 405:300–308

Padidam M, Sawyer S, Fauquet CM (1999) Possible emergence of new geminiviruses by frequent recombination. Virology 265:218–225

Patel HP, Mahatma L (2022) Disease rating scale for the resistant screening against begomoviruses causing yellow mosaic disease in different crops. In: Geminivirus: detection, diagnosis and management. Academic Press, Cambridge, MA, pp 561–574

Patil MN, Jagadeesh KS, Krishnaraj PU, Patil MS, Vastrad AS (2011) Plant growth promoting rhizobacteria (PGPR) mediated protection in bhindi against yellow vein mosaic virus. Indian J Plant Prot 39:48–53

Pérez-Padilla V, Fortes IM, Romero-Rodríguez B, Arroyo-Mateos M, Castillo AG, Moyano C, De León L, Moriones E (2020) Revisiting seed transmission of the type strain of Tomato yellow leaf curl virus in tomato plants. Phytopathology 110(1):121–129

Pun KB, Doraiswamy S (1999) Effect of age of okra plants on susceptibility to Okra yellow vein mosaic virus. Indian J Virol 15:57–58

Rahman M, Hussain D, Malik TA, Zafar Y (2005) Genetics of resistance to cotton leaf curl disease in *Gossypium hirsutum*. Plant Pathol 54:764–772

Rajamanickam S, Karthikeyan G (2014) Sequence diversity analysis of Tobacco streak virus infecting okra (Abelmoschus esculentus L.) in India. J Pure Appl Microbiol 8(6):4759–4767
Renukadevi P, Sangeetha B, Malathi VG, Nakkeeran S, Satya VK (2022) Enigmatic emergence of seed transmission of geminiviruses. In: Geminivirus: detection, diagnosis and management. Academic Press, Cambridge, MA, pp 285–306
Rishi N (2009) Significant plant virus diseases in India and a glimpse of modern disease management technology. J Gen Plant Pathol 75:1–18
Samarajeewa PK, Rathnayaka RMUSK (2004) Disease resistance and genetic variation of wild relatives of okra (*Abelmoschus esculentus* L.). Ann Sri Lanka Dep Agric 6:167–176
Sangeetha B, Malathi V, Alice D, Suganthy M, Renukadevi PA (2018) A distinct seed-transmissible strain of tomato leaf curl New Delhi virus infecting Chayote in India. Virus Res 258:81–91
Sanwal SK, Singh M, Singh B, Naik PS (2014) Resistance to yellow vein mosaic virus and okra enation leaf curl virus: challenges and future strategies. Curr Sci 106:470–1471
Sanwal SK, Venkataravanappa V, Singh B (2016) Resistance to bhendi yellow vein mosaic disease: a review. Indian J Agr Sci 86:835–843
Sastry KS (2013) Seed-borne plant virus diseases. Springer Publishers, p 327
Sastry KSM, Singh SJ (1973a) Field evaluation of insecticides for control of whitefly Bemisia tabaci in relation to the incidence of yellow vein mosaic of okra (*Abelmoschus esculentus*). Indian Phytopathol 26:129–138
Sastry KSM, Singh SJ (1973b) Restriction of yellow vein mosaic virus spread in okra through the control of vector, whitefly Bemisia tabaci. Indian J Mycol Plant Pathol 3:76–80
Sastry KSM, Singh SJ (1974) Effect of yellow vein mosaic virus infection on growth and yield of okra crop. Indian Phytopathol 27:294–297
Sattar MN (2012) Diversity and interactions of begomoviruses and their associated DNA-satellites. Doctoral thesis. Swedish University of Agricultural Sciences, Uppsala, p 74
Saunders K, Bedford ID, Briddon RW, Markham PG, Wong SM, Stanley J (2000) A unique virus complex causes Ageratum yellow vein disease. Proc Natl Acad Sci USA 97:6890–6895
Sayed SS, Rana D, Krishna G, Reddy PS, Bhattacharya PS (2014) Association of *Begomovirus* with okra (*Abelmoschus esculentus L.*) leaf curl virus disease in southern India. SAJ Biotechnol 1:102
Schafleitner R, Kumar S, Lin CY, Hegde SG, Ebert A (2013) The okra (*Abelmoschus esculentus*) transcriptome as a source for gene sequence information and molecular markers for diversity analysis. Gene 517:27–36
Seal SE, Bosch FVD, Jeger MJ (2006) Factors influencing begomovirus evolution and their increasing global significance: implications for sustainable control. Crit Rev Plant Sci 25:23–46
Seka K, Ouattara A, Assiri KP, Kra KD, Hoareau M, Lefeuvre P, Atta Diallo H, Lett JM (2016) First reports of Cotton leaf curl Gezira virus and Okra yellow crinkle virus associated with Okra leaf curl disease in Cote d'Ivoire. New Dis Rep 34:8
Serfraz S, Amin I, Akhtar KP, Mansoor S (2014) Recombination among begomoviruses on malvaceous plants leads to the evolution of okra enation leaf curl virus in Pakistan. J Phytopathol 163:764–776
Shetty AA, Singh JP, Singh D (2013) Resistance to yellow vein mosaic virus in okra: a review. Biol Agric Hortic 29:159–164
Shih SL, Green SK, Tsai WS, Lee LM, Levasseur V (2007) First report of a distinct begomovirus associated with okra yellow crinkle disease in Mali. Plant Pathol 56:718
Shih SL, Kumar S, Tsai WS, Lee LM, Green SK (2009) Complete nucleotide sequences of okra isolates of Cotton leaf curl Gezira virus and their associated DNA-beta from Niger. Arch Virol 154:369–372
Shruti BAS, Nargund VB, Dharmatti PR (2016) Survey, symptomatology and molecular characterization of Okra enation leaf curl virus (OELCuV) in North Karnataka. In: International conference (VIROCON 2016), Bangalore, p 143
Sindhumole P, Manju P (2013) Association of okra (*Abelomoschus esculentus* (L.) Moench) yellow vein mosaic virus incidence with population of its vectors under Kerala conditions. Entomon 38:131–138

Singh SJ (1990) Etiology and epidemiology of whitefly-transmitted virus diseases of okra in India. Plant Dis Res 5:64–70

Singh SJ, Dutta OP (1986) Enation leaf curl of okra – a new virus disease. Indian J Virol 2:114–117

Singh B, Rai M, Kalloo G, Satpathy S, Pandey KK (2007) Wild taxa of okra (*Abelmoschus* species): reservoir of genes for resistance to biotic stresses. Acta Hortic 752:323–328

Singh B, Sanwal S, Venkataravanappa V, Halder J (2013) Breeding strategies for biotic stresses of okra: prospects and potential. In: Abstract book of national symposium on abiotic and biotic stress management in vegetable crops, Varanasi, pp 32–33

Sisodia P, Mahatma L (2020) Detection of Bhendi yellow vein mosaic virus (BYVMV) from the different parts of bhendi [*Abelmoschus esculentus* (L.) Moench] plant, flower and seed. Int J Curr Microbiol App Sci 9:389–395

Sohrab SS, Mirza Z, Karim S, Rana D, Abuzenadah AM, Chaudhary AG, Bhattacharya PS (2013) Detection of begomovirus associated with okra leaf curl disease. Arch Phytopathol Plant Protect 46:1047–1053

Sohrab SS, Yasir M, El-Kafrawy SA, Mirza Z, AlZahrani HSM, Bhattacharya P, Rana D (2015) Recombination analysis and in silico structural characterization of ßC1 protein gene from Okra leaf curl betasatellite. Plant Omics J 8:565–571

Stephan D, Siddiqua M, Hoang AT, Engelmann J, Winter S, Maiss E (2008) Complete nucleotide sequence and experimental host range of Okra mosaic virus. Virus Genes 36:231–240

Suruthi V, Nakkeeran S, Renukadevi P, Malathi VG, Rajasree V (2018) Evidence of seed transmission of dolichos yellow mosaic virus, a begomovirus infecting lablab-bean in India. Virus Dis 29:506–512

Swanson MM, Harrison BD (1993) Serological relationships and epitope profiles of isolates of okra leaf curl geminivirus from Africa and the Middle East. Biochimie 75:707–711

Tharmila CJ, Jeyaseelan EC, Ihsan U, Wetten AC, De Costa DM, Shaw MW (2017) First report on association of Okra yellow vein mosaic virus with yellow vein mosaic disease of okra (*Abelmoschus esculentus*) in Sri Lanka. Plant Dis 101:1335

Thresh JM (2006) Control of tropical plant virus diseases. Adv Virus Res 67:245–295

Tiendrebeogo F, Lefeuvre P, Hoareau M, Villemot J, Konaté G, Traoré AS, Barro N, Traoré VS, Reynaud B, Traoré O, Lett JM (2010) Molecular diversity of Cotton leaf curl Gezira virus isolates and their satellite DNAs associated with okra leaf curl disease in Burkina Faso. Virol J 7:48

Tsai WS, Shih SL, Lee LM, Wang JT, Duangsong U, Kenyon L (2013) First report of Bhendi yellow vein mosaic virus associated with yellow vein mosaic of okra (*Abelmoschus esculentus*) in Thailand. Plant Dis 97:291

Uppal BN, Varma PM, Capoor SP (1940) Yellow vein mosaic of bhendi. Curr Sci 9:227–228

Usha R (2008) Bhendi yellow vein mosaic virus. In: Rao GP, Kumar PL, Holguin-Pena RJ (eds) Characterization, diagnosis and management of plant viruses, vol 3. Studium Press, Houston, pp 387–392

Varma PM (1952) Studies on the relationship of the Bhendi yellow vein mosaic virus and its vector, the whitefly (*Bemisia tabaci* Gen.). Indian J Agric Sci 22:75–91

Vashisht V, Sharma B, Dhillon G (2001) Genetics of resistance to yellow vein mosaic virus in Okra. Crop Improv 28:218–225

Venkataravanappa V (2008) Molecular characterization of okra yellow vein mosaic virus. Ph.D. thesis. GKVK, Bengaluru

Venkataravanappa V, Reddy MK (2013) Begomovirus characterization, and development of phenotypic and dna-based diagnostics for screening of okra genotype resistance against bhendi yellow vein mosaic virus. Biotech 3:461–470

Venkataravanappa V, Lakshminarayana Reddy CN, Swaranalatha P, Jalali S, Briddon R, Krishna Reddy M (2011) Diversity and phylogeography of begomovirus-associated beta satellites of okra in India. Virol J 8:555

Venkataravanappa V, Lakshminarayana Reddy CN, Jalali S, Krishna Reddy M (2012a) Molecular characterization of distinct bipartite Begomovirus infecting bhendi (Abelmoschus esculentus L.) in India. Virus Genes 44:522–535

Venkataravanappa V, Lakshminarayana Reddy CN, Swarnalatha P, Devaraju JS, Krishna Reddy M (2012b) Molecular evidence for association of Cotton leaf curl Alabad virus with yellow vein mosaic disease of okra in North India. Arch Phytopathol Plant Protect 45:2095–2113

Venkataravanappa V, Lakshminarayana Reddy CN, Jalali S, Krishna Reddy M (2013a) Molecular characterization of a new species of Begomovirus associated with yellow vein mosaic of bhendi (Okra) in Bhubhaneswar, India. Eur J Plant Pathol 136:811–822

Venkataravanappa V, Lakshminarayana Reddy CN, Devaraju A, Jalali S, Krishna Reddy M (2013b) Association of a recombinant Cotton leaf curl Bangalore virus with yellow vein and leaf curl disease of okra in India. Indian J Virol 24:188–198

Venkataravanappa V, Lakshminarayana RC, Jalali S, Krishna RM (2014a) Association of Tomato leaf curl New Delhi virus DNA-B with bhendi yellow vein mosaic virus in okra showing yellow vein mosaic disease symptoms. Acta Virol 59:125–139

Venkataravanappa V, Prasanna HC, Reddy CNL, Reddy MK (2014b) Evidence for two predominant viral lineages, recombination and subpopulation structure in begomoviruses associated with yellow vein mosaic disease of okra in India. Plant Pathol 64:508–518

Venkataravanappa V, Lakshminarayana Reddy CN, Jalali S, Briddon RW, Krishna Reddy M (2015a) Molecular identification and biological characterisation of a Begomovirus associated with okra enation leaf curl disease in India. Eur J Plant Pathol 141:217–235

Venkataravanappa V, Lakshminarayana Reddy CN, Jalali S, Krishna Reddy M (2015b) Association of tomato leaf curl New Delhi virus DNA-B with bhendi yellow vein mosaic virus in okra showing yellow vein mosaic disease symptoms. Acta Virol 59:125–139

Venkataravanappa V, Kodandaram MH, Reddy CNL, Shankarappa KS, Krishna Reddy M (2017) Comparative transmission of Bhendi yellow vein mosaic virus by two cryptic species of the whitefly, Bemisia tabaci (Hemiptera: Aleyrodidae). Biotech 7:331

Venkataravanappa V, Lakshminarayana Reddy CN, Saha S, Krishna Reddy M (2018) Recombitant tomato leaf curl New Delhi virus is associated with with yellow vein mosaic disease of okra in India. Physiol Mol Plant Pathol 104:108–118

Wang Y, Ning Z, Hu Y, Chen J, Zhao R, Chen H, Ai N, Guo W, Zhang T (2015) Molecular mapping of restriction-site associated DNA markers in allotetraploid upland cotton. PLoS One 10(4):e0124781

Wilson FD, Brown JK (1991) Inheritance of response to cotton leaf crumple virus infection in cotton. J Hered 82:508–509

Xie K, Cai JH, Hu DM, Wei X, Jia Q, Qin BX, Chen BS, Meng JR, Liu YL (2010) First report of Okra leaf curl disease in China. J Plant Pathol 92:S409

Zerbini FM, Briddon RW, Idris A, Martin DP, Moriones E, Navas-Castillo J, Rivera-Bustamante R, Roumagnac P, Varsani A, ICTV Report Consortium (2017) ICTV virus taxonomy profile: Geminiviridae. J Gen Virol 98:131–133

Zhou X (2013) Advances in understanding Begomovirus satellites. Annu Rev Phytopathol 51:357–381

Zhou X, Liu Y, Robinson DJ, Harrison BD (1998) Four variants among Pakistani isolates of cotton leaf curl virus and their affinities to DNA-A of geminivirus isolates from Okra. J Gen Virol 79:915–923

Zia-Ur-Rehman M, Hameed U, Ali CA, Haider MS, Brown JK (2017) First report of Chickpea chlorotic dwarf virus infecting okra in Pakistan. Plant Dis 101:1336

7 Okra Breeding Overview in Nigeria and Challenges Ahead

Dorcas Olubunmi Ibitoye

7.1 Introduction

The widely cultivated okra, *Abelmoschus esculentus* (L.) Moench, belongs to the Malvaceae (mallow) family, and it is an important autogamous species grown in the tropical, subtropical, and warmer portions of the temperate region of the world (Patil et al. 2015; Ibitoye and Kolawole 2022). Petropoulos et al. (2018) described okra fruits as sources of primary and secondary metabolites, which are rich in dietary fiber, carbohydrates, vitamins (A, B, and C), and minerals (sodium, potassium, magnesium, calcium, zinc, iron, manganese, and nickel). The seeds contain oil (13–22%) and protein (20–24%) with balanced amounts of lysine and tryptophan (Hughes 2009). Thus, okra is a multipurpose fruit vegetable, highly nutritive and medicinal, playing a substantial role in averting malnutrition and alleviating food insecurity.

Global okra production is about 10.6 million metric tons, with India and Nigeria being the first and second largest producers, contributing about 60% and 17% of the total global production, respectively. In addition, Nigeria contributes significant role in okra production in Africa as the largest producer, contributing about 54% of the total production on the continent (FAOSTAT 2020). In spite of the high production quantity recorded, okra production in Nigeria has been limited with lots of complex interactions of biotic and abiotic stresses, which have contributed to the decline in fruit yield over the years.

Okra is one of the mandate crops of NIHORT, Nigeria, but it still remains an understudied crop in terms of genetic improvement in Nigeria. Developing an efficient product profile that will meet the demands of different okra stakeholders along

D. O. Ibitoye (✉)
Genetic Resources Unit, National Horticultural Research Institute (NIHORT), Ibadan, Nigeria
e-mail: Ibitoye.Dorcas@nihort.gov.ng

S. K. Tikoo et al. (eds.), *Okra: Status, Challenges and Opportunities*,
https://doi.org/10.1007/978-981-97-9963-3_7

the value chain and increase productivity in Nigeria requires a systematic demand-breeding approach that will increase the confidence of the stakeholders in the breeding program so that the final product will be readily adopted even before official release and registration. NIHORT has commenced efforts in okra breeding, putting into use the product profile approach, which integrates and considers the needs of all actors along the value chain in our breeding program.

Among the challenges facing the okra breeding program at NIHORT include the unavailability of modern breeding technologies, such as molecular marker techniques and genomics, which would enhance the efficiency of our breeding program. The institute is open to collaboration with public and private organizations with respect to germplasm sharing and regional/international varietal trials to make our breeding program more robust and facilitate the availability of improved okra varieties.

7.2 Okra Breeding Efforts in NIHORT

(i) *Participatory Rural Appraisal Study*

Past efforts between 1973 and 1985 had led to the release and registration of three (3) okra and one (1) released in 2016. These were selections from available landraces and introduced germplasm. Concerted breeding efforts, however, commenced in 2019 with an appraisal study in locations with high okra production in order to develop improved okra varieties that meet stakeholders' demands, following a pipeline that will lead to well-adopted varieties with preferred traits for stakeholders (Fig. 7.1). Preferences differ across value chain actors. High yield (100%) was scored as a very important trait by farmers, fol-

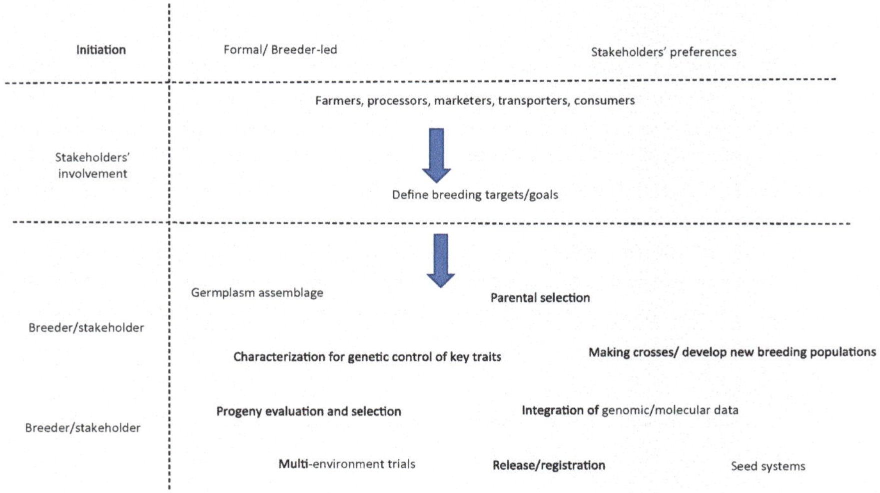

Fig. 7.1 Okra breeding pipeline framework in NIHORT, Nigeria

Table 7.1 Trait preference in okra by farmers

Traits	Very important	Important	Not important
High yield	120(100)	–	–
High viscosity	70(58.3)	41(31.2)	9(7.5)
Resistance to pests and diseases	108(90)	11(9.2)	1(0.8)
Earliness	99(82.5)	15(12.5)	6(5)
Spineless	74(61.7)	36(30)	10(8.3)
High market acceptability	102(85)	15(12.5)	3(2.5)
Drought tolerance	89(74.2)	18(15)	13(10.8)
Large pods	65(54.2)	29(24.2)	26(21.7)

[a]*N* = 120, percentages in parentheses. Table showing preferences of respondents in novel okra varieties

Table 7.2 Trait preference in okra by consumers and marketers in surveyed locations

Traits	Sobi Barracks	Ogundele	Aigoro	Balla	Yeregi	Mean derived rank score
High viscosity	1	1	2	2	1	1.4
Spineless	2	2	1	1	2	1.6
Large pods	3	3	3	3	3	3
High yield	5	5	5	4	4	4.6
High market acceptability	7	6	4	5	6	5.6
Earliness	4	4	8	6	8	6
Resistance to pests and diseases	6	8	6	7	5	6.4
Drought tolerance	8	7	7	8	7	7.4

[a]*N* = 120, percentages in parentheses. Table showing preferences of respondents in novel okra varieties

lowed by resistance to pests and diseases (Table 7.1), while high viscosity followed by spineless were ranked as the first and second preferred traits by consumers and marketers, respectively (Table 7.2). Diseases (75%) and insect pests (74%) were ranked first and second, respectively, as production constraints with great extent of damage (Table 7.3).

(ii) *Germplasm Assemblage and Characterizations*

Germplasm assemblage and characterization of okra followed the appraisal study. This was done in order to identify okra varieties with potential sources of genes for the identified trait preferences across the okra value chain actors. Over 100 different okra accessions were collected across different growing regions in Nigeria. Seed regeneration and characterization followed the acquisition. Controlled pollination (Fig. 7.2) was carried out on each accession in order to maintain the genetic integrity within and between accessions, respectively. Observed morphological and agronomic diversities among the accessions were recorded and documented for further use (Fig. 7.3).

Table 7.3 Production constraints as ranked during focus group discussions across communities

Production constraints	[a]Great extent	Some extent	Little extent
Lack of improved varieties	80(66.7)	39(32.5)	1(0.8)
Drought	85(70.8)	17(14.2)	18(15)
Diseases	90(75)	21(17.5)	9(7.5)
Insect pests	89(74.2)	19(15.8)	12(10)
Cost of land rent/lease	58(48.3)	17(14.2)	45(37.5)
Gender with respect to land ownership	3(2.5)	14(11.7)	103(85.8)
Poor soil fertility	63(52.5)	25(20.8)	32(26.7)

[a]$N = 120$, percentages in parentheses

Fig. 7.2 Okra seed regeneration plot with floral isolation for each accession

Fig. 7.3 Observed diversity in fruit and leaf shapes and color

7.3 Conclusion

Inclusion of all stakeholders along the okra value chain at the very beginning of variety improvement holds great prospects for the adoption of novel okra varieties since the development of these varieties is demand-driven. The differences in the selected trait preferences among different stakeholders justify the need to harness information from all stakeholders along the value chain in order to drive the adoption and acceptability of improved novel varieties. The observed variability in the characterization of the germplasm provides a strong basis for selecting parents that possess genes for identified farmers' preferred traits.

References

FAOSTAT (2020). https://www.fao.org/faostat/en/#data/QCL. Accessed on the 20 Sept 2022 at 4.30pm

Hughes J (2009) Just famine foods? What contributions can under-utilized plants make to food security? Acta Hortic 806:39–48

Ibitoye DO, Kolawole OA (2022) Farmers' appraisal on okra [Abelmoschus esculentus (L.)] production and phenotypic characterization: a synergistic approach for improvement. Front Plant Sci 13:787577. https://doi.org/10.3389/fpls.2022.78757

Patil P, Sutar S, Joseph JK, Malik S, Rao S, Yadav S et al (2015) A systematic review of the genus Abelmoschus (Malvaceae). Rheedea 1:14–30

Petropoulos S, Fernandes A, Barros L, Ferreira ICFR (2018) Chemical composition, nutritional value and antioxidant properties of Mediterranean okra genotypes in relation to harvest stage. Food Chem 242:466–474

Okra Germplasm Collection, Conservation and Utilization

8

K. Joseph John, A. Suma, K. V. Bhat, M. Latha,
and Chitra Devi Pandey

8.1 Introduction

Okra has medicinal properties and good nutritional value, making it a viable choice for the nutraceutical business (Elkhalifa et al. 2021). Immature pods are used as a vegetable, while the extract of the pods serves as a thickening agent in numerous recipes, such as soups and sauces, to augment their viscosity (Dhaliwal 2010; Kumar et al. 2013). K, Na, Mg and Ca are the key components in the pods (Kumar et al. 2010). The mucilage found in the pod walls is made up of chemically linked polysaccharides that are rich in minerals and proteins (Woolfe et al. 1977) and are used in food and medicine (Kumar et al. 2010). Because of consumer preference, high nutritive value, round-the-year cultivation prospects and export potential, okra has gained an important position in India's horticulture scenario.

The African origin of this crop was proposed by De Candolle (1883), Chevalier (1940) and Vavilov (1951) due to the abundance of wild relatives in the Nile Valley, Ghana and Ethiopia. From Ethiopia, it is thought to have moved to Arabia (Dhaliwal 2017) and was further probably taken to Egypt by Arabs (Muimba-Kankolongo, 2018). The Asiatic origin was contested by Masters (1875), Bates (1968) and Grubben (1977) because of the existence of wild progenitors *A. ficulneus* (L.) Wight & Arn. and *A. tuberculatus* Pal & Har B. Singh in South and Southeast Asia. As proposed by Joshi and Hardas (1976), *A. esculentus* might have originally been present in Africa and Asia as a polyphyletic species. Vredebregt (1990) also supported the Asiatic origin due to the presence of high genotypic diversity of *A. tuberculatus* and *A. ficulneus*, which are thought to be the progenitors of *A. esculentus*.

K. J. John (✉) · A. Suma · M. Latha
ICAR-National Bureau of Plant Genetic Resources, Regional Station, Thrissur, Kerala, India

K. V. Bhat · C. D. Pandey
ICAR-National Bureau of Plant Genetic Resources, Pusa Campus, New Delhi, India

S. K. Tikoo et al. (eds.), *Okra: Status, Challenges and Opportunities*,
https://doi.org/10.1007/978-981-97-9963-3_8

Okra cultivation is largely confined to tropical and subtropical areas of the world. Apart from the cultivated okra, there are three more domesticated species, i.e. *A. caillei* (A. Chev.) Stevels, known as West African okra or African Okra or Guinean Okra, cultivated in West, Central and some parts of East Africa (Dhaliwal 2017) and India; *A. manihot* (L.) Medik., cultivated for its leaves used as a vegetable in Papua New Guinea (Hamon and van Sloten 1995), Solomon Islands and other South Pacific Islands (Keatinge 2009); and *A. moschatus* Medik., cultivated in India for its seeds used for extracting ambrette seed oil. Okra is mainly grown in India, Nigeria, Sudan, Pakistan, Ghana, Egypt, Benin, Saudi Arabia, Mexico and Cameroon, with India being the leading producer of okra, followed by Nigeria (https://www.atlasbig.com/en-in/countries-by-okra-production, accessed on 11 Jan 2023).

Taxonomically, okra is placed in the family Malvaceae. The genus name *Abelmoschus* was derived from the Arabic word '*Kaab-el-misk*', meaning father of musk (Sivarajan and Pradeep 1996), and the type species is *A. moschatus* Medik. (John et al. 2013). Earlier, okra was treated under the genus *Hibiscus*, even though the genus *Abelmoschus* was originally established by Medikus (1787) and accepted by Gaertner (1791) and Moench (1794). Later, when Hochreutiner (1924) clearly defined the taxonomic traits of *Abelmoschus*, a general consensus for acceptance of *Abelmoschus* as a separate genus was arrived upon. The genus is characterized by an asymmetrical, spathaceous, deciduous calyx that falls off as a unit as against the campanulate or cupular regular, accrescent calyx in *Hibiscus* (John et al. 2013). The plant is an annual herb, 1.0–3.00 m tall, with a deep tap root system.

8.2 Plant Genetic Resources Management in Okra

8.2.1 Collection and Conservation

Okra genetic resources are conserved in various genebanks across the world. The World Vegetable Centre, Taiwan (1438 collections) (WorldVegGRSU: Accession list accessed on 13 Jan 2023) and the National Bureau of Plant Genetic Resources (NBPGR), New Delhi, under ICAR (Indian Council of Agricultural Research) (2926 accn.), are the leading centres with notable collections of okra. The number of cultivated okra germplasm housed in the Indian National Genebank, ICAR-NBPGR, New Delhi, is reasonably high, with a significant exotic collection (568) introduced from other countries. Besides, a large number of indigenous germplasm collections from Northeastern states (Meghalaya, Manipur, Arunachal Pradesh, Mizoram, Tripura, Nagaland, Assam), Andaman and Nicobar Islands, and from other states, namely Kerala, Himachal Pradesh, Uttar Pradesh, Delhi, Telangana, Bihar, West Bengal, Gujarat, Rajasthan, Odisha, Goa, Haryana, Uttarakhand, Jammu & Kashmir, Andhra Pradesh, Jharkhand, Karnataka, Punjab, Madhya Pradesh, Maharashtra, Chhattisgarh and Tamil Nadu (ICAR-NBPGR Seed GeneBank | Dashboard (ernet.in) accessed on 11 Jan 2023), are also represented. Figure 8.1 depicts the collection of wild *Abelmoschus* germplasm from Mizoram,

Fig. 8.1 Collecting wild *Abelmoschus* germplasm from Mizoram, India

India. However, the representation of wild *Abelmoschus* species is inadequate; hence, future missions have to be planned for collecting the less represented taxa without neglecting the natural hybrids between the cultivated and the wild species of *Abelmoschus*. The value of Indian collections lies in their wealth of various wild species of *Abelmoschus* occurring in India (Charrier 1984).

8.2.2 Characterization/Evaluation

Okra breeders are challenged to develop new varieties suited to varying climatic, edaphic and biotic constraints in the major okra-growing regions, befitting the needs of the exporters and meeting quality and consumer preferences. In addition to yield, other economic traits, such as earliness, prolonged flowering and fruiting periods, desirable plant types (viz. dwarf with shorter internodes for protected cultivation), genotypes resistant to various biotic and abiotic stresses, less mucilage content, among others are also desirable. A large number of okra germplasm have been characterized, and wide variability has been documented for agro-morphological traits, which resulted in the identification of trait-specific and elite okra genotypes (Table 8.1) (Bisht et al. (1995); Binalfew and Alemu (2016); Ansari et al. (2020). Still, a large number of germplasm collections maintained in the gene banks are not fully characterized/evaluated.

Maximum morphological variability in cultivated okra was observed on the west coast of India, from Kanyakumari to the Konkan region. Fruit variability for colour ranged from cream to light green to dark green, light purple to dark purple, red, their

Table 8.1 Trait-specific germplasm of okra

Trait	Accessions	References
Superior for pod yield and its components	IC128037, IC128045, IC117018, IC117095, IC117212, IC128061, IC128069, IC128071, IC128119, EC329372, EC329375	Bisht et al. (1995)
	IC43735, IC43736, IC43720, IC128088, IC39137A, IC111500 and IC265147	Nizar et al. (2004)
	EC305771 (*A. caillei*)	Karuppaiyan (2006)
	IC45802, EC329386, EC305609	Nizar et al. (2007)
	IC282248, IC27826-A, IC29119-B, IC31398-A, IC45732, IC89819, IC89976, IC90107, IC99716 and IC111443	Reddy et al. (2012)
	1006, 1113, Parbani Kranti, Sel 1 (fruits per plants) IC43750, Sel 6, IC218877, IC69286, Sel-87 (fruit length); EC329406, 1097, KA079, 1096 (average fruit weight); 1097, 1108, 1096, IC265147 (fruit girth); 1097, IC265147, Hosahalli, IC69286-Sel-87 (branches per plant)	Kumar et al. (2013)
	245161-A, 240587-A and 245162-B (from Ethiopia)	Binalfew and Alemu (2016)
	Ajeet-121	Ansari et al. (2020)
	IC282293, EC169363, IC108235, IC111014, IC111507, MDU-1, Parbhani Kranti and Varsha Uphar	Sindhumole and Manju (2014)
	IC-18530, Elephant Tusk and SK-BS-11	Fayaz et al. (2022)
Early flowering	IC128062, IC128065, IC128072, IC128088, IC128096, IC11479, IC117218, EC325356, EC329364 and EC329370	Bisht et al. (1995)
Germplasm showing resistance/tolerance to YVMV	Ok No. 6, LORM 1, VRO3 and Punjab 7	Batra and Singh (2000)
	IC69286, IC21887, EC305619	Nizar et al. (2004)
	Nun 1145 and Nun 1144 (moderate resistance reactions under both glass house and field conditions)	Venkataravanappa et al. (2013)
	VRO 109, VRO 104, VROB 178, 307 10-1 and No. 315 (confirmation through molecular techniques)	Sanwal et al. (2014)
	VRO-5 (immune), IC117216, IC140934 and Parbhani Kranti (moderately resistant)	Kumar and Tayde (2018)
	IC15027	Narayanan and Muthiah (2017)
	AE 64, AE 65 and AE 66	Rynjah et al. (2018)

Wild species resistant to YVMV	*A. manihot*, *A. pungens*	Ugale et al. (1976), Karuppaiyan (2006)
	A. caillei cv. Susthira	Gopalakrishnan et al. (2004)
	A. angulosus (Completely free, *A. tetraphyllus*, *A. moschatus*, *A. caillei* and *A. tetraphyllus*)	Prabu et al. 2007
	A. enbeepeegearense	John et al. (2013)
	IC582757 (*A. enbeepeegeerensis*), NIC5952 (*A. moschatus*), Jpn/N-2176 (*A. manihot*) IC90340 (*A. tuberculatus*) and IIVR-Tube-1 (*A. tuberculatus*)	Sanwal et al. (2014)
Resistance to Okra Enation Leaf Curl Virus (OELCV)	VRO-5, VRO-6, KS-410, VRO-4, 111-8-9 (Red), 131-8-1 (Red) and IIVR-10	Singh et al. (2009)
Resistance/tolerance to shoot and fruit borer	IC15027	Narayanan and Muthiah (2017)
Resistance/tolerance to Aphids	EC305656, EC305664, EC305696 IC128092, IC128095, EC305743, EC306737, IC031850, IC105742 and IC117235; GOH-3, GAO-5	Bisht et al. (1995) and Jayabal et al. (2019)
Resistance/tolerance to leaf hopper (Jassid)	BBSR-37, BBSR-36 and BBSR-57	Yogeshkumar and Padhi (2022)
Resistance/tolerance to mites	GAO-5 (highly resistant to red spotted mite) Arka Anamika, GO-2, Arka Abhay, GJO-3 and JOL-10-18 (resistant to red spider mite)	Siddhapara et al. (2018)
	Kasturi Bhendi (against red spider mite)	Sheeba et al. (2011)
Resistance/tolerance to nematodes (*Meloidogyne* spp.)	Sanam, Dikshah, Arka Anamika, Ikra-1 and Ikra-2 (moderately resistant)	Mukhtar et al. (2014)
	A. angulosus var. *grandiflorus* (IC203863) and *A. tetraphyllus* (IC470742) (moderately resistant)	Holajjer et al. (2022)
Resistance /tolerance to fruit rot (*Choanephora cucurbitarum*)	EC329424, EC329361, EC329359	Nizar et al. (2004)

combinations, and reddish/ purplish streaks on a pale yellow background were observed. Fruit lengths varied from 12.00–13.00 cm to 35.00–40.00 cm, five-ridged to multi-ridged and soft pubescent to dense hairy types were also found. Besides, very tall genotypes with a perennating tendency for more than one season were also observed during collection missions. The tradition of homestead cultivation of vegetables, including okra, in the West Coast is one of the main reasons for the occurrence of high variability. Ethnic diversity linked to culinary traditions and variation in edaphic and climatic conditions in the hills and valleys of the Western Ghats probably might have contributed to the evolution of this diversity. Coloured okra fetches a premium price in the local vegetable markets across the Western Ghats. Even though commercial cultivation is restricted to very few locations, landraces are still prevalent; however, productivity is low. Figure 8.2a, b illustrates a section of fruit variability in okra conserved in the National Genebank, India.

8.3 Genetic Stocks Identified and Registered with Germplasm Registration System, ICAR-NBPGR, New Delhi

A total of five unique accessions under *Abelmoschus* taxa have been registered as genetic stock. The ICAR-Indian Institute of Vegetable Research, Varanasi, had registered two genotypes viz., a dwarf (12–15 cm) and bushy okra genotype (IC526802) bearing 2–3 effective branches (INGR No. 5026) and a thin (0.9 cm) and long-fruited genotype (IC565533) (INGR09125). IC0523737, a GMS line (INGR10156) and IC0189926, with completely ridge-less fruit, long stalk, easy to snap, shorter internode length, suitable for both fresh market and processing (INGR10157), were registered by ICAR-Indian Institute of Horticultural Research, Bengaluru, India. Further, an infra-specific hybrid between *Abelmoschus moschatus* subsp. *moschatus* (IC140985) and *A. moschatus* subsp. *tuberosus* (IC470750) was registered for its showy red flowers having ornamental value (INGR16036) (http://www.nbpgr.ernet.in:8080/registration/InventoryofGermplasm.aspx accessed on 11 Jan 2023).

8.4 Wild Genetic Resources

Despite numerous studies, authors differ as to the number of species included in the genus *Abelmoschus*. According to Charrier (1984), over 50 species have been described throughout the world, which included synonyms and misidentifications (Vredebregt 1991). van Borssum Waalkes (1966), in his revision of the genus, considered only six species: *A. moschatus*, *A. Manihot*, *A. crinitus*, *A. ficulneus*, *A. esculentus* and *A. angulosus*. Of late, Patil et al. (2015) supported the view of Sutar et al. (2013) that there exist 11 species, 3 subspecies and 4 varieties under the genus *Abelmoschus* in India. The species treatment of Sutar et al. (2013) is given in

Fig. 8.2 (**a**, **b**) Fruit variability in okra

Fig. 8.2 (continued)

Table 8.2. However, John et al. (2013) contradicted the natural occurrence of *A. manihot* in India. Recently, Misra et al. (2018) described a new taxon under *A. angulosus* at the variety level, i.e. *A. angulosus* var. *mahendragiriensis*, sourced from the Mahendragiri hills of Odisha in the Eastern Ghats of India, and John et al. (2020) upgraded variety *pungens* to the species level and described a new variety under *A. pungens*, i.e. *A. pungens* var. *mizoramensis*.

A fairly diverse gene pool exists for *Abelmoschus*. The primary gene pool consists of genetically diverse *A. esculentus* cultivars/landraces that are easily crossable. A cultivar of *A. esculentus* named '*Anakomban*' (meaning 'like elephant's tusk') is a very popular long-fruited (30–60 cm) landrace. '*Aarumasavendai*' (six-months bhendi) and '*Thamaravenda*' (lotus-like broad leaves) are profuse pod-bearing cultivars, but their potential is confined either to a specific climate or season or soil type. '*Maravenda*' (tree-type bhendi) is a landrace of *A. caillei*). A few lesser-known landraces, namely *Anjilavendai* (flowering starts at the five-leaf stage) and Palvendai (bearing white fruit), were reported by Karuppaiyan (2006). It is important to assess how much residual heterozygosity these conventional cultivars still contain, as they serve as donors for incorporating useful traits.

A. tuberculatus as the closest relative of *A. esculentus* is widely accepted. However, the sterility of direct and reciprocal F_1 hybrids reconfirms its distinctiveness as a different biological species, as reported by Patil et al. (2015). In *Abelmoschus* taxa, *A. manihot*, *A. tetraphyllus*, *A. moschatus*, *A. angulosus* and

Table 8.2 The taxonomic treatment of *Abelmoschus* in India (Sutar et al. 2013) and recent revisions

Genus	Species	Subspecies	Variety	References
Abelmoschus	*crinitus* Wall.			Sutar et al. (2013)
	ficulneus (L.) Wight & Arn.			
	angulosus Wall. ex Wight & Arn.		*grandiflorus*	
			angulosus	
			purpureus	
	moschatus Medik.			
	manihot (L.) Medik.			
	tetraphyllus (Roxb. ex Hornem.) Wall.		*tetraphyllus*	
			pungens	
	esculentus (L.) Moench.			
	tuberculatus Pal and Singh			
	caillei (A. Chev.) Stevels			
	palianus Sutar, K.V. Bhat & S.R. Yadav			
	enbeepeegeearense K.J. John, Scariah, Nissar, K.V. Bhat & S.R. Yadav			
	angulosus		*mahendragiriensis* R.C. Misra	Misra et al. (2018)
	pungens		*mizoramensis*	John et al. (2020)

A. caillei are included in GP_3, whereas *A. ficulneus* and *A. tuberculatus* are kept under GP_2 (GOI 2011). The crossability relationships of Indian taxa of *Abelmoschus* were worked out at ICAR-NBPGR, Regional Station, Thrissur and various interspecific hybrids were developed (John et al. 2013) (Fig. 8.3). Development of F_1 hybrids with various wild species was possible, but the generations could not be advanced due to the sterility of the F_1 hybrids. This problem of sterility was overcome by treating the F_1 seedlings with 0.1% colchicine for 4 days, for 4 hours from 7.00 am to 11.00 am, with less than 50% seedling recovery. Amphidiploids developed in various interspecific cross combinations were characterized for their morphological characters by Suma et al. (2023), and the variability in fruits of amphidiploids is depicted in Fig. 8.4.

Abelmoschus x *tuberculatus* is reported to be the amphidiploid okra developed by Kuwada (1966) with its closest progenitor, *Abelmoschus tuberculatus*. In a later investigation, F_1 hybrids of *A. tuberculatus* with *A. manihot* were also developed (Kuwada 1974). Similar crosses between cultivated okra and *A. tuberculatus* were also attempted by Pal et al. (1952). Akhond et al. (2000) and Samarajeewa (2003) described the success of wide hybridization attempts involving okra with *A.*

Fig. 8.3 Various interspecific crosses in *Abelmoschus* genus. (**a**) *Abelmoschus esculentus* x *A. tetraphyllus*; (**b**) *Abelmoschus esculentus* x *A. angulosus* var.*grandiflorus*; (**c**) *Abelmoschus caillei* x *A. angulosus*; (**d**) *Abelmoschus caillei* x *A. pungens var. mizoramensis*

Fig. 8.4 Fruit variability in *Abelmoschus* amphidiploids

moschatus and *A. angulosus*, respectively. Further extensive crossability studies were carried out by Bhat et al. (2014) under the National Agricultural Innovation Project of ICAR (Indian Council of Agricultural Research) during 2009–13. In their study, cross combinations were made between *A. esculentus* x *A. angulosus* var. *grandiflorus*, *A. esculentus* x *A. tetraphyllus*, *A. esculentus x A. caillei, A. ficulneus* x *A. tuberculatus*, *A. tuberculatus* x *A. ficulneus*, *A. moschatus* subsp. *moschatus* x *A. moschatus* subsp. *tuberosus*, *A. tetraphyllus x A. tetraphyllus* var. *pungens, A. pungens* x *A. angulosus* var. *grandiflorus*, *A. ficulneus x A. esculentus*, A. *tetraphyllus* x *A. esculentus* and *A. manihot* x *A. esculentus*. Crossability behaviour among various species was dealt by John et al. (2013) and Suma et al. (2019). Patel et al. (2021) reported a high crossability index of around 80% in the *A. esculentus* and *A. tetraphyllus* cross, when *A. esculentus* was used as the female parent. Further, Sandeep et al. (2022) observed that the crosses were compatible only when cultivated *A. esculentus* was used as the female parent in crosses between *A. esculentus* (cv. Arka Anamika) and the wild species *A. manihot, A. ficulneus* and *A. moschatus*.

Various taxa under *Abelmoschus* bear very attractive flowers; hence, there exists scope for selecting germplasm with aesthetic value, though flowering is transient, anthesis in the dawn and withering during the afternoon. At ICAR-NBPGR Regional Station, Thrissur, under the NAIP on Biosystematics project, several interspecific and intraspecific hybrids of ornamental value were also developed using red-flowered *A. moschatus* subsp. *tuberosus* as one parent with *A. crinitus*, *A. enbeepeegearense* and *A. moschatus* subsp. *moschatus* as the other parent (Bhat et al. 2014; John et al. 2013). F_2 (backcross and selfed progeny) of *A. moschatus* subsp. *moschatus* and *A. moschatus* subsp. *tuberosus* showed wide floral colour variability and patterns having ornamental value.

8.5 Distribution of Various *Abelmoschus* Species in India

A. esculentus is cultivated worldwide, ranging from the Mediterranean to the Equatorial regions. The wild species' habitats are very diverse. John et al. (2013) opined that the Western Ghats, Northeastern India and Chota Nagpur plateau are important areas for the conservation of the wild *Abelmoschus* gene pool in its natural habitat. Bisht and Bhat (2006); Yadav et al. (2014); Misra et al. (2018) and John et al. (2013, 2020) gave detailed information regarding the distribution of *Abelmoschus* spp. in India and worldwide. *A. angulosus* var. *angulosus* is reported to be niche-specific, occurring in high-altitude areas of Kerala and Tamil Nadu states of India, whereas var. *grandiflorus* is reported from the Western Ghats (covering Karnataka, Kerala, Maharashtra, Tamil Nadu, Goa (occasional). *A. angulosus* var. *purpureus* is rare and found in the high ranges of the Western Ghats, and var. *mahendragiriensis* is reported only from the Mahendragiri hills of Odisha. *A. crinitus* is distributed in eastern (Andhra Pradesh) as well as Western India (Gujarat, Maharashtra) and also in states like Chhattisgarh, Jharkhand, Bihar, Orissa, Madhya Pradesh, Uttar Pradesh and Uttarakhand. *A. caillei* is an introduced and cultivated species in Kerala, Karnataka and Tamil Nadu, to a limited extent as a homestead vegetable. *A. enbeepeegeearenis* is yet another endemic species from the low-elevation Western Ghats. *A. ficulneus* (L.) Wight & Arn. ex Wight is distributed in Andhra Pradesh, Tamil Nadu, Karnataka (Deccan), Madhya Pradesh, Maharashtra, Rajasthan, Gujarat, Uttar Pradesh, West Bengal, Jammu and Kashmir, India. *A. manihot* (L.) Medikus is an introduced species from Southeast Asia. *A. moschatus* subsp. *moschatus* var. *moschatus* is occasional in Andhra Pradesh, Tamil Nadu, Karnataka (Deccan), Maharashtra, Rajasthan, Gujarat, Kerala, Assam, Tripura, Arunachal Pradesh and Meghalaya; however, subsp. *biakensis* is limited to the Nicobar Islands and subsp. *tuberosus* to Northeastern India (Meghalaya—Khasi & Jaintia Hills and Assam). *A. palianus* is described from Chhattisgarh, and *A. pungens* occurs throughout the Northeastern Hill states and Uttarakhand. Recently, *A. pungens* var. *mizoramensis* was reported from Mizoram.

8.6 Chromosome and Ploidy Level of *Abelmoschus* spp.

Different species in the *Abelmoschus* genus possess varying levels of ploidy and chromosome numbers (GOI 2011). The lowest chromosome number of an *Abelmoschus* taxa is attributed to $2n = 56$ for *A. angulosus* (Ford 1938), while the highest is for *A. caillei* syn. *A. manihot* var. *caillei* (close to $2n = 200$) (Singh and Bhatnagar 1975; Siemonsma 1982a, b). Ugale et al. (1976) reported $2n = 130$ and Joshi and Hardas (1976) $2n = 138$ for *A. manihot* subsp. *tetraphyllus*. Similarly, Tyagi (2002) reported that the 2n number for *A. manihot* does not exceed 66, based on chromosome counting from flower buds and root tips from 7 cultivars. Based on the cytogenetic reports, Bisht and Bhat (2006) observed some degree of homology between *A. esculentus*, *A. tuberculatus* and *A. ficulneus*, wherein 29 out of the 65 chromosomes of *A. esculentus* showed complete homology with 29 of *A. tuberculatus* and the remaining 36 had incomplete homology with *A. ficulneus* and between *A. esculentus*, *A. tetraphyllus* and *A. pungens*. Recently, Merita et al. (2012) studied the chromosome counts of *Abelmoschus* spp. (*A. angulosus* var. *grandiflorus*, *A. moschatus* subsp. *moschatus* and *A. moschatus* subsp. *tuberosus* and *A. esculentus*) in India.

8.7 Utilization of Genetic Resources

Besides yield and its components, the major breeding objectives are resistance/tolerance to biotic and abiotic stresses. Varieties suited to different agro-climatic zones of India, released by various research organizations in India, are given in Table 8.3. There are several breeding methods adopted by plant breeders to develop superior varieties, with the major approach being selection followed by recombination breeding. The germplasm maintained in the genebank is featured by heterogeneity within the collection, which can be observed during preliminary characterization. Therefore, in order to enhance the utilization of okra genetic resources, activities should be planned for purifying the collections, followed by their systematic

Table 8.3 Varieties released in India

Name	Released by	Characters
Pusa Sawani	ICAR, Indian Agricultural Research Institute, New Delhi	First variety to have developed anywhere in the world possessing resistance (symptomless carrier type) to yellow vein mosaic (YVM) virus Selection from a cross between IC1542 x Pusa Makhmali Fruits slender, dark green, five-ridged, attractive and 10–12 cm long at marketable maturity Average fruit yield: 75 quintals per hectare
Punjab Padmini	Punjab Agricultural University, Ludhiana	Selection from *A. esculentus* (Reshmi x Pusa Sawani) x *A. manihot* subsp. *manihot* (EC31830) Suitable for spring and rainy season Average yield: 115 quintals per hectare Possesses field tolerance to YVM virus, with low infection at later stages
Punjab No. 7	-do-	Selection from an interspecific cross *A. manihot* subsp. *manihot* × *A. esculentus* cv. Pusa Sawani. Suitable for cultivation both for spring and rainy seasons Fruits medium long (15–20 cm), green, tender and ridged Average yield: 110 quintals per hectare Highly resistant to YVM virus
Punjab 8	-do-	Developed through mutation breeding by treating seeds of the variety Pusa Sawani with the chemical EMS (Ethyl Methane Sulphonate) @ 1% Suitable for cultivation both for spring and rainy seasons Fruits are thin, long, dark green and five-ridged Average yield: 140 quintals per hectare Highly resistant to YVM virus
Pusa A-4	IARI, New Delhi	Selection from an intervarietal cross Suitable for both spring and rainy season cultivation Fruits are five-ridged, dark green, 12–15 cm long, having excellent shelf life Average fruit yield 100–120 quintals per hectare Resistant to YVM virus and tolerant to Jassid
Arka Anamika	ICAR-IIHR, Bangaluru	Developed from an interspecific cross *A. esculentus* (IIHR 20–31) × *A. tetraphyllus* Suitable for cultivation both for spring and rainy seasons Fruits are dark green and five-ridged Average yield: 125 quintals per hectare Resistant to YVM virus

(continued)

Table 8.3 (continued)

Name	Released by	Characters
Arka Abhay	-do-	Sister line of Arka Anamika (same cross) Fruits medium long, dark green, having 5–6 ridges Average yield: 100 quintals per hectare Resistant to yellow vein mosaic virus
Arka Nikita	-do-	Hybrid between GMS-4 X IIHR-299-14-11-585 Yield 21–24 t/ha in 125–130 days duration Developed through genetic male sterile line
Parbhani Kranti	MAU, Parbhani	Developed from an interspecific cross *A. esculentus* cv. Pusa Sawani × *A. manihot* introduced from Japan Suitable for cultivation both for spring and rainy seasons Fruits are 8–9 cm long, smooth, dark green, tender, five-ridged, with a long beak Average yield is 85–90 quintals per hectare during spring and 115 quintals per hectare during rainy season Resistant to YVM virus
Hisar Unnat	CCS HAU, Hisar)	Developed by selection from the cross Selection 2-2 × Parbhani Kranti suitable for cultivation both for spring and rainy seasons Fruits are green, attractive, five-ridged and 15–16 cm long Average yield: 100 quintals per hectare Highly resistant to Yellow Vein Mosaic Virus
Varsha Uphar	-do-	Developed by selection from the cross, Lam Selection 1 × Parbhani Kranti Suitable for cultivation both for spring and rainy seasons Fruits five-ridged, dark green and long (18–20 cm) Average yield: 100 quintals per hectare Resistant to YVMV
Kashi Lila	ICAR-IIVR, Varanasi	Developed by selection from the cross (NIC 9305 × HRB 9-2) × IIVR-50 Early maturing (35–40 days) and suitable for spring and rainy seasons Fruits dark green and 10–13 cm long, with average fruit weight of 9–11 g Average fruit yield: 150–190 quintals per hectare Resistant to yellow vein mosaic and leaf curl virus
Kashi Shristi (VROH-12)	-do-	F_1 hybrid High yield potential (18–19 t/ha) Medium tall and short internodes Suitable for both summer and Kharif season Fruits dark green, medium fruit length Tolerant to YVMV Recommended for cultivation in Uttar Pradesh

(continued)

Table 8.3 (continued)

Name	Released by	Characters
Kashi Lalima (VROR-157)	-do-	Redish purple fruits tolerant to YVMV and OLCV Yield of 14–15 t/ha Rich in anthocyanin and phenolics Suitable for both summer and Kharif season Recommended for cultivation in Uttar Pradesh
Kashi Chaman VRO-109	-do-	Fruits dark green with fruit length of 11–14 cm Resistant to YVMV and OLECV under field conditions Yield 150–160 q/ha Suitable for both summer and rainy season
Kashi Vardaan	-do-	Suitable for both summer and rainy seasons Yield potential around 140–150 q/ha Resistant to both YVMV and OLCV Fruits medium, dark-green, easily cooked, good for both table and export purposes
Shitla Jyoti	-do-	Hybrid suitable for warm humid climate with relatively long day length Fruit green, 12–14 cm Yield 180–200 q/ha Resistant to YVMV and OLCV
Shitla Uphar	-do-	Fruits green, 11–13 cm Yield 150–170 q/ha Resistant to yellow vein mosaic virus and OLCV
Kashi Satdhari	-do-	Bears 18–25 fruits per plant with seven ridges, 13–15 cm long Yield 110–140 q/ha Resistant to YVMV under field conditions
Kashi Vibhuti	-do-	Dwarf growth habit, plant height 60–70 cm during rainy and 45–50 cm during summer seasons Bears 18–22 fruits per plant with 8–10 cm length Yield 170–180 q/ha Resistant to YVMV and OLCV
Kashi Mangali	-do-	Developed through pure line selection Yield 130–150 q/ha Resistant to YVMV and OLCV under field conditions
Kashi Mohini	-do-	Yield of 130–150 q/ha Tolerate high temperatures during the summer season and resistant to YVMV under field conditions
Hybrid—Kashi Bhairav	-do-	Yield 200–220 q/ha Fruits dark green with 10–12 cm Resistant to YVMV and OLCV under field conditions

(continued)

Table 8.3 (continued)

Name	Released by	Characters
CO-3, Bhendi hybrid COBh H 1	TNAU	Cross between Varsha Uphar selection and PA 4 (T) Resistant to YVMV disease with high level of market preference The fruits are dark green, slender, less fibrous and sparsely pubescent Yields about 22.1 /ha
CO-2 (1987)	-do-	F_1 hybrid between A.E. 180 x Pusa Sawani Fruits very long with 7–8 ridges. Yield potential of 15–16 tonnes/ha
COBh H 3	-do-	Slender, dark green fruit Yield 25.7 t ha-1, 25–29 fruits per plant Resistant to Yellow Vein Mosaic Virus (YVMV) disease
CO. 3 (1991)	-do-	F1 hybrid between Parbhani Kranti x MDU.1 Yields about 16–18 tonnes/ha Moderate resistance to Yellow Vein Mosaic disease
Salkeerthi	KAU	Re-selection from Average fruit length 27 cm with average single fruit weight of 28 g Average yield: 16.2 t/ha
Kiran	-do-	Selection from Kilichundan Local Pods light green colour
Aruna	-do-	Pureline selection from IC218870 Red fruited and resistant to YVMV Average yield: 15.8 t/ha
Susthira	-do-	Perennial type Average yield: 18 t/ha Fruits long and green, 22 cm long, weighing 24.5 g Resistant to YVMV
Anjitha	-do-	*A. esculentus* var. Kiran and *A. manihot* and further inducing mutations by gamma irradiation Resistant to YVM disease and tolerant to fruit and shoot borer attack Yield 14.6 t/ha
Manjima	-do-	Hybrid variety between Gowreesapattom local x IC282257 Green-fruited Moderately tolerant to shoot and fruit borer. Yield 16 t/ha
Pusa Bhindi-5	IARI, New Delhi	Developed through pedigree method High yielding (18.0 t/ha) Resistant to Bhendi Yellow Vein Mosaic Virus Plants straight with 2–3 branches with completely dark green stem and petioles

(continued)

Table 8.3 (continued)

Name	Released by	Characters
Pusa Okra Hybrid-1	do-	High yielding hybrid (263 q/ha), prolific bearer with shorter internodes Resistant to YVMV and Enation Leaf Curl Virus (ELCV) diseases and tolerant to leafhopper and red spider mites Suitable for kharif as well as spring summer season
Pusa Lal Bhindi-1:	do-	Developed through pedigree method Fruits reddish purple in colour and all plant parts, including stems and petioles, are reddish purple Moderately resistant to YVMV and ELCV diseases Yield 15 t/ha on short internodes Rich in anthocyanin (130 μg/g) and antioxidant content (357 μg/g GAE)

Source: Dhaliwal (2017) https://iivr.icar.gov.in/kashi-shristi-vroh-12-f1-hybrid; https://kau.in/basic-page/varieties-released; https://nhb.gov.in/pdf/vegetable/okra/okr011.pdf; https://iihr.res.in/varieties-crop-name/okra; https://agritech.tnau.ac.in/horticulture/horti_vegetables_bhendi_Varieties.html

characterization. Germplasm field days/diversity days/breeders meet can be organized by the germplasm curators to display the available variability in the field during characterization so that researchers can effectively utilize the diverse germplasm.

8.8 Future Prospects

Okra is a crop with vast nutritional properties and export potential. One of the major prospects in okra breeding lies with the genetic enhancement of existing germplasm collections by incorporating genes from secondary gene pools, especially for transferring genes for resistance to yellow vein mosaic disease and enation leaf curl virus disease. Morphological, cytological and molecular studies need to go hand in hand to effectively widen the okra genetic base. Entire okra collections, including wild species, should be evaluated in YVMV hotspots. The fertile amphidiploids developed should be screened for YVMV resistance at hotspots, and the promising amphidiploids should be advanced through backcrossing with the cultivated okra. In addition, attempts should be evaluated to develop difficult-to-achieve cross combinations involving *A. enbeepeegeearense* and *A. moschatus*. Niche-specific wild taxa like *A. angulosus* var. *purpureus* and *A. angulosus* var. *angulosus* need to be utilized for wide hybridization. As a sizable number of central Indian collections are morphologically similar, efforts should be focused on removing the duplicates for effective handling. Exploration and collection of wild species and landrace diversity from remote and unvisited tribal tracts in the Western Ghats, Northeastern region, Jharkhand and Chhattisgarh, and collecting niche-specific species and taxa like *A. angulosus* var. *angulosus*, var. *purpureus*, var. *mahendragiriensis*, and their conservation in the genebank, are the other priority areas. Attention should also be given to in-situ conservation of wild species of *Abelmoschus to* minimize the risks of genetic erosion in the wild gene pool (Charrier 1984).

Acknowledgements We gratefully acknowledge the support and facilities provided by the Director, ICAR-National Bureau of Plant Genetic Resources, New Delhi and the Heads, Division of Germplasm Evaluation, ICAR-NBPGR.

References

Akhond YAM, Molla MAH, Islam MO, Ali M (2000) Cross compatibility between *Abelmoschus esculentus* and *A. moschatus*. Euphytica 114(3):175–180

Ansari AM, Ahmad E, Nazrussalam, Singh DN (2020) Evaluation of okra (*Abelmoschus esculentus* L. Moench) genotypes for important quantitative characters. Int J Appl Agric 1(1):1–5

Bates DM (1968) Notes on the cultivated Malvaceae. 2. Abelmoschus Baileya 16:99–112

Batra VK, Singh J (2000) Screening of okra varieties to yellow vein mosaic virus under field conditions. Veg Sci 27:192–193

Bhat KV, Yadav SR, Rao SR, Joseph JK, Latha M, Malik SK, Negi KS (2014) NAIP project report on biosystematics of the genera *Vigna, Cucumis* and *Abelmoschus*. National Bureau of Plant Genetic Resources, New Delhi

Binalfew T, Alemu Y (2016) Characterization of okra (*Abelmoschus esculentus* L. Moench) germplasms collected from Western Ethiopia. Int J Res Agric For 3:11–17

Bisht IS, Bhat KV (2006) Okra (*Abelmochus* species). In: Singh RJ (ed) Genetic resources, chromosome engineering, and crop improvement. Vegetable crops. CRC Press, Boca Raton, pp 148–162

Bisht IS, Mahajan RK, Rana RS (1995) Genetic diversity in South Asian okra (*Abelmoschus esculentus*) germplasm collection. Ann Appl Biol 126:539–550

Charrier A (1984) Genetic resources of the genus *Abelmoschus* Med. IBPGR, Rome, 61 p

Chevalier A (1940) L'origine, la culture et les usages de cinq Hibiscus de la section *Abelmoschus*. Rev Bot Appl 20:319–328

De Candolle AP (1883) Origine des Plantes Cultivees, Paris, pp 150–151

Dhaliwal MS (2010) Okra (*Abelmoschus esculentus*) L. (Moench). Kalyani Publishers, New Delhi

Dhaliwal MS (2017) Okra (*Abelmoschus esculentus*) L (Moench). In: Dhaliwal MS (ed) Handbook of vegetable crops, 3rd edn. Kalyani Publishers, New Delhi, pp 279–310

Elkhalifa A, Alshammari E, Adnan M, Alcantara JC, Awadelkareem AM, Eltoum NE, Mehmood K, Panda BP, Ashraf SA (2021) Okra *(Abelmoschus esculentus*) as a potential dietary medicine with nutraceutical importance for sustainable health applications. Molecules (Basel, Switzerland) 26(3):696. https://doi.org/10.3390/molecules26030696

Fayaz Z, Masoodi UH, Afroza B, Nazir G, Wani AS, Jan U, Rashid M (2022) Evaluation of growth and yield parameters of okra [*Abelmoschus esculentus* (L.) Moench] genotypes under temperate conditions of Kashmir valley. Environ Ecol 40(3A):1368–1372

Ford CE (1938) A contribution to a cytogenetical survey of the Malvaceae. *Genetica, 20*, 431–452

Gaertner J (1791) De Fructibus et Seminibus *Plantarum* 2:249–253

GOI (2011) Series of crop specific biology documents – biology of *Abelmoschus esculentus* L. (Okra). Dept of Biotechnology and Ministry of Science and Technology & Ministry of Environment and Forests, Govt. of India, 35p

Gopalakrishnan TR, Gokulapalan C, Mini C, Krishnakumary K, Babu SKV, Mathew KS, Sadhankumar PG (2004) Three decades of vegetable research in Kerala Agricultural University, Directorate of Extension, KAU, Mannuthy, 153p

Grubben JH (1977) Tropical vegetables and their genetic resources. IBPGR, Rome, pp 111–114

Hamon S, van Sloten DH (1995) Okra. In: Smartt J, Simmonds NW (eds) Evolution of crop plants. John Wiley & Sons, New York, pp 350–357

Hochreutiner BPG (1924) Genres nouveaux et genres discutes de la famille des Malvaces. Candollea 2:79–90

Holajjer P, John KJ, Pandravada SR, Sivaraj N, Kodaru A (2022) Screening of *Abelmoschus* gene pools for sources of resistance to root knot nematode, *Meloidogyne incognita*. Indian J Nematol 52(1):16–20

Jayabal TD, Chinniah C, Kalyasundaram M (2019) Screening of okra, *Abelmoschus esculentus* L. (Moench) germplasm collections against two spotted spider mite, *Tetranychus urticae* (Koch) based on damage grading index. Electron J Plant Breed 11(2):425–429

John JK, Nissar VAM, Latha M, Sutar S, Patil P, Malik SK, Negi KS, Keisham M, Yadav SR, Bhat KV (2013) Genetic resources and crossability relationships among various species of *Abelmoschus*. Curr Hort 1(1):35–46

John JK, Krishnaraj MV, Pradheep K, Patil P, Harish GD, Bhat KV (2020) A new variety of *Abelmoschus pungens* (Malvaceae) from Indo-Burma Biodiversity Hotspot. Rheedea 30(2):270–277

Joshi AB, Hardas MW (1976) Okra. In: Simmonds NW (ed) Evolution of crop plants. Longman, London, pp 194–195

Karuppaiyan R (2006) Breeding for resistance to shoot and fruit borer (*Earias vittella*) in okra (*Abelmoschus esculentus*). Ph.D thesis, Kerala Agricultural University, Vellanikkara, 178p

Keatinge JDH (2009) Ensuring slippery cabbage won't slip away. AVRDC-The World Vegetable Centre Newsletter, December 31, pp 1–2

Kumar KI, Tayde AR (2018) Screening of okra genotypes against yellow vein mosaic virus disease (OYVMV) under field conditions in Allahabad. J Pharmacogn Phytochem 7(1):660–662

Kumar S, Dagnoko S, Haougui S, Ratnadass A, Pasternak D, Kouame C (2010) "Okra" (*Abelmoschus* spp.) in West and Central Africa: potential and progress on its improvement. Afr J Agric Res 5:3590–3598

Kumar A, Kumar P, Nadendla R (2013) A review on *Abelmoschus esculentus* (Okra). Int Res J Pharm Appl Sci 3:129–132

Kuwada H (1966) The new amphidiploid plant named "*Abelmoschus tubercular-esculentus*," obtained from the progeny of the reciprocal crossing between *A. tuberculatus* and *A. esculentus*. Jpn J Breed 16:21–30

Kuwada H (1974) F_1 hybrids of *Abelmoschus tuberculatus* and *A. manihot* with reference to the genome relationship. Jpn J Breed 24:207–210

Masters MT (1875). In: Hooker JD (ed) Flora of British India. Ashford Kent 1, pp 320–348

Medikus (1787) Uebereinigekinstlichegeschlechteraus der Malvenfamilie, denn derklasse der Monadelphien, Mannheim, pp 45–46

Merita K, Joseph JK, Yadav SR, Bhat KV, Rao SR (2012) Chromosome counts in wild and cultivated species of *Abelmoschus* Medikus. From the Indian sub-continent. J Hortic Sci Biotechnol 87(6):593–599. https://doi.org/10.1080/14620316.2012.11512917

Misra RC, Pani DR, Bharathi LK, Ahlawat SP (2018) *Abelmoschus angulosus* var. *mahendragiriensis* (Malvaceae): a new taxonomic variety of wild okra from Eastern Ghats of India. Genet Resour Crop Evol 65:993–1002

Moench C (1794) Methodus Plantas Horti Botanici et Agri Marburgensis, a Staminum Situ Describendi

Muimba-Kankolongo A (2018) Vegetable production. In A. Muimba-Kankolongo (Ed.), Food crop production by smallholder farmers in Southern Africa (pp. 205–274). Oxford: Academic Press. https://doi.org/10.1016/B978-0-12-814383-4.00011-6

Mukhtar T, Hussain MA, Kayani MZ, Aslam MN (2014) Evaluation of resistance to root-knot nematode (*Meloidogyne incognita*) in okra cultivars. Crop Prot 56:25–30

Narayanan US, Muthiah C (2017) In vivo screening of okra (Abelmoschus esculentus L.) germplasm collections against sucking pests. Electron J Plant Breed 8(1):187–192

Nizar MA, John J, Karuppaiyan R (2004) Evaluation of okra germplasm for fruit yield, quality and field resistance to yellow vein mosaic virus. Indian J Plant Genet Resour 17(3):241–244

Nizar MA, John J, Karuppaiyan R (2007) Selection response for fruit yield in okra (*Abelmoschus esculentus* (L.) Moench). J Plant Genet Resour 20(1):43–47

Pal BP, Singh HB, Swarup V (1952) Taxonomic relationships and breeding possibilities of species of *Abelmoschus* related to okra (*A. esculentus*). Bot Gaz 113:455–464

Patel BN, Hegde GK, Manu TG (2021) Interspecific hybridization as a way of resistance transfer against viruses in okra: hindrances and way forward. Plant Genet Resour 19:357–362

Patil P, Sutar S, John JK, Malik SK, Rao SS, Yadav SR, Bhat KV (2015) A systematic review of the genus *Abelmoschus* (Malvaceae). Rheedea 25(1):14–30

Prabu T, Warade SD, Ghante PH (2007) Resistance to okra yellow mosaic virus in Maharashtra. Veg Sci 34(2):119–122

Reddy MT, Haribabu K, Ganesh M, Reddy KC, Begum H, Subbarama R, Reddy K, Babu JD (2012) Genetic analysis for yield and its components in okra (*Abelmoschus esculentus* (L.) Moench). J Agric Technol 34(2):133–141

Rynjah S, Sathiyamurthy VA, Saraswathi T, Harish S (2018) Field screening of okra (*Abelmoschus esculentus* L. Moench) genotypes against okra yellow vein mosaic virus disease. J Pharmacogn Phytochem 7(5):1852–1854

Samarajeewa PK (2003). Wild *Abelmoschus* species in the improvement of okra. In: Jayasuriya AHM, Vaughan DA (eds) Conservation and use of crop wild relatives. Proceedings of the Joint Department of Agriculture, Sri Lanka and National Institute of Agrobiological Sciences, Japan, workshop held on 3 Feb 2003, pp 97–108

Sandeep N, Dushyanthakumar BM, Sridhara S, Dasaiah L, Mahadevappa SK, El-Shehawi AM, Althaqafi MM, Aloufi S, Sharma H, Alaklabi A, Elansary HO (2022) Characterization of okra species, their hybrids and crossability relationships among *Abelmoschus* species of the Western Ghats Region. Horticulturae 8:587. https://doi.org/10.3390/horticulturae8070587

Sanwal SK., Venkataravanappa V, Chauhan NS (2014) Screening of wild and cultivated okra germplasm against yellow vein mosaic disease. Paper presented on international conference on horticulture for nutritional livelihood and environmental security in hills: opportunity and challenges, Kalimpongm Darjeeling, India during 22–24 May 2014

Sheeba S, Roseleen J, Ramaraju K (2011) Effect of okra entries on biology of two spotted spider mite *Tetranychus urticae* (Koch.). Ann Plant Prot Sci 19(1):67–70

Siddhapara MR, Virani VR, Bhadani DJ (2018) Screening of okra cultivars/genotypes against two spotted red spider mite, *Tetranychus urticae* Koch (Acari: Tetranychidae). J Entomol Zool Stud 6(2):1799–1803

Siemonsma JS (1982a) West-African okra – morphological and cytogenetical indications for the existence of a natural amphidiploid of *Abelmoschus esculentus* (L.) Moench and *A. manihot* (L.) Medikus. Euphytica 31:241–252

Siemonsma JS (1982b) La Culture du Gombo (*Abelmoschus* spp.), Legume-Fruit Tropical (avec Reference Speciale à la Cote d'Ivoire). Ph.D. Thesis, Wageningen Agricultural University, Wageningen, The Netherlands, 279 pp

Sindhumole P, Manju P (2014) Evaluation of okra [*Abelmoschus esculentus* (L.) Moench] germplasm for yield and quality of fruits. Indian J Plant Genet Resour 27(1):28–35

Singh HB, Bhatnagar A (1975) Chromosome number in okra from Ghana. Indian J Genet Plant Breed 36:26–27

Singh B, Sanwal SK, Rai M, Rai AB (2009) Sources of biotic stress resistance in vegetable crops: a review. Veg Sci 36(2):133–146

Sivarajan VV and Pradeep AK (1996) Malvaceae of Southern Peninsular India: A Taxonomic Monograph. Daya Publishing House, New Delhi, India. 312 pp

Suma A, Joseph John K, Latha M, Bhat KV (2019) Wide hybridization and development of amphidiploids incorporating yellow vein mosaic virus tolerance in okra. In: Viswanath S (ed)

Abstracts of papers of national seminar on forestry, plant genetics and improvement. 3–4 December. Kerala Forest Research Institute, Kerala, p 94

Suma A, John JK, Bhat KV, Latha M, Lakshmi CJ, Pitchaimuthu M, Nissar VAM, Thirumalaisamy PP, Pandey CD, Pandey S, Kumar A, Gautam RK, Singh GP (2023) Genetic enhancement of okra [*Abelmoschus esculentus* (L.) Moench] germplasm through wide hybridization. Front Plant Sci. https://doi.org/10.3389/fpls.2023.1284070

Sutar SP, Patil P, Aitawade M, John J, Malik S, Rao S, Yadav S, Bhat KV (2013) A new species of *Abelmoschus* Medik. (Malvaceae) from Chhattisgarh, India. Genet Resour Crop Evol 60:1953–1958

Tyagi AP (2002) Cytogenetics and reproductive biology of some BELE (*Abelmoschus manihot* Linn., Medic subsp. *manihot*) cultivars. S Pac J Nat Sci 20:4–8

Ugale SD, Patil RC, Khupse SS (1976) Cytogenetic studies in the cross between *Abelmoschus esculentus* and *A. tetraphyllus*. J Maharashtra Agric Univ 1:106–110

van Borssum Waalkes J (1966) Malesian Malvaceae (revised). Blumea 14(1):1–25

Vavilov NI (1951) The origin, variation, immunity and breeding of cultivated plants (trans: Cheaster KS). Chrom Bot 13:1–366

Venkataravanappa V, Lakshminarayana Reddy CN, Krishna Reddy M (2013) Begomovirus characterization, and development of phenotypic and DNA-based diagnostics for screening of okra genotype resistance against Bhendi yellow vein mosaic virus. 3 Biotech 3(6):461–470. https://doi.org/10.1007/s13205-012-0107-z

Vredebregt JH (1990) Taxonomic and ecological observations on species of *Abelmoschus* Medik. Report of an international workshop on okra genetic resources. NBPGR Int Crop Netw Ser 5:69–76

Vredebregt JH (1991) Taxonomic and ecological observations on species of *Abelmoschus* Medik. In: Report of an international workshop on Okra genetic resources held at the NBPGR, New Delhi, India, 8–12 Oct 1990, pp 69–76

Woolfe ML, Martin F, Otchere CG (1977) Studies on the mucilages extracted from okra fruits (Hibiscus esculentus L.) and baobab leaves (Adansonia digitata L.) J Sci Food Agric 28: 519–529

Yadav SR, Bhat KV, Latha M, John KJ, Aitwade M, Rao SR, Scariah S, Nissar M, Umdale S, Patil P, Krishnan G, Khedsana R (2014) An illustrated guide for the identification of *Vigna* Savi, *Cucumis* L. and *Abelmoschus* Medik. species in India. NBPGR, New Delhi; Shivaji University, Kolhapur and North Eastern Hill University, Shillong

Yogeshkumar HD, Padhi J (2022) Biophysical basis of resistance in Okra to Jassids, *Amrasca biguttula biguttula* (Ishida). Indian J Entomol 84(4):906–911. https://doi.org/10.55446/IJE.2021.350

Part III

Okra Breeding

9 Okra Breeding: History, Key Milestones and Challenges: Indian Perspective

R. K. Yadav, N. D. Vinay, Mahesh Badiger, Manjsh Saakre, B. S. Tomar, Harshwardhan Choudhary, and Suman Lata

9.1 Introduction

Okra or lady's finger, locally known as *bhindi*, is one of the important vegetables grown in the tropical and subtropical regions of the world. A temperature around 25 °C with high relative humidity (65–85%) is ideal for plant growth and fruiting. Lower temperatures (<18 °C) reduce germination, while temperatures above 42 °C affect plant growth and reduce fruit yield due to desiccation and the dropping of flower buds.

The immature fruit of okra is used in fresh or processed (dehydrated and frozen) forms. Okra pods are good in nutritive value as they contain 10.4 g dry matter, 3100 calorie energy, 1.8 g protein, 90 mg calcium, 1.0 mg iron, 0.1 mg carotene, 0.07 mg thiamine, 0.08 mg riboflavin and niacin, and 18 mg vitamin C in a 100 g consumable portion (Grubben 1977). Dry seed kernels of okra are a rich source of protein (around 18%) and oil (around 20%) and may be used as a substitute for edible oil. Besides its food value, the by-products of okra (fibre, mucilage and oil) have extensive applications in the textile, paper and medicine industries (BeMiller et al. 1993). World-wise okra is produced at around 8.7 million tonnes, and out of which Indian production accounts for more than 63%. In India, the area under okra cultivation is 530 thousand hectares, and total production is 6350 thousand tonnes, with a productivity level of 12.0 t/ha (NHB 2020).

R. K. Yadav (✉) · N. D. Vinay · M. Badiger · B. S. Tomar · H. Choudhary · S. Lata
Division of Vegetable Science, ICAR-Indian Agricultural Research Institute, New Delhi, India

M. Saakre
ICAR-National Institute of Plant Biotechnology, New Delhi, India

S. K. Tikoo et al. (eds.), *Okra: Status, Challenges and Opportunities*,
https://doi.org/10.1007/978-981-97-9963-3_9

9.2 Crop Biology/Botany and Breeding Behaviour

Okra is a tall-growing annual with an erect single or branched stem having a well-developed tap root system. Flowers are borne solitary in leaf axils, which start appearing normally from the 3rd to the 7th node of the base and continue along with the growing stems. Flower calyx is generally non-lobed and tubular. Corolla is having five large yellow petals. The staminal column is tubular, with numerous stamens attached to the corolla base. The stigma is 5–10 lobed, having a velvety red/purplish colour. The fruit has a 5–10 chambered, beaked capsule, with or without sutures/ridges.

Okra is protogynous in nature, in which the stigma is receptive 2 hours before anthesis and remains receptive till 3–4 hours after anthesis, with varying degrees of receptivity. The flower is hermaphrodite, and self-pollination is predominant; however, cross-pollination may take place from 4% to 19% through insects. Hence, it is classified as an often cross-pollinated vegetable crop. The extent of cross-pollination depends upon the cultivar, competitive flora, insect population and season, etc. The inbreeding depression is very less in okra, unlike in the case of other self-pollinated crops, due to self-fertilization. It generally takes about 40–50 days from sowing to flowering and fruit set. The fruits develop to marketable (unripe harvest) maturity in 5–6 days after anthesis and seed harvest in 30–32 days from anthesis.

Okra belongs to the Malvaceae family and the genus *Abelmoschus*.

The chromosome number ($2n$) of cultivated okra (*A. esculentus* L. (Moench)) has been variably reported by different workers. The common somatic chromosome numbers ($2n$) reported are 72, 108, 120, 132 and 144 in a regular series of polyploids with $n = 12$. However, the most common somatic chromosome number ($2n$) of 130 is found in cultivated okra, and it is considered to be an amphidiploid between *Abelmoschus tuberculatus* ($2n = 58$) and *A. ficulenus* ($2n = 72$) (Datta and Naug 1968).

9.3 Breeding

Okra breeding in India started with the efforts of Dr. Harbhajan Singh, who introduced high-yielding varieties and biologically important wild relatives of okra. Later, researchers applied conventional breeding methods such as pure line selection and hybridization for cultivar development. Pusa Makhmali, Gujarat Bhindi 1 and Co 1 have been developed through pure line selection. The pedigree method is the most widely followed method in okra improvement, while in the case of interspecific hybridization, "backcross pedigree" is preferred. Famous cultivars such as Pusa Sawani, Hisar Unnat and Varsha Uphar are derived from intervarietal hybridization. Interspecific hybridization is the go-to method in okra specifically to transfer YVMV resistance from wild relatives, through which several YVMV-resistant cultivars have been developed. Punjab Padmini, Punjab-7, Parbhani Kranti were developed using *Abelmoschus manihot* as a gene source, while Arka Anamika and Arka Abhay were derived from a *Abelmoschus esculentus* x *A. tetraphyllus* cross.

In okra, mutation breeding is being used to generate novel desirable variability. Through gamma irradiation, two cultivars, namely MDU-1 and Parbhani Tillu, were developed, whereas EMS treatment was done to evolve Punjab-8(EMS-8), a YVMV-resistant and fruit borer-tolerant cultivar. Heterosis is exploited to evolve hybrids that are very much popular due to their uniformity, high yield, earliness, stress tolerance and wide adaptability.

Trait-targeted breeding in okra has been successful, with the identification of several lines and the development of varieties for processing, export and resistant to stresses. Despite the high economic importance of okra and decades of breeding efforts, its genetic improvement has been slow compared to other vegetable crops, primarily due to the limited application of advanced biotechnological tools. The complex genome architecture, due to its polyploid nature, is a major limiting factor in the utilization of genomics and molecular breeding in okra. Other challenges in okra breeding include the frequent breakdown of *Begomovirus* resistance and the existence of pre- and post-zygotic barriers in wide hybridization, which need serious attention.

Breeding objectives in okra:

- Early maturing and high yield
- Development of suitable ideotype/plant architecture: i.e. short internodal length with a greater number of nodes, having deeply cut leaves at 45° from the stem
- Fruit quality: medium-long, smooth, 5-ridged pods, dark green colour, fruit shape and size, tenderness, Pods devoid of conspicuous hairs and better shelf life
- Processing cultivars suitable for canning, freezing, dehydration and edible oil extraction
- Higher nutrition content: vitamins and minerals
- Pods suitable for export markets and the processing industry (8–10 cm long)
- Resistance to disease (YVMV), okra leaf curl virus/enation leaf curl virus (ELCV), powdery mildew and cercospora leaf spot
- Resistance to insects like jassids/leaf hoppers, fruit and shoot borers, whitefly and red spider mites
- Development of multiple disease- and pest-resistant varieties/hybrids
- Resistance/tolerance to abiotic stresses such as low and high temperature, moisture stress, salinity, etc.
- Development of an economical method of hybrid seed production by genetic mechanisms

9.4 Breeding Methods and History and Achievements of Okra Breeding in India

Okra is an often cross-pollinated crop and does not show inbreeding depression. The common breeding methods applicable to any autogamous crop are employed in okra improvement. Okra breeding mainly relies on conventional methodologies.

Various breeding approaches, such as introduction, pure line selection, pedigree selection, distant hybridization, backcrossing and heterosis breeding, have been applied in okra improvement. Advanced techniques such as mutation breeding and polyploidy are also employed to address specific problems.

9.4.1 Varietal Development Using Traditional Breeding Methods

(a) *Plant Introduction*

It is the simplest and earliest method employed for crop improvement. Plant introduction division, IARI, New Delhi, introduced several cultivars of okra from the USA, such as Perkins Long Green, Clemson Spineless and Louisiana, into India, and these were grown widely across the country for fresh market and canning purposes. Furthermore, introduced germplasm has played a significant role in okra improvement in India. For instance, a wild relative, *Abelmoschus manihot* ssp. *manihot*, was introduced from Africa (Ghana) and has been successfully used as a source of resistance to YVMV.

(b) *Pure Line Selection*

It is the easiest method, which involves identification of a superior plant from a mixed/heterogeneous population. Landraces/cultivars collected from farmers' fields are used as source material. The first improved variety of okra, Pusa Makhmali, in India was developed through pure line selection using local germplasm collected from West Bengal. Co-1 was developed by single plant selection from Red Wonder. Similarly, Gujarat Bhindi-1 is a product of pure line selection.

(c) *Hybridization and Selection*

Intraspecific Hybridization

Intraspecific or intervarietal hybridization followed by pedigree selection was the most widely followed breeding method in okra during the period 1960–1990, resulting in several high-yielding varieties, including the most widely cultivated, high-yielding, yellow vein mosaic virus (YVMV) resistant cultivar Pusa Sawani (Singh et al. 1962). Similarly, Selection-2 is a derivative of a multiple intervarietal crossing project.

Interspecific Hybridization

Wild relatives of okra exhibit varying levels of resistance to major biotic stresses, namely Yellow Vein Mosaic Virus (*Abelmoschus manihot*, *A. manihot* spp. *tetraphyllus*, *A. tetraphyllus* var. *pungens*, *A. moschatus*, *A. crinitus*, *A. angulosus*, *A.*

caillei), Jassids (*A. manihot* spp. *manihot*) and Fruit & Shoot borers (*A. manihot*). Some of the wild species of okra have been successfully used over the years to transfer resistance to cultivated okra.

Martin (1982) first time inferred that backcrossing is useful for the transfer of the traits of disease resistance along with other desirable traits, namely vigour, woodiness, perennialism, branched inflorescence and a high number of seeds per pod from West African species (*Abelmoschus manihot*). In India, limited studies have been carried out on interspecific hybridization in the genus *Abelmoschus* with varying degrees of success (Pal et al. 1952; Joshi and Hardas 1956; Gadwal et al. 1968; Jambhale and Nerkar 1986; Martin 1982; Siemonsma 1982). Pre- and post-zygotic barriers are major hindrances in the success of wide crosses in okra. To overcome these obstacles, the help of special techniques such as colchiploidization, repeated crossing, pollen treatment, reciprocal crossing, ovule culture, embryo culture and polyploidy is applied. For instance, successful hybrids of the interspecific cross *A. esculentus* × *A. tuberculatus* can be obtained but exhibit a high degree of pollen sterility in F_1 plants and onwards, resulting in low or no seed set (Gadwal et al. 1968). Colchicine-induced polyploidization of F_1 has been widely practiced to restore fertility.

In the beginning, *Abelmoschus manihot* was being extensively involved in the breeding program of okra for yellow vein mosaic virus disease resistance. *A. esculentus* ($2n = 130$) is crossable with *A. manihot* ($2n = 68$) and produces amphiploids ($2n = 198$) with the help of colchicine treatment. Badiger et al. (2024) reported that crossability studies of *Abelmoschus manihot* var. tetraphyllus (IC90476-1) and *Abelmoschus* variety Pusa A-4 and *moschatus* (IC141055) with commercially grown *A. esculentus* found no serious pre- and post-zygotic barriers in *Abelmoschus manihot* var. *tetraphyllus*. *A. esculentus* is crossable with *A. moschatus* and *A. ficulneus*, but F_1 shows pollen sterility, which can be advanced with special aids such as colchicine treatment and embryo and ovule cultures. *A. esculentus* ($2n = 130$) is crossable in both directions with *A. tuberculatus*. NBPGR Regional Station, Trissur has got success in interspecific crosses of almost all wild okra species except *A. crinitus*, following colchicine treatment. Further improvement for the development of desirable traits is in progress.

Martin reported the development of a new variety, "Winter Bush", in 1982 by interspecific hybridization between Guinean okra and *A. esculentus*, followed by backcrossing to Clemson Spineless. Research workers in India applied interspecific hybridization followed by backcrossing in the development of YVMV-resistant varieties, namely, Parbhani Kranti (Jambhale and Nerkar 1979; Jambhale and Nerkar 1986), P-7 (Thakur 1976) and Punjab Padmini (Dhillon and Sharma 1982), using *A. manihot* as the gene source, while Arka Anamika and Akra Abhay were derived from *A. esculentus* and *A. tetraphyllus* crosses (Dutta 1984; Dutta 1991). These workers employed the backcross pedigree method in varietal development.

(d) *Heterosis/Hybrid Breeding*

Okra has been intensively studied, and heterosis has been reported for various economic traits, viz. early flowering, plant height, pod number, weight and size of pods, marketable and total pod yield. Furthermore, since the hybrid vigour is realized in the first generation (F_1) of crossing, superior combinations are selected in the F_1 generation itself, which saves a lot of time as compared to pedigree breeding. A typical heterosis breeding involves the following steps: (1) production of inbred lines, (2) testing of the combining ability of inbreds, (3) improvement of inbred lines and (4) production of F_1 hybrids. Commercially, hand emasculation and pollination methods are used for hybrid production. However, Genic male sterility (GMS) has been developed, which could be used to economize the hybrid seed cost (Dutta 1991).

In okra, hybrids show superior performance over open-pollinated varieties (OPV) with respect to earliness, uniformity, high yield, resistance to one or more pests (jassids, aphids, borers), diseases (YVMV, ELCV and wilt) and better fruit quality. However, a number of hybrids developed by both the public and private sectors and tested in the All India Coordinated Vegetable Improvement Project trials have not shown consistency in their performance over locations and years, barring a few. Two hybrids, namely Co. 2 (AE 180 × Pusa Sawani) and Co. 3 (Parbhani Kranti × MDU-1), were developed by Tamil Nadu Agricultural University, Coimbatore. Pusa Okra Hybrid-1 from the Indian Agricultural Research Institute, New Delhi, and Arka Nikita from the Indian Horticultural Research Institute, Bengaluru, India, have been released and notified for cultivation in various agroclimatic zones/conditions. Similarly, several other hybrids have been identified for different agroclimatic zones (Chauhan et al, 2019).

(e) *Mutation Breeding*

Mutation breeding is the quickest way to develop new value-added traits in an already favoured/adopted cultivar with an elite genetic background. In okra, mutation breeding has been employed by various workers to create desired genotypes. Till now only a few mutants of commercial importance have been isolated in India, viz. Punjab-8 (EMS-8), an EMS (1%) induced mutant of Pusa Sawani, which exhibited yield improvement (107% increased) and enhanced YVMV resistance and tolerance to the fruit borer. Additionally, this mutant was also recommended for canning and dehydration (Sharma and Arora 1991). Apart from this, most of the mutants in okra are produced through gamma irradiation, which is considered the most effective mutagen, especially at 60–80 kRad dosage (Abraham and Bhatia 1984). MDU-1, developed by Tamil Nadu Agricultural University, Coimbatore in 1978, is a gamma-ray-induced mutant of Pusa Sawani; and Parbhani Tillu was developed through induced mutation that was found suitable for processing also (Kulkarni and Nerkar 1992). Other attempts with significant results include the YVMV-resistant mutant plant (B-21) from 400Gy gamma-radiated Okura variety in the M4 generation. Three dwarf mutants and two genic male-sterile mutants were isolated from the mutated population of Hisar Unnat generated by gamma rays.

(f) *Polyploidy Breeding*

Polyploidy has also played a significant role in the evolution of okra as well as its genetic improvement. *Abelmoschus caillei* is known to be a natural amphidiploid of *A. esculentus* and *A. manihot* (Siemonsma 1982). It has been established that *A. esculentus* ($2n = 130$) is an amphidiploid of *A. tuberculatus* ($2n = 58$) and an unknown species with $2n = 72$ (Joshi and Hardas 1976). Naturally occurring alloploids such as *A. esculentus* and West African okra (*A. caillei*) are widely cultivated, which showed considerable outcrossing in nature even at the species level. In okra, efforts to induce polyploidy have yielded limited success; colchicine-induced polyploids exhibited retarded growth and abnormal flowering and fruiting (Patil 1963). In okra, colchicine-induced polyploidy of F_1 crosses is widely used to overcome post-zygotic barriers, i.e. sterility of interspecific hybrids during gene transfer from wild taxa to cultivated okra. Kuwada (1966) obtained a fertile amphidiploid by duplication of the F_1 between *A. esculentus* and *A. manihot*, called Nori-Asa. Later, synthesized an amphidiploid between *A. tuberculatus* and *A. ficulneus*. An amphidiploid between *A. esculentus* and *A. manihot* ssp. *manihot* was produced by Jambhale and Nerkar (1986). It showed 81.8% seed fertility, against 7.07% in the F_1 and exhibited resistance to yellow vein mosaic virus. Amphidiploids from crosses between *A. esculentus* and *A. manihot* ssp. *tetraphyllus* were also developed.

9.5 Trait-Targeted Breeding

(a) *Breeding for Fruit Quality and Processing Traits*

Okra is a hot-season crop; hence, its availability is season-bound in temperate regions where its pods can be preserved and processed for later use in the off-season. Mostly, fresh pods of okra are preserved through dehydration, freezing and canning. For processing purposes, fruits should preferably be dark green, tender, with less fibre, less mucilage, high in protein, vitamins, dry matter and minerals, with a high dehydration/rehydration ratio for dehydration. Dehydration is the simplest method of okra preservation. Whole small pods can be sun-dried, while large pods are cut lengthwise in halves or quarters, or crosswise into circular pieces before drying. "Pusa Sawani" was reported to be the best for dehydration, followed by "Sel. 2" for its pod colour, texture, taste, flavour, dehydration and rehydration ratio and organoleptic qualities. Dehydrated okra contains comparatively higher amounts of carbohydrates, fat and ash but contains less protein and ascorbic acid than fresh okra. However, vacuum-dehydrated produce may retain higher ascorbic acid, colour pigment and mucilage at each of the dehydration temperatures than hot-air oven-dried okra.

For canning and freezing, okra fruits should be smaller in size, tender, dark green, with low dry matter and rich in nutrients. Pusa Sawani, Dwarf Green Smooth and Vaishali Vadhu were found to have better quality fruits suitable for canning. EMS 8 and Parbhani Tillu were also found to be promising cultivars for canning. US cultivars Emerald and Clemson Spineless have been recommended for canning. The quality attributes required for canning of okra fruits are also suitable for freezing. Clemson Spineless is a principal variety recommended for freezing. "Sel. 2" has also been found to be suitable for freezing. The cultivars Louisiana Market, Louisiana Green Velvet and Gold Coast, having short pods, were also found suitable for freezing. Fruit quality attributes in *A. manihot* are considered as good for processing. These can be transferred to cultivated okra background through interspecific hybridization. In India, so far, no direct breeding work has been carried out focusing on the quality and processing attributes of okra.

(b) *Breeding for Pest Resistance*

Spotted bollworm or fruit and shoot borer (*Earias* spp.) and jassids are major pests of okra. Crimson Smooth Long, IC-7194, IC-8899 and *A. manihot* var. *Ghana* were found to be resistant to Jassid. In cultivated okra accessions, a low degree of resistance against spotted bollworm has been observed. Narnaul Special, Perkins Long Green, Clemson Spineless, White Snow and Red Round were found to be less infested by insects and had less than 10% infestation against shoot and fruit borers. Resistance to leaf hopper (*Amrasca bigutula*), especially at early growth of the plant, was classified on the basis of insect numbers and the rate of damage in the field. Resistance to leaf hopper was observed in AE-11.

(c) *Breeding for Begomovirus Resistance in Okra*

Okra is infected by *Begomoviruses*, which are members of the family Geminiviridae, possessing circular ssDNA encapsidated in twinned quasiisometric particles (Venkataravanappa et al. 2015), transmitted by whitefly (*Bemicia tabaci*). Okra is reported to be infected by at least 27 begomoviruses, of which Bhendi Yellow Vein Mosaic Virus (YVMV) and Okra Enation Leaf Curl Virus (ELCV) diseases are the predominant ones that significantly affect fruit yield and quality (Mahesh and Yadav 2019; Badiger 2018; Yadav et al. 2019; Puneeth et al. 2022b; Santhiya et al. 2022; Sanwal et al. 2014). As of now, the New World *Begomoviruses* mostly have bipartite genomes (having DNA-A and DNA-B components), whereas in the Old World, the presence of both bipartite and monopartite (having DNA-A homolog only) genomes has been reported. The monopartite *Begomoviruses* are more prevalent than the bipartite types and are often associated with a class of ssDNA molecules known as satellite molecules.

Begomoviruses pose a serious threat to okra production worldwide. Due to quick spread of the disease and high infection rates in the field, management through physical, biological and chemical means is difficult. Hence, resistance breeding is the most appropriate strategy for disease management. Continuous efforts are being made in begomovirus resistance breeding, and several resistant varieties and hybrids have been released for cultivation.

Since in the cultivated gene pool, no durable resistance source is available, alternative sources of resistance reported in wild species, like *A. manihot*, *A. manihot* subsp. *manihot* and *A. tetraphyllus*, have been exploited through interspecific hybridization to evolve YVMV-resistant varieties (Dhankar et al. 2005). Similar to YVMV, for another newly reported viral disease, ELCV, which is appearing aggressively all over the World, there is no durable resistance source against the disease available in cultivated taxa of okra. Hence, breeders rely on crop wild relatives such as Abelmoschus *moschatus, A. crinitus*, *A. ficulneus*, *A. angulosus* and *A. manihot* for stable and reliable sources of resistance to ELCV (Santhiya et al. 2022). Further, reduced incidence and almost zero incidence of ELCV are reported in *A. tetraphyllus* (IC90476-1), *A. moschatus* (IC 141055) and *A. caillei* (Sikkim) genotypes (Santhiya et al. 2022). However, gene transfer through wide hybridization is curtailed by pre- and post-zygotic barriers due to genetic and cytological differences (Badiger et al. 2024).

Okra breeders have applied several tools, such as ovule culture, embryo culture and polyploidy, to address the cross-incompatibility issues (Gadwal et al. 1968) and evolved several viral-resistant varieties of okra. Backcrossing followed by the pedigree method of selection has been employed for transferring YVMV resistance genes from the wild species into a cultivated okra background. As of now, A. *manihot*, *A. caillei* and *A. tetraphyllus* have been the major sources of resistance. Varieties Punjab-7 and Prabhani Kranti were the resistant varieties developed using *A. manihot* ssp. *manihot*, while using *A. tetraphyllus*, varieties like Arka Anamika and Arka Abhay were developed at IIHR, Bangalore.

Although continuously several biotic stress (YVMV and ELCV) resistant varieties and hybrids have been developed in India by both public and private sectors, however, at the same time, frequent breakdown of resistance due to the rapid emergence of new or novel viral strains and sometimes outbreaks of polyphagous viruliferous whitefly vectors have been reported (Venkataravanappa et al. 2015). An alternative approach to controlling YVMV and ELCV infection in okra is through managing whitefly vector populations by insecticides, physical barriers, or by removal of symptomatic virus-infected plants. However, due to the continuous use of pesticides, whiteflies are showing resistance to several insecticides. Presently, virus resistance breeding in okra mainly relies on conventional methods. However, for quick outcomes, current conventional breeding should be complemented with molecular and genomic approaches for the development of pre-breeding lines resistant to viral diseases (Sanwal et al. 2014; Puneeth et al. 2023).

9.6 Biotechnology and Genomic Tools Implications in Okra Improvement

(a) *Molecular Breeding in Okra*

Okra is a tricky crop to deal with in molecular biology; even today, limited protocols are available in laboratories across the world for good-quality DNA isolation that can be utilized efficiently for advanced genomic studies. The presence of significant amounts of mucilaginous polysaccharides is the primary barrier, which makes it tedious and time-consuming for the isolation and purification of DNA from leaf tissue (Daher et al. 2016). These polysaccharides and polyphenols bind to nucleic acids during cell lysis, causing a reduction in the quantity and quality of isolated DNA. In addition, these bound impurities also inhibit various molecular activities like restriction enzyme digestion, polymerase chain reaction (PCR), or in vitro labelling. Besides, a number of researchers have optimized DNA isolation protocols in okra to isolate good-quality DNA.

9.7 Genetic Diversity and Germplasm Characterization Using Molecular Markers

Okra is an economically valuable crop, even though very little attention has been given to measure the genetic diversity in okra at the DNA level. Various studies have observed ample variability at the phenotypic stage among cultivated okra (*A. esculentus*), which requires confirmation at the DNA level using molecular markers. Only a small number of reports on genetic diversity studies using molecular markers are available in okra (Lata et al. 2021).

Okra genetic diversity between different accessions was initially studied using random amplified polymorphic DNA (RAPD) markers. RAPD markers showed a very high level of genetic diversity among *A. esculentus* and *A. caillei* (Aladele et al. 2008). Additionally, RAPD is also successfully utilized to measure genetic variation in cultivated taxa comprising 44 okra genotypes from India, 39 genotypes from Pakistan and 29 Nigerian okra genotypes (Prakash et al. 2011). Amplified Fragment Length Polymorphism (AFLP) marker analysis showed a correlation between the germplasm collection site and 21 Jordanian landraces. In a different AFLP study, 50 okra genotypes from various nations were discovered to exhibit diverse geographic distributions and modest levels of polymorphism (12%). Younis et al. found similar outcomes in their analysis of Egyptian okra using 42 ISSR and 5 AFLP primers. Also, 25 CAAT box-derived polymorphism (CBDP) markers and 36 start codon-targeted (SCoT) markers were used to investigate the genetic diversity among 89 accessions of okra, comprising of cultivated and wild species (Puneeth et al. 2022a, b). In this work, the CBDP markers outperformed the SCoT markers in terms of evaluating genetic diversity between wild and cultivated accessions of okra.

The Simple Sequence Repeat (SSR) markers, being multi-allelic in nature and having co-dominant inheritance with wide genome coverage, make them a crucial marker tool for plant breeding applications. For the first time in okra, genomic SSRs were generated using next-generation sequencing (NGS) technology, Roche 454 Titanium pyrosequencing. A total of 2708 microsatellites were derived from assembled reads in okra, and finally, 50 SSR primers were selected from 402 microsatellites for PCR (Ravishankar et al. 2018). Further, a total of 106,224 EST-SSRs were identified using transcriptome sequencing of contrasting okra varieties. Among them 183 SSR primers were validated in okra accessions (Puneeth et al. 2023). Anjan et al. (2022) utilized 65 SSR markers to study genetic diversity in 96 okra accessions comprising of cultivated and okra genotypes. Further, to maximize the use of global okra germplasm in enduring breeding programs around the world, intensive biodiversity studies are required.

9.8 Marker-Assisted Selection (MAS)

Molecular markers are very effective for the early identification of genotypes by confirming the presence or absence of a particular locus in genotypic data without waiting until phenotypic expression. However, very little molecular marker information is available for use in okra breeding. Because of its polyploid nature and large genome size, limited work has been done on marker discovery, genetic mapping/linkage map construction (Mishra et al. 2017). Additionally, due to large transcriptome arising from polyploidy also creates complexity in the annotation of de novo assembly of short sequence reads in genomic/transcriptomic studies (Mishra et al. 2017). Therefore, most genetic diversity-related work is reported in okra. However, a few scientific reports have now demonstrated the utility of marker-assisted selection in YVMV resistance breeding. To confirm virus infection, molecular markers can be used to amplify viral DNA from plants to detect YVMV-resistant genotypes.

(b) *Application of Tissue Culture and Genetic Transformation in Okra*

In vegetable breeding, the backcross method is commonly employed for transferring individual genes from wild relatives to the cultivated genetic background. In okra, it is extensively used for transferring YVMV resistance from its wild relatives. It involves extensive crossing and selection procedures, which generally take 10–15 years to develop a new variety. Moreover, sexual gene transfer mainly relies on cross-compatibility between wild and cultivated taxa. Unfortunately, in okra, most of the wide crosses are hampered by pre- and post-zygotic barriers. To overcome these problems, advanced technologies, like genetic transformation, are applied. A number of transgenic plants have been developed to improve yield, quality, enhance nutritional value, and reduce the use of pesticides and fertilizers.

Efficient transformation and in vitro regeneration protocols are crucial for the successful application of genetic engineering in okra improvement. Despite its economic value and decades of effort, progress in okra genetic engineering has been hampered by the absence of an efficient transformation protocol. Okra is a recalcitrant species, which is difficult for tissue culture regeneration, and only a few plant regeneration protocols are reported. For example, direct shooting from cotyledon and cotyledonary node explants, while the regeneration of okra plants from callus tissue using cotyledonary node as an explant was attained by Roy and Mangat. Further, a plant regeneration protocol based on embryogenesis was developed from suspension culture. Meristem culture techniques was used for the regeneration of disease-free plants. Effective protocol for the indirect shoot organogenesis of okra from leaf-disc and hypocotyl through a callus phase was used. Recently, successful regeneration using epicotyl and embryo explants was reported, where wild species, interspecific hybrids, along with cultivated okra genotypes, were used for regeneration in okra (Kiran et al. 2024).

Agrobacterium-mediated gene transfer technique is the most commonly followed transformation in crop plants. However, okra is found to be highly recalcitrant to *Agrobacterium*-mediated genetic transformation and regeneration. The standardized protocols for *Agrobacterium*-mediated transformation in okra have been reported by a few workers using various plant parts of okra as explants, which include cotyledonary axil, Hypocotyls, Shoot tips, Zygotic embryos, etc. (Mishra et al. 2017). *Agrobacterium*-mediated transformation using zygotic embryos as explants and confirmed the presence of the *cry1Ac* gene (conferring resistance to fruit borer) was done in okra. Okra seeds as explants for transformation was developed, which was found to be simple, less time-consuming and cost-effective and having the least chances of somaclonal variation (Manickavasagam et al. 2015). Further, the research focus is required to develop transgenics with coat protein (CP) genes using novel approaches such as RNAi and CRISPR technology that confer resistance to diseases, namely YVMV and OELCV (Mishra et al. 2017).

RNA interference (RNAi) is a naturally occurring phenomenon of sequence-specific and homology-dependent gene silencing induced by double-stranded (ds) RNA, which leads to target mRNA degradation in the presence of a protein complex. Host-delivered RNA interference (HD-RNAi) has been utilized to suppress targeted viral genes, which leads to a durable resistance strategy to combat various okra viruses using genetic engineering tools. Recently, a group of researchers developed YVMV-resistant lines using the HD-RNAi strategy. A hairpin-loop RNA (hpRNA) gene construct was designed and developed targeting the β-satellite βC1 ORF of the begomovirus. The transgenic okra plants developed through the in planta transformation strategy displayed resistance against the target virus and remained healthy during the crop span.

The CRISPR/Cas9 genome editing technology can be utilized for conferring resistance as well as examining gene function in the host. The CRISPR/Cas9 system was developed to target not only against the virus resistance but also to enhance yield-related parameters and plant architecture in numerous vegetable crops. However, researchers have not yet explored similar resistance approaches against viruses that infect okra.

9.9 Doubled Haploids

Homozygosis lines are prerequisites for breeding programs. In the hybridization as well as heterosis breeding, completely homozygous inbred lines are used as parents. Traditional breeding methodologies are time-taking, particularly to attain the homozygosity of the new inbred lines, which require around six to seven generations of selfing to acquire 99.22% homozygosity. Moreover, the methods are labour intensive, tedious and involve hand self-pollination every year. However, the development of 100% homozygous inbred lines in a shorter span can be achieved by generation of haploid plants through in vitro pollen or ovule culture followed by the chromosomes doubling (creation of doubled haploids). Haploid plants are a perfect model for genetic studies and provide opportunities to inspect hemizygous qualities, to study the effect of recessive genes and for easy detection of novel mutations. Completely homozygous doubled haploids are important genetic material for heterosis breeding and genetic mapping studies.

The introduction of in vitro techniques such as andro- and gynogenensis in the early 1980s led to a surge of interest in haploid crops for crop improvement. Among the different media tested, MS medium was found to be the most responsive, inducing 95% of calluses on average. The highest percentage (92.5%) of callus and root development was recorded after 28 days of incubation in dark conditions. However, root development could not be achieved even after a number of treatments and further subculturing in okra. Hence, in order to maximize the use of doubled haploids for accelerated breeding, there is a need to develop an effective method for androgenesis/gynogenesis in okra.

(c) *Next Generation Sequencing (NGS): Transcriptomics and Proteomics Studies in Okra*

In the field of molecular breeding, next-generation sequencing (NGS) has revolutionized the process of identifying and developing SSR and SNP markers. NGS techniques have the advantages of cost efficiency and the ability to identify a large number of SSRs in a small amount of time (Ravishankar et al. 2018). However, a limited number of studies using genomic approaches have been conducted on okra. Transcriptome analysis of contrasting traits can be a prospective move towards the identification of transcript/gene sequences in okra, where limited genome sequence information is available, and it is still considered an orphan crop. Several workers have started transcriptome studies in okra since the last decade (Lata et al. 2021). The first study on leaf and pod in okra was reported by RNA sequencing, and from short read assembly, more than 150,000 unigenes have been identified, and 935 SSRs were derived from this. SSR markers developed through unigenes obtained from transcriptome assembly were used by several workers in genetic analysis of okra and realized usefulness more often for diversity studies (Schafleitner et al. 2013). Similarly, in okra, Ravishankar et al. (2018), through the Roche 454 Titanium pyrosequencing platform, generated data isolating the genomic SSR markers. A total of 402 microsatellites have been designed, and 50 arbitrarily chosen SSR primers have been standardized for PCR studies in okra.

Regulation of gene expression at post-transcriptional levels is mediated by endogenous microRNAs (miRNAs). An investigation of miRNAs was conducted using next-generation sequencing. The database miRPlant was used to analyse the data, and 128 known and 845 novel miRNA candidates were found. They discovered 18 new miRNAs in addition to the 57 recognized miRNAs from 15 families. Furthermore, in 2018, Zhang and colleagues have discovered 293,971 distinct unigene sequences and 931 unigenes connected to enzymes involved in the production of flavonoids. The results revealed that 13 essential enzymes, which are responsible for the metabolism of fructose and mannose, were probably encoded by 691 unigenes. These transcriptome data will be helpful to identify and analyse genes involved in the synthesis of bioactive substances. Transcriptome data of *A. esculentus* leaves was generated, where 16,307 unigenes, 76 transcription factors and 9578 putative SSRs were discovered.

Particularly in fewer studied species like okra, transcriptome analysis has become a potent method for obtaining gene sequences and creating molecular markers. For instance, a pooled transcriptome (leaf and pod) of *A. esculentus* derived from RNA sequencing produced about 46 million base pairs of data and over 150,000 unigenes. From this, 935 SSR motifs were derived; out of these, 199 were tested, and about 161 were revealed to be highly polymorphic. However, the sequencing depth was insufficient to fully reflect the transcriptome due to the small number of full-length cDNAs and the comparatively small size of unigenes. Thus, in order to eliminate redundancy in the unigene annotation and enrich the unigene collection with full-length transcripts, deeper transcriptome sequencing is required. However, the allopolyploidy of *Abelmoschus*, which causes the transcriptome to comprise transcripts from many genomes with marginally different sequences, could possibly be the cause of some of the annotation redundancy. In addition, due to the increased size of the transcriptome, efficient de novo assembly of short sequence reads of polyploidy organisms like okra is highly challenging (Gruenheit et al., 2012). Although sequencing RNA is quite inexpensive, extensive sequencing of the transcriptome is constrained by the lack of funding for exploration of genomically orphan crops like okra, which culminates in short unigenes that only represent partial gene sequences. For okra research in the future, long unigenes from deep sequencing that represent full-length gene sequences will therefore be more advantageous.

Proteomics analysis has made protein expression studies easier at a global level. Proteomics analysis offers a great deal of information about the function of individual proteins involved in many biological processes. Soil salinity is also an important abiotic stress that adversely affects plant growth and agricultural output. It will be easier to produce crops that can withstand salt stress if researchers decode the mechanisms by which plants defend themselves in certain cases. Proteomic studies are more suitable for understanding the molecular players that are responsible for tolerating various stress conditions. Zhan et al. (2019) examined the proteins that were differently expressed between the NaCl-treated and control seedlings using a

TMT-based proteomic method. A total of 7179 proteins have been identified by proteomic data. Of them, 317 were differentially expressed proteins (DEPs), of which the expression of 165 proteins increased and the expression of 152 proteins decreased when treated with NaCl.

References

Abraham V, Bhatia CR (1984) Induced mutations in okra (*A. esculentus*). Mutation breed. Newsletter 24:1–3

Aladele SE, Ariyo OJ, de Lapena R (2008) Genetic relationships among West African Okra (*Abelmoschus caillei*) and Asian genotypes (*Abelmoschus esculentus*) using RAPD. Afr J Biotechnol 7:1426–1431

Anjan D, Kumar YR, Rakesh B, Harshawardhan C, Khade YP (2022) Study of heterotic potential for important antioxidant and nutritional traits in okra [*Abelmoschus esculentus* (L.) Moench.]. Indian J Agric Sci 98(8):986–990

Badiger MK (2018) Ph.D. thesis on Studies on genetic diversity and interspecific crosses between *Abelmoschus esculentus* and its wild relatives for resistance to BYVMV. IARI, New Delhi

Badiger M, Yadav RK, Sharma BB, Bhat KV, Tomar BS, Lata S, Das A (2024) Pollen germination, pollen–pistil interaction and crossability studies in interspecific and induced colchiploid population of Abelmoschus species. Genet Resour Crop Evol 71(1):107–127

BeMiller JN, Whistler RL, Barkalow DG, Chen CC (1993) Aloea, chia,flax seed, okra, psyllium seed, quince seed, and tamarin gums. In: Whistler RL, BeMiller JN eds Industrial gums. Academic Press, New York, pp 227–256

Chauhan VBS, Kumar R, Behera TK, Yadav RK, Verma AK (2019) Inheritance of fruit weight and mode of gene action for yield contributing traits in tomato. Res J Biotechnol 14(4):73–78

Daher RK, Stewart G, Boissinot M, Bergeron MG (2016) Recombinase polymerase amplification for diagnostic applications. Clin Chem 62(7):947–958

Datta PC, Naug A (1968) A few strains of *Abelmoschus esculentus* (L.) Moench their karyological in relation to phylogeny and organ development. Beitr Biol Pflanzen 45:113–126

Dhankar SK, Dhankar BS, Yadava RK (2005) Inheritance of resistance to yellow vein mosaic virus in an interspecific cross of okra. Indian J Agric Sci 75:287–290

Dhillon TS, Sharma BR (1982) Interspecific hybridization in okra (*Abelmoschus sp.*). Genet Agrar 36:247–256

Dutta OP (1984) Breeding okra for resistance to yellow vein mosaic virus and enation leaf curl virus. Annu. Rep. IIHR, Bangalore, p 43

Dutta OP (1991) Okra germplasm utilization at IIHR, Bangalore. International workshop on okra genetic resources. NBPGR, New Delhi, pp 114–116

Gadwal VR, Joshi AB, Iyar RD (1968) Interspecific hybrids in *Abelmoschus* through ovule and embryo culture. Indian J Genet 28:269–274

Grubben GJH (1977) Tropical vegetables and their genetic resources. HD Tindall & JT Williams (eds.) IBPGR. Rome. 197 p.

Jambhale ND, Nerkar YS (1979) Inheritance of chlorine and induced chlorophyll mutant in okra (*Abelmoschus esculentus* (L.) Moench). J Maharashtra Agric Univ 4(3):316

Jambhale ND, Nerkar YS (1986) Parbhani Kranti, a yellow vein mosaic-resistant okra. Hort Sci 21:1470–1471

Joshi AB, Hardas MW (1956) Alloploid nature of okra, *Abelmoschus esculentus* (L.) Moench. Nature 178:1190

Joshi AB, Hardas MW (1976) Okra. In: Simmonds NW (ed) Evolution of crop plants. Longman, London, pp 194–195

Kiran BS, Yadav RK, Lata S, Sharma BB, Talukdar A, Ghosh A et al (2024) In vitro studies of epicotyl and embryo culture in cultivated, wild and inter-specific hybrids of okra (*Abelmoschus* spp.). Plant Cell Tissue Organ Cult (PCTOC) 156(3):80

Kulkarni UG, Nerkar YS (1992) Parbhani Tillu-an induced bhindi mutant suitable for processing. J Maharashtra Agric Univ 17(3):496

Kuwada H (1966) The new amphidipioid plant named Abelmoschus tubercular esculentus obtained from the progeny of the reciprocal crossing between A. tuberculatus and A. esculentus. Jpn J Breed 16: 21–30

Lata S, Kumar Yadav R, Tomar BS (2021) Genomic tools to accelerate improvement in Okra (*Abelmoschus esculentus*). Intech Open. https://doi.org/10.5772/intechopen.97005

Mahesh B, Yadav RK (2019) Screening of germplasm of *Abelmoschus* against biotic stresses. Indian J Agric Sci 89(12):2085–2090

Martin FW (1982) Okra, potential multiple-purpose crop for the temperate zones and tropics. Econ Bot 36(3):340–345

Mishra GP, Singh B, Seth T, Singh AK, Halder J, Krishnan N, Singh PM (2017) Biotechnological advancements and begomovirus management in okra (*Abelmoschus esculentus* L.): status and perspectives. Frontiers in plant science, 8, 360. https://doi.org/10.3389/fpls.2017.00360

NHB (2020) National Horticulture Database, 2017, Ministry of Agriculture. Govt. of India, Gurgaon

Pal BP, Singh HB, Swarup V (1952) Taxonomic relationships and breeding possibilities of species of *Abelmoschus* related to okra *(A. esculentus)*. Bot Gaz 113:455–464

Patil JA (1963) Polyploidy in vegetable crops. "Okra" *(Abelmoschus esculentus)*. Poona Agric Coll Mag 54:20–25

Prakash K, Pitchaimuthu M, Ravishankar K (2011) Assessment of genetic relatedness among okra genotypes [*Abelmoschus esculentus* (L) Moench] using RAPD markers. Electron J Plant Breed 2:80–86

Puneeth PV, Suman L, Yadav RK, Wankhede Dhammaprakash P, Tomar BS, Choudary H, Avinash T, Bidaramali V, Talukdar (2022a) Exploring the genetic diversity using CAAT box-derived polymorphism (CBDP) and start codon targeted (SCoT) markers in cultivated and wild species of okra (*Abelmoschus esculentus* (L.) Moench). Genet Resour Crop Evol. https://doi.org/10.1007/s10722-022-01458-8

Puneeth PV, Yadav RK, Suman L, Amalendhu G, Choudary H, Tomar BS, Bidaramali V, Krishnan GB, Anjan D, Avinash T (2022b) Vulnerability studies of okra [*Abelmoschus esculentus* (L.) Moench.] genotypes to bhendi yellow vein mosaic virus (BYVMV) disease. Indian J Hortic 79(2):186–193

Puneeth PV, Yadav RK, Lata S, Asif M, Talukdar A, Choudhary H, Tomar A (2023) Development of EST-SSR markers using transcriptome data in okra (*Abelmoschus esculentus* L.): a genetically orphan crop. Indian J Genet Plant Breed 83(01):95–102. https://doi.org/10.31742/ISGPB.83.1.12

Ravishankar KV, Muthaiah G, Mottaiyan P, Gundale SK (2018) Identification of novel microsatellite markers in okra (Abelmoschus esculentus (L.) Moench) through next-generation sequencing and their utilization in analysis of genetic relatedness studies and cross-species transferability. J Genet 97(1):39–47

Roy MK, Mangat BS (1989) Regeneration of plants from callus tissue of okra (*Abelmoschus esculentus*). Plant Sci 60:77–81. https://doi.org/10.1016/0168-9452(89)90046

Santhiya S, Yadav RK, Suman L, Sharma BB, Talukdar A, Amlendu G, Vaishali C, Janani J (2022) Screening of wild okra (*Abelmoshus esculentus*) against yellow vein mosaic and enation leaf curl diseases. Indian J Agric Sci 92(10):1268–1273

Sanwal SK, Singh M, Singh B, Naik PS (2014) Resistance to Yellow Vein Mosaic Virus and Okra Enation Leaf Curl Virus: challenges and future strategies. Curr Sci 106(11):1470–1471

Schafleitner R, Kumar S, Lin CY, Hegde SG, Ebert A (2013) The okra (Abelmoschus esculentus) transcriptome as a source for gene sequence information and molecular markers for diversity analysis. Gene 517(1):27–36

Sharma BR, Arora SK (1991) Mutation breeding in okra In: Indian society of Genetics and plant Breeding. Ed. Golden Jubilee symposium on Genetic Research and Education.Current Trends and the next fifty years, New Delhi, IN. Indian Agricultural Research Institute, 734P.

Siemonsma JS (1982) West African okra – morphological and cytogenetical indicators for the existence of a natural amphidiploid of *Abelmoschus esculentus* (L.) and *A. manihot* (L.) Mediks. Euphytica 31:241–252

Singh HB, Joshi BS, Khanna PP, Gupta PS (1962) Breeding for field resistant to yellow vein mosaic in bhindi. Indian J Genet Plant Breed 22(2):137–144

Thakur MR (1976) Inheritance of yellow vein mosaic in a cross of okra species, *Abelmoschus esculentus* × *A. manihot ssp. manihot*. SABRAO J 8:69–73

Venkataravanappa V, Reddy CNL, Jalali S, Briddon RW, Reddy MK (2015) Molecular identification and biological characterization of a begomovirus associated with okra enation leaf curl disease in India. Eur J Plant Pathol 141:217–235. https://doi.org/10.1007/s10658-014-0463-0

Yadav RK, Mahesh B, Sumanlata (2019) Studies of genetic variability and association of yield traits and YVMV disease in *Abelmoschus* species. Indian J Agric Sci 89(4):688–694

Zhan Y, Wu Q, Chen Y, Tang M, Sun C, Sun J, Yu C (2019) Comparative proteomic analysis of okra (*Abelmoschus esculentus* L.) seedlings under salt stress. BMC Genomics 20:381

Utilization of Male Sterility in Hybrid Seed Production of Okra [*Abelmoschus esculentus* (L.) Moench]

10

M. Pitchaimuthu and O. P. Dutta

10.1 Introduction

Okra [*Abelmoschus esculentus* (L.) Moench] is one of the popular traditional and commercial vegetable crops cultivated in tropical, subtropical and warmer regions of the world. The total area and production of okra are reported to be around two million ha and ten million tonnes, respectively. It is mainly grown in India, Nigeria, Sudan, Pakistan, Ghana, Egypt, Benin, Saudi Arabia, Mexico and Cameroon. The largest area and production are in India, followed by Nigeria. The highest average productivity is reported from Egypt (13.3 tonnes/ha), followed by Saudi Arabia (12.5 tonnes/ha). India ranks first in the world, with 5227.6 thousand ha area and production of about 64,163.3 thousand tonnes (around 63% share of the total world production) of okra. In India, the okra seed market is projected to grow from $87.12 million in 2022 to $148.45 million in 2029 at a CAGR of 7.91%. Fortune business insights (2022) reports that approximately 4800 MT of seeds are sold in India today, with over 90% being hybrid seeds. The average productivity is very low in India (11.70 t/ha), and the crop is prone to damage by various fungi, nematodes and mainly viruses like YVMV and ELCV. Marked heterosis (3.8–71%) has been reported in okra for yield and its components. Commercial F_1 hybrid seed production is a time-consuming and cumbersome process involving hand emasculation and hand pollination, developing a GMS system-based hybrid seed production system of great consequence. Recently, several hybrids released both from the public and private sectors, such as Cobh-4, Kashi Bhairav and Pusa Okra Hybrid 1, Radhika,

*Paper presented at the Global Okra Round Table-1 10–12th October 2022 at ICAR-IARI, New Delhi

M. Pitchaimuthu (✉) · O. P. Dutta
Division of Vegetable Crops, ICAR-Indian Institute of Horticultural Research, Bengaluru, Karnataka, India
e-mail: m.pitchaimuthu@icar.gov.i

S. K. Tikoo et al. (eds.), *Okra: Status, Challenges and Opportunities*,
https://doi.org/10.1007/978-981-97-9963-3_10

NS-862, Singam, Mahyco10, OH-2324 and Lavanya, are covering a considerable area under commercial cultivation, and the GMS system would economise on seed production costs and improve purity.

10.2 Development of Genetic Male Sterility (GMS) in Okra

The occurrence of male sterility in okra in nature is very rare. However, it has been induced artificially by gamma radiation at ICAR, IIHR, Bengaluru by the authors. This GMS in okra is conditioned by a single recessive gene when present in a homozygous (ms1 ms1) condition (Dutta 1983). The gene was found to be stable and not influenced by environmental factors. The anthesis was normal, but anther dehiscence was partial. The microsporogenesis was normal up to the tetrad stage, and thereafter, microsporogenesis degenerated (Figs. 10.1 and 10.2). In earlier days, male sterility was considered a disadvantage, but now it has a place of pride in the breeding of both cross- and self-pollinated crops.

The GMS line (deliberate exclusion of pollen in this kind of situation) has solved many problems, such as: (a) elimination of tedious emasculation, (b) providing correct combinations of parents, (c) providing more flexibility to breeding and (d) facilitating the quick incorporation of diverse genes for disease resistance in a hybrid. The male sterile gene was incorporated into a YVMV-tolerant line (Pitchaimuthu et al. 2012).

Field designs for maximizing hybrid seed yield using male sterile lines under natural cross-pollination have been standardized, and the combining ability of the parents using male sterile lines has been worked out (Pitchaimuthu and Dutta 2002). A field design where alternate planting of two rows of the plants was done, keeping a ratio of 2:1 and maintaining a plant population of 4166 male sterile plants per hectare, gave a hybrid seed yield of 5.66 quintals per hectare, which was 52.56% higher seed yield in comparison with their fertile control (Dutta 1983).

Fig. 10.1 Male fertile flower with yellow-coloured pollen grains

Fig. 10.2 Male sterile flowers showing no pollen grains

In the systematic okra improvement programme at ICAR-IIHR, Bengaluru, led by a team of scientists, 20 F_1 hybrids were developed using GMS lines. Based on combining ability, six top-ranking hybrids were tested along with commercial checks, Shakti and Sarika, for yield, quality and resistance to YVMV at IIHR, Bengaluru and a hotspot area of Guntur, Andhra Pradesh, during the Summer and Kharif seasons of 2014–2016 (Fig. 10.3). The results of mean pooled data revealed that during the summer season, OKMSH-3 (Arka Nikita) was found to be significantly superior with 20.82 t/ha and a 34.32% yield increase over check Shakti and a 35.37% yield increase over Sarika (Table 10.1), followed by OKMSH-7 (19.11 t/ha) (Fig. 10.4). During the Kharif season, among the hybrids evaluated, OKMSH-3 recorded the highest mean fruit yield of 23.69 t/ha, which recorded a 39.27% yield increase over check Shakti and a 45.24% over Sarika, followed by OKMSH-2 (20.12 t/ha) (Table 10.2).

Evaluation of okra GMS hybrids for YVMV incidence A total of six GMS-based F_1 hybrids, including the selected F1 hybrid named Arka Nikita by the institute, along with two commercial checks, were screened at ICAR-IIHR, Bengaluru, during 2014–2015 and KVK Guntur (A.P.) during 2016. At ICAR-IIHR, Bengaluru, Arka Nikita did not show any symptoms of YVMV in both the years (0% incidence). However, at the hotspot area of Guntur during 2016, Arka Nikita had an incidence of 26.83% YVMV disease, whereas the commercial check Shakti showed a mean incidence of 25.90%, and Sarika recorded 27.60%, respectively (Table 10.3).

10.3 Mean Performance of Okra Hybrid

The mean squares due to crosses were highly significant for all the characters studied except earliness (days to flower opening). The range of mean performance for the six crosses and two commercial checks is presented in Table 10.4. Days to first

Fig. 10.3 Flow chart of GMS-based F_1 Arka Nikita development

fruit harvest among the crosses ranged from 39 days to 51 days. The number of branches per plant ranged from 2.33 to 4.33 among the crosses. Plant height ranged from 122.0 to 182.33 cm. Fruit length varied from 12.93 to 16.17 cm. Fruit diameter varied from 1.233 to 1.86 cm. Fruit weight varied from 25 to 28 cm. Among the crosses, OKMSH-3 (Arka Nikita) was found to be significantly superior to the

Table 10.1 Mean performance of the okra GMS hybrids for yield (t/ha) during summer seasons of 2014–16

Estimated yield (t/ha) over the seasons					Percent increase over checks	
Genotype	2014	2015	2016	Mean	Shakti (C1)	Sarika (C2)
OKMSH-1	12.07	18.23	15.76	15.35		
OKMSH-2	10.87	19.35	18.25	16.15		
OKMSH-3 (Arka Nikita)	**17.67**	**23.37**	**21.43**	**20.82**	**34.32**	**35.37**
OKMSH-4	14.56	19.60	15.65	16.60		
OKMSH-7	15.05	21.49	20.85	19.11		
OKMSH-9	14.00	20.00	15.15	16.38		
Sarika (C1)	11.66	19.43	15.06	15.38		
Shakti (C2)	12.50	19.65	14.36	15.50		
Mean	13.54	20.14	17.06	16.91		
Significance	0.5%	0.5%	0.5%	0.5%		
C.D./SEm±	3.22	2.49	1.92	2.54		
CV %	9.45	7.08	6.43	7.65		

Fig. 10.4 Dark green fruits of Arka Nikita

Salient Characteristics of Arka Nikita GMS-based F_1 hybrid

- Yield 25 t/ha
- Fruits are dark green, smooth and thin
- Tender fruits (10–12 cm length)
- Early fruiting: 39–43 days
- Green stem and unique petal base (Purple colour in inner petals)
- Suitable for Kharif, rabi and summer seasons
- Seed production is easy and saves 70% labour cost by avoiding emasculation
- Rich in iodine: 33.31 µg/kg
- Rich in mucilaginous substance: 1.08% (FW)
- Rich in edible fibre: 8.85% (DW)
- Rich in potassium: 3.7%

Table 10.2 Mean performance of the okra GMS hybrids for yield (t/ha) during Kharif seasons of 2014–16

Estimated yield (t/ha) over the seasons					Percent increase over checks	
Genotype	2014	2015	2016	Mean	Shakti (C1)	Sarika (C2)
OKMSH-1	14.85	16.33	14.84	15.34		
OKMSH-2	15.48	23.16	21.73	20.12		
OKMSH-3 Arka Nikita	**23.00**	**22.33**	**25.75**	**23.69**	**39.27**	**45.24**
OKMSH-4	12.66	18.33	20.75	17.24		
OKMSH-7	19.66	18.50	18.76	18.97		
OKMSH-9	14.20	11.50	24.85	13.01		
Sarika (C1)	15.70	12.33	20.91	16.31		
Shakti (C2)	18.58	17.00	17.91	17.01		
Mean	16.14	17.43	20.68	17.71		
Significance	0.5%	0.5%	0.5%	0.5%		
C.D./SEm±	2.504	1.781	3.63	2.638		
CV %	8.855	6.344	10.89	8.696		

Table 10.3 Mean incidence of YVMV disease in okra GMS hybrids during summer seasons of 2014–16

Incidence of YVMV over the seasons				
Genotype	2014 IIHR	2015 IIHR	2016 A.P. Guntur	Mean
OKMSH-1	1.58	0	33.32	34.9
OKMSH-2	0	0	16.33	16.33
OKMSH-3 Arka Nikita	**0**	**0**	**26.83**	**26.83**
OKMSH-4	0	0	33.61	33.61
OKMSH-7	1.46	0	23.33	24.79
OKMSH-9	1.19	6.53	30.86	38.58
Sarika (C1)	0.70	0	26.90	27.60
Shakti (C2)	2.00	0	23.90	25.90
Mean	6.93	6.53	31.88	45.34
Significance	0.5%	0.5%	0.5%	–
C.D./SEm±	–	–	8.09	–
CV %	–	–	13.59	–

Table 10.4 Advantages of GMS-based F_1 hybrid seed production over conventional method of hybrid seed production in okra

SL.No	Parameters	Advantages (0.5%)	Conventional method
1.	Hand emasculation	Not required	Required
2.	Produce 1 kg of hybrid seeds No. of flowers to be crossed Time consumed	274 2 hours 53 minutes	274 9 hours 37 minutes
3.	Seed rate (kg /ha)	6 kgs	4 kgs
4.	Spacing (cm)	60 × 15	60 × 30
5.	Removal of fertile plants	Required	Not required
6	Cost benefit ratio	1: 4.36	1: 2.68

GMS based hybrid seed production achievement: 70% saving in time and manual labour

commercial checks Sarika and Shakti. The mean performance of OKMSH-3 had 33 days for earliness, 39 days to first fruit harvest, number of branches 4.333, plant height of 169 cm, fruit length of 12.93 cm, fruit weight of 26.5 g, fruit colour dark green and tolerance to YVMV. However, it was susceptible to powdery mildew and wilt during the rabi/rainy season, which can be managed by systematic spraying of fungicides.

Okra hybrid seed production demands highly skilled labour. The okra flowers, being hermaphrodite, need hand emasculation and pollination, which are laborious, time-consuming and expensive. Further during the processes of hand emasculation, the fingertips of the labourers are seriously injured due to the presence of certain chemicals in the stamen and calyx of the flower bud, causing a lot of pain. Okra leaves are covered with glandular spines, which cause a lot of itching and irritation to the skin during the processes of emasculation and pollination. In male sterile lines, emasculation is not required. This provides a lot of relief from hand injury and skin irritation. Further, as compared to fertile lines, there is a 70% reduction in time and labour requirements for seed production. Adoption of the okra hybrid Arka Nikita by the growers gave an average net return of Rs. 40,000 per hectare due to 25% increased yield over other commercial hybrids and Rs. 85,000 net returns from commercial hybrid seed production. Presently, the okra hybrid Arka Nikita is being tested under the All India Co-ordinated Vegetable Improvement Project (AICRPVC) at 15 centres all over India. In Karnataka, this hybrid has been introduced in farm trials for frontline demonstration at 12 locations under the control of UHS, Bagalkot and UAHS, Shivamogga. A successful demonstration of Arka Nikita has been conducted at KVK, Hirehalli (Table 10.5).

Hybrid seed production in okra is labour-intensive. It requires 180 man-days per acre for the production of hybrid seeds by emasculation and pollination. To produce 1 kg of okra hybrid seeds by conventional methods, it takes 9 hours and 52 minutes to complete the emasculation and hand pollination of 277 flowers, whereas, by using genetic male sterile lines, it requires only 3 hours to complete the same task

Table 10.5 Economics of hybrid seed production using GMS line in okra (per hectare)

SL	Particulars	Geneic male sterile line-based hybrid (Rs/ha)	Normal fertile line-based hybrid (Rs/ha)
1	Land preparation	3000.00	3000.00
2	Seeds & sowing	2350.00	2350.00
3	FYM + Fertilizer & application cost	16,535.00	16,535.00
4	Intercultural operations	2800.00	2800.00
5	Emasculation & pollination	22,230.00[a]	44,460.00
6	Plant protection	6200.00	6200.00
7	Irrigation	7500.00	7500.00
8	Harvesting & cleaning	1200.00	1200.00
9	Total cost cultivation (Rs)	61,815.00	83,695.00
Seed yield (Q/ha):	5.99		5.00
Gross return (Rs):	269,550		225,000
Procurement Price (Rs):	500/kg		150/kg
Benefit Cost ratio (Rs):	4.36		2.68

[a]No emasculation, only pollination

(Table 10.6, Dutta 1983). Thus, there is a 70% saving in time. Further, the use of male sterile lines protects the okra labour force from physical hand injury and skin irritation, thereby increasing their efficiency and reducing their physical drudgery. Keeping in view the above advantages of male sterile lines, the corporate sector has shown a lot of interest in these lines for incorporating their own breeding efforts. A total of 16 National and International companies have licensed the seeds of male sterile lines developed by the authors at ICAR-IIHR for their hybrid seed development projects.

Genetic male sterility systems have been utilized for commercial hybrid seed production in okra. The female and male lines are grown in a 4:1 ratio. However, to maintain a good plant population in the female rows, it is suggested that the seed parent should be sown with three to four seeds per hill. The male sterile line is maintained in the heterozygous form by crossing with a maintainer line under adequate isolation distance or under cover. The okra GMS-based hybrid Arka Nikita, with a high yield potential (25 t/ha), can be a very useful addition to the national programme. The use of the GMS-based hybrid will reduce the cost of hybrid seed production and hence increase profitability (Table 10.5, Pitchaimuthu et al. 2012). The GMS-based hybrid also possesses high mineral and fibre content, making it a good choice from a nutritional standpoint (Tables 10.7 and 10.8).

Table 10.6 Mean performance of various parameters of okra hybrids during the summer seasons of 2014–2016

Hybrid	1 Earliness (Days)	2 Days to first harvest	3 No. of branches	4 Plant height (cm)	5 Fruit length (cm)	6 Fruit diameter (cm)	7 Fruit weight (g)	9 Fruit colour	Reaction to diseases YVMV	PM	Wilt
OKMSH-1	33.66	40.66	3.000	122.000	13.767	1.667	25.6	Dark Green	T	S	S
OKMSH-2	35.33	41.33	2.333	154.333	13.533	1.867	26.0	Dark Green	MR	S	S
OKMSH-3	**33.00**	**39.00**	**4.333**	**169.000**	**12.933**	**1.700**	**26.5**	**Dark Green**	**T**	**S**	S
OKMSH-4	33.66	39.66	2.667	160.000	15.133	1.233	26.5	Dark Green	T	S	S
OKMSH-7	35.66	41.66	3.667	182.333	13.867	1.900	27.0	Dark Green	MR	S	S
OKMSH-9	37.50	43.50	3.000	155.667	13.200	1.733	28.0	Dark Green	T	S	S
Sarika (C1)	42.00	48.00	3.667	165.000	16.167	1.833	25.0	Lush Green	T	S	S
Shakti (C2)	45.00	51.00	4.667	182.333	15.767	1.633	26.0	Dark Green	MR	S	S
Mean	37.00	48.66	3.416	161.33	14.295	1.695	33.6	–	31.88	S	S
Significance	NS	0.5%	0.5%	0.5%	0.5%	0.5%	NS	–	0.5%		
C.D.5% SEm±	–	3.56	1.114	18.513	1.165	0.388	–	–	8.09	–	–
CV %	–	10.68	18.621	6.552	4.652	9.412	–	–	13.59	–	–

Significant at $p < 0.05$ level respectively

Table 10.7 Estimation of mineral content in okra pre-released hybrids during summer 2015

Hybrids	N(%)	P(%)	K(%)	Ca(%)	Mg(%)	Cu(ppm)	Zn(ppm)	Fe (ppm)	Mn (ppm)
GMS4x285	3.70	0.52	3.10	2.76	0.98	14.2	28.13	51.1	12.36
GMS4x551	3.16	0.60	3.00	2.92	0.98	14.46	31.13	45.33	10.66
GMS4x294	3.11	0.61	2.96	2.18	0.98	7.80	26.6	15.43	5.30
GMS4x299	**2.53**	**0.75**	**3.70**	**3.41**	**1.00**	**16.00**	**43.34**	**21.26**	**5.53**
GMS4x875	3.30	0.67	3.33	4.04	0.98	14.66	25.13	30.13	8.06
299 (P)	2.16	0.62	3.30	3.11	1.00	18.80	38.40	28.00	11.30
GMS4 (P)	2.81	0.54	2.40	3.37	0.99	17.00	30.20	34.50	5.40
Sakthi©	2.34	0.54	2.70	3.41	0.98	18.20	34.86	29.10	7.50

Table 10.8 Estimation of mucilage, crude fibre and Iodine content of okra of okra pre- released hybrids during summer 2015

Sl. No	Hybrids	% Mucilage (Fresh Weight)	% Fibre (Dry wt)	Iodine content (μg/kg) Dry Wt
1	**GMS4 x 299**	**1.08**	**8.85**	**33.31**
2	GMS4x 875	0.69	6.8	31.21
3	GMS4x 285	0.79	7.2	21.85
4	GMS4x 551	0.79	8.57	30.00
5	GMS4x 294	0.83	8.57	26.50
6	Sarika (C)	0.71	6.80	27.00
7	Sakthi (C)	0.83	8.30	30.50
8.	A. Anamika	0.98	6.89	25.50

10.4 Conclusion

Keeping in view the higher percentage of seed setting and elimination of the labour-oriented process of emasculation by using a genetic male sterile line (GMS), the production of hybrid seed could be made considerably cheaper in okra. In addition, the seed of high genetic purity will be produced with the use of GMS lines, simply because the selfed seed of the female parent is not produced at all. Shroff et al. (1988) made a very extensive investigation and reported that the elimination of emasculation helped in reducing the labour cost by at least 50%, and hybrid seed of cotton could be produced very economically.

The results obtained on the basis of the present investigation revealed that the hybrid seed production of okra could be feasible, easy and cheaper with the employment of a genetic male sterile line as the female parent of a hybrid. Efforts are in progress to identify male sterile and male fertile plants before flowering through cost-effective molecular markers for easy removal of fertile plants in the large-scale seed production plot and maintain the total plant population to get more seed yield.

Arka Nikita has an export potential to many international markets. It will contribute to foreign exchange and cut down imports from other countries. This GMS-based hybrid is tolerant to YVMV, and there will be a drastic reduction in the use of pesticides, which in turn reduces the costs of seed production, and lower pesticide use makes it a more enviro-friendly option too. This hybrid can very well adapt to organic farming and fetch higher market prices besides the high yield and better quality of the produce. It creates employment opportunities for farm youth and farm women during commercial production as well as hybrid seed production. As parents of public bred hybrid Arka Nikita are accessible for licensing on a non-exclusive basis to any entrepreneur, it encourages individual entrepreneurs to set up their economically viable hybrid seed production programmes successfully.

10.5 The Way Forward

- The development of practically feasible molecular markers may provide an appropriate cost-effective selection strategy to discard 50% of male fertile sister plants at the seedling stage, which may open the way to exploit monogenic recessive male sterile lines (Kumar and Singh 2004).
- The development of cytoplasmic genetic male sterility with a usable fertility restorer system is required. Work to discover this is ongoing at IIHR.
- Conversion of RAPD markers to SCAR markers to identify homozygous recessive gene msms at an early stage (before flowering) to rogue out fertile plants in the seed production plot.

References

Dutta OP (1983) Male sterility in okra [*Abelmoschus esculentus* (L.) Moench.] and bottle gourd (*Lagnaria scieraria* (Mol. (Standl) and utilization in hybrid seed production. Ph.D thesis, UAS, Bangalore

Fortune Business Insights (2022) The India okra seed market is projected to grow from $87.12 million in 2022 to $148.45 million at a CAGR of 7.91% in forecast period, 2022–2029. https://www.fortunebusinessinsights.com/india-okra-seed-market-106721

Kumar S, Singh PK (2004) Mechanisms for hybrid development in vegetables. J New Seeds 6(4):300–407

Pitchaimuthu M, Dutta OP (2002) Combing ability using genetic male sterile lines in okra. International conference on vegetable crops. November 11–14, 2002, Bangalore, India, p 109

Pitchaimuthu M, Dutta OP, Swamy KRM (2012) Studies on inheritance of Geneic Male Sterility (GMS) and hybrid seed production in okra [*Abelmoschus esculentus* (L.) Moench.]. J Hortic Sci 7(2):199–202

Shroff VN, Mandloi KC, Pandey SC, Nigam (1988) Use of Asiatic hybrid seed production in Cotton. J Cotton Res Dev 2:56–61

11 Okra Improvement in Egypt and North Africa: Current Status and Opportunities

Ehab A. Ibrahim and Hamdino M. I. Ahmed

11.1 Introduction

Okra (*Abelmoschus esculentus*) is an old vegetable crop that originated in Ethiopia, was grown by the ancient Egyptians, and is now cultivated worldwide (Kumar et al. 2013). Its cultivation became widespread over North Africa (Lamont 1999). Until recently, researchers paid little attention to the improvement of okra because it was considered a minor crop (Düzyaman 2005). The edible okra fruits, or pods, have a high nutritional value, as they are an excellent source of vitamins, fiber, carbs, proteins, and amino acids (Gemede et al. 2015; Liu et al. 2021; Petropoulos et al. 2018). Thus, it offers considerable medicinal advantages in addition to being an excellent nutritional source (Elkhalifa et al. 2021; Gemede et al. 2015). Okra's elevated position as a functional food with considerable nutraceutical benefits has led to an increase in consumption (Durazzo et al. 2018). Okra is a conventional crop that requires little agronomic inputs and can significantly boost sustainable agriculture production in countries like Egypt and Sudan. This underutilized crop can be crucial in producing income and alleviating poverty for the rapidly rising population by assisting in securing food, nutrition, and health security.

Okra belongs to the *Malvaceae* family and has polyploid chromosomes with $2n = 72–144$. There are several species around the globe, but *Abelmoschus esculentus* ($2n = 130$) is known to be the most common and widely cultivated species (Joshi and Hardas 1956). Extensive biodiversity supports the widespread subsistence farming system by providing food, nutritional security, health benefits, and revenue diversification. Therefore, it is essential to increase the genetic potential of locally grown vegetables, like okra species, for crop yield and nutritional quality. Breeders must first assess and define their germplasm in order to uncover sources of genes for

E. A. Ibrahim (✉) · H. M. I. Ahmed
Vegetable Crops Research Department, Horticulture Research Institute, Agricultural Research Center, Giza, Egypt

S. K. Tikoo et al. (eds.), *Okra: Status, Challenges and Opportunities*,
https://doi.org/10.1007/978-981-97-9963-3_11

unique traits such as greater and higher quality, yield, tolerance to biotic and abiotic challenges, and broader adaptability. Characterization of genetic resources is defined as identifying, differentiating, or separating accessions based on their morphological and/or biochemical quality traits (Ali et al. 2014; Mohammed et al. 2022). The genetic characterization and evaluation of this crop's genetic resources in North Africa are currently poorly understood. Okra varies both within and across species in different phytogeographic zones of Egypt and Sudan (Helmy and Ragheb 2019; Mohamed 2018). Most of the types cultivated over time in various locations in these countries are landraces. As a result of adopting low-yielding native landraces and poor agronomic management techniques, okra production and productivity have decreased (Eltigani et al. 2021). Despite the importance of the okra crop, no significant advances have been made in this crop, and Egyptian and Sudanese farmers still only grow open-pollinated or local cultivars. Farmers typically choose their cultivars during harvest based on morphological traits like plant size, color, and pod size. The seeds from the vigorous plants are saved for the subsequent season. Therefore, plant breeders must create okra varieties unique to their area.

Traditional farmers regularly utilize selection to achieve various objectives and eventually mix them up. These conventional farmers grow a wide variety of okra cultivars (landraces) adapted to different environments (Ibrahim et al. 2018; Shalan et al. 2011). Standardizing the properties present in these landraces is essential to satisfy the end uses and markets as okra gains popularity as an export commodity. Currently, there are a few exotic varieties that can be produced locally to meet export markets. Furthermore, North African nations have few production technologies, new cultivar developments, and okra management techniques. Fewer improved varieties of okra are available for farmers to grow in Egypt and Sudan because there are few reports of breeders striving to improve the crop by creating core collections for breeding to increase yield or quality. Therefore, local okra genetic resources must be assembled, characterized, assessed, and preserved to maximize and enable their application in any improvement program for the crop. This will allow for the harnessing and utilization of valuable traits (Helmy and Ragheb 2019; Masoud et al. 2007; Mohamed 2018).

Unfortunately, many local landraces risk going extinct because of their small populations. Low production is a major factor in the decline of small landraces due to the replacement of small landraces with new, high-output landraces. The long-term existence of small local landraces depends on selection to improve their productivity (Ibrahim et al. 2018; Shalan et al. 2011). Climate change may make plant species previously neglected or thought to be of little significance more important (Massawe et al. 2015). Utilizing landraces is one method for ensuring a good fit between crops and environmental constraints brought on by climate change. Characterizing the okra genotypes now grown in North Africa will help breeders create cultivars that can survive in harsh climates and expand the use of the crop to increase food security (Mohammed et al. 2022). However, it has been determined that the lack of improved genotypes is the most urgent issue that demands priority attention as a first step toward enhancing okra in Egypt and Sudan (Abed et al. 2020; Helmy and Ragheb 2019; Mohamed 2018).

The difficulties and possibilities of employing genetic improvement for regional okra genotypes in North Africa are discussed in the present review. To increase output and tolerance to salinity and drought, this review documents the current state and prospects of okra genetic diversity for key morphological and physiological features. The review provides insightful analyses of how genetic diversity is used in selection programs for salinity, drought, and economic aspects and how genetic diversity is investigated using conventional and genomic methodologies. Herein, the challenges and prospects of improving local okra genotypes in North Africa will be discussed. We will highlight the genetic diversity, breeding methods, substantial achievements in abiotic stress tolerance, and modern molecular technologies to enable okra genetic improvement in North Africa. The review also provides a foundation of knowledge for future okra breeding and cultivar development in places with limited freshwater and high salinity.

11.2 Production Status of Okra in North Africa

Okra is produced annually on an estimated 2.5 million acres of land around the world, totaling 10.5 million tons (FAOSTAT 2022). Okra is grown in Africa, where the continent produces 33% of the world's supply, with Northern African nations accounting for over 10.9% of the continent's overall production. In comparison to East Africa (5.4 t/ha) and West Africa (1.6 t/ha), North Africa has relatively high average okra yields (11.1 t/ha) (FAOSTAT 2022). Sudan produces the most okra in Northern Africa, accounting for 315,812 tons annually, or 3.05% of the world's total production, followed by Egypt, which produces 63,362 tons annually, or 0.61% (FAOSTAT 2022).

11.3 Genetic Diversity

To guarantee future global food security, plant genetic resources are essential. Thus, elite germplasm comprising desirable traits with high genetic and phenotypic diversity must be selected for subsequent breeding programs (Yadav et al. 2018). Okra development through an effective selection program can only be accomplished with reliable data on the correlation and genetic variability of important traits, being fully aware that crop improvement depends on the population's genetic diversity (Helaly et al. 2017; Nwangburuka et al. 2012; Singh et al. 2017). Evolutionary biologists have long sought to characterize and quantify genetic diversity to provide breeders with the crucial data to make informed decisions about exploiting plant genetic resources between closely related crop varieties and within them. For the improvement of okra in Egypt and Sudan, among other things, collection, characterization, and genetic diversity research are crucial first steps that lay the groundwork for maximizing the crop's potential and use. The quantity and level of variation in an accession's genetic makeup determine the value of germplasm (Aladele et al. 2008; Helaly et al. 2017). The efficacy of crop improvement programs is influenced mainly by genetic variability, genetic advancement, direct and indirect influences on yield and yield characteristics, and character association (Singh et al. 2017). Diversity limitation directly impacts the lack of materials for new breeding projects. To develop high-yielding okra cultivars, the genetic diversity of okra landraces will be essential to breeding programs. The possibility of okra improvement through the study of its genetic resources has been highlighted by several researchers (Helmy and Ragheb 2019; Ibrahim et al. 2013, 2018). This issue is essential for selecting and breeding desired plant landraces (Alemu and Mohammed 2022). However, there isn't much information available about how diverse okra is regarding significant agronomic features in North Africa. Understanding the genetic diversity and interactions among the germplasm might help breeding efforts to improve okra quality and boost tolerance to biotic and abiotic problems substantially. Interspecific hybridization is possible among *Abelmoschus* species (Reddy 2010). This might increase genetic diversity and widen the gene pool for breeding projects.

There is evidence of genetic diversity in the okra accessions from North Africa, including cultivated genetic resources and wild species for quantitative phenotypic traits (Helmy and Ragheb 2019; Ibrahim et al. 2018; Mohamed 2018). Okra's genetic improvement would benefit from a thorough grasp of the genetic diversity in its various traits. This will make it easier to identify helpful genes and their behavior to support hybridization programs. Characterizing a range of genetic variation among genotypes is crucial for the preservation and subsequent acquisition of germplasm resources, even though accessions from multiple origins are required as parent stocks to create better varieties (Govindaraj et al. 2015). Farmers' varieties or cultivated native varieties that have been adapted to a particular agroecological and farming system without using a scientific type of selection are commonly referred to as landraces. Landraces are an excellent source of genetic variation for okra breeding operations since they are invariably heterogeneous. Among the valuable genes discovered in landraces are those for early maturity, yield potential, abiotic tolerance, and other desirable characteristics. Landraces are most prevalent in regions with the most diversity (Alemu and Mohammed 2022; Mir et al. 2020). The local okra cultivars are subjected to outcrossing since they are open-pollinated. Therefore, the populations generated from these cultivars exhibit a vast degree of genetic diversity. Farmers typically produce their own seeds without sufficient isolation. As a result, some of the traits are undesirable, and the plants' growth and yield are likely to decrease (Helmy and Ragheb 2019; Ibrahim et al. 2018; Ragheb and Helmy 2022).

Numerous local okra cultivars, especially those produced in Egypt and Sudan, exhibit significant variation in a wide range of traits, including vegetative growth, earliness, pod quality, and yield (Helmy and Ragheb 2019; Ibrahim et al. 2013; Mohamed 2018; Ragheb and Helmy 2022). They are a rich source of variation since they can be employed as the primary selection material in breeding projects to enhance the okra crop's attributes (Figs. 11.1, 11.2, and 11.3). Changes in the genetic makeup of the germplasm can be inferred from the observed phenotypic variations in these studies. Although only a few of this research revealed that certain traits have strong heritability and might be selected throughout development efforts,

Fig. 11.1 Vegetative growth of some Egyptian okra genotypes

more research is required to determine how characteristics in okra are related. Hence, it is essential to use efficient breeding strategies based on selection evaluations for okra improvement (Abed et al. 2020; Ibrahim et al. 2013, 2018). The diversity between two parents is necessary to achieve heterosis and produce transgressive segregates. Therefore, finding a suitable donor parent among the cultivated species of okra is necessary for okra development (Abed et al. 2020).

To develop new varieties with high-yielding traits, plant breeders must understand whether two characters are positively or negatively correlated (Priyanka et al. 2018). Correlations among the characters are crucial for crop improvement and boosting selection efficiency (Falconer and Mackay 1996). For a crop to make rapid progress in selection and genetic improvement, knowledge of the strength and direction of the relationship between yield and attributes associated with yielding is vital (Mohammed et al. 2022; Nwangburuka et al. 2012). This will show how two or more plant features affect yield, providing indirectly valuable tools for choosing yield (Alam et al. 2021). In certain studies, there were significant positive and/or negative correlations among various characteristics of okra (Abed et al. 2020; Ibrahim et al. 2013, 2018; Ragheb and Helmy 2022; Rambabu et al. 2019). This demonstrates that some characteristics, particularly those expensive to assess or easily influenced by the environment, may be indirectly addressed in the selection criteria (Ibitoye and Kolawole 2022). There were positive and desirable correlations

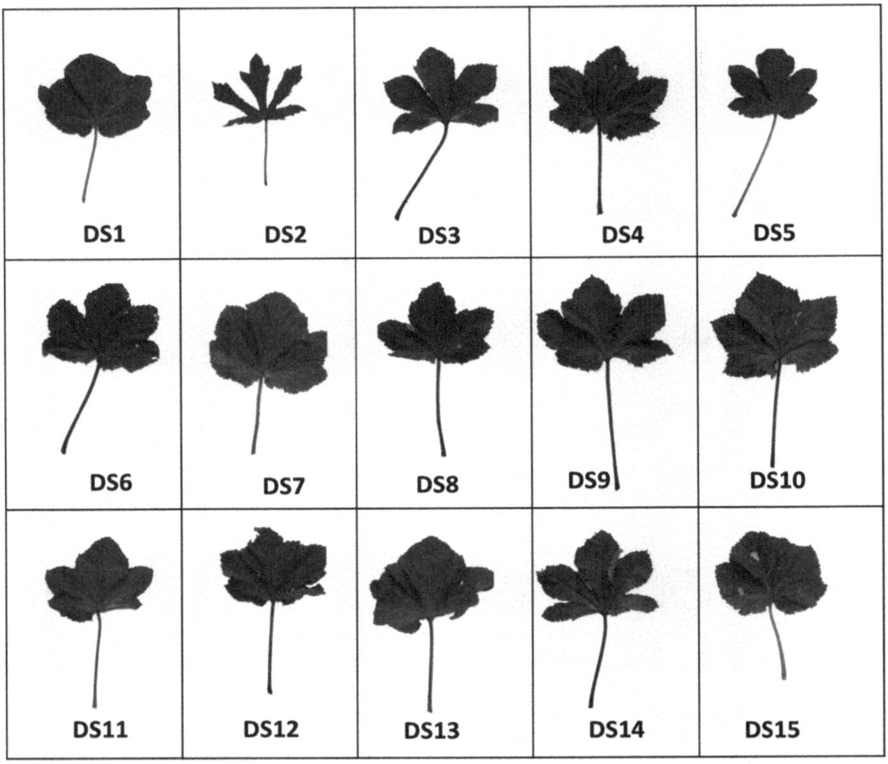

Fig. 11.2 Leaf shape in some Egyptian okra genotypes

between earliness traits (days to the first flower, days to the first pod, days to 50% flowering, and days to the first harvest) and between them and plant height and early yield. However, there were negative correlations between earliness traits and the fruiting period (Ragheb and Helmy 2022).

11.4 Application of Traditional Breeding Methodologies

Although traditional breeding methods are time-constrained and may require many years for varietal development and are less efficient due to their high labor demands and susceptibility to human mistakes, they are used to create new types with desirable traits (Ahmar et al. 2020). In order to investigate genetic diversity and use hybrid vigor in okra germplasms for productivity, nutritional quality, and adaptability, some writers strongly suggested adopting traditional breeding methods (El-Gendy et al. 2012; El-Sherbeny et al. 2018). Genetic variation makes it simpler to detect different genotypes, even though conventional breeding techniques have

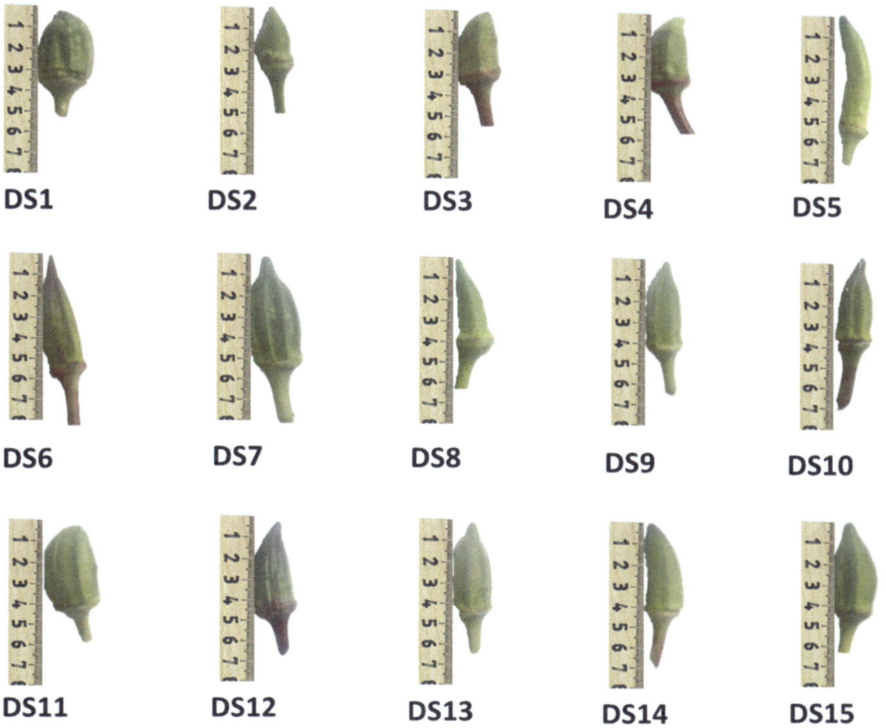

Fig. 11.3 Pod shape and color in some Egyptian okra genotypes

always been a response to improving the desired qualitative and quantitative traits in plant species.

11.4.1 Inbreeding with Selection

The breeder's main goal is to develop superior genotypes from available landraces or varieties. This crop's variety offers lots of opportunities for genetic advancement through selection. Selection is a good strategy when there is much diversity, especially when the quantitative traits are inherited by additive gene activity. Utilizing inbreeding creatively can improve the response to selection. Genetic variation in a population is redistributed by inbreeding; within an inbred line, it declines or vanishes, whereas, between groups of lines, it rises. Additionally, as relatives become more genetically similar due to inbreeding, the covariance between them frequently increases (Walsh and Lynch 1998). It was possible to recover desirable families from Egyptian okra by inbreeding and selection. In this regard, the biological and economic characteristics of various produced families due to utilizing individual plant selection on certain okra populations have been well investigated under

Egyptian conditions (Ibrahim et al. 2013, 2018; Ragheb and Helmy 2022). In all studied characters, they discovered a substantial amount of diversity. They came to the conclusion that selecting individuals with these features should be satisfying and practicable and that breeding programs can use selected families to create new cultivars.

11.4.2 Hybridization

Although hybrid seeds are more costly than varieties, they are widely used because they offer one of the quickest ways to boost production and produce higher income. Despite hybrid okra plants having greater genetic potential, seed production is still challenging (Medagam et al. 2012). In order to examine genetic diversity and utilize hybrid vigor in okra germplasms in terms of productivity, nutritional quality, and adaptability, several writers recommended mainly depending on conventional breeding methods (Abed et al. 2020; El-Sherbeny et al. 2018; Elgendy et al. 2012; Nagesh et al. 2014). A hybridization-based breeding technique would be desirable to get beyond the yield limitations in the current open-pollinated okra types. Intra- and inter-specific hybridization among locally adapted cultivars may greatly aid in producing a broad-based segregating population. Among these, recurrent selection could be applied to produce hybrids that combine high yield and abiotic and biotic resistance with other desirable traits, like high nutritional content and fruit characteristics required for export (Abed et al. 2020; Amiteye et al. 2019; Sandeep et al. 2022).

Heterosis breeding is the most effective method for boosting the production of cross-pollinated vegetable crops, including okra. The exploitation of heterosis is a widely used strategy for commercial okra production (Nagesh et al. 2014). Furthermore, it is commonly known that cross-combinations with large degrees of parental germline divergence reveal heterosis (Patel and Patel 2016). Heterosis breeding is a method for improving okra quality and productivity (Abed et al. 2020; Verma and Sood 2015). Traditional heterosis analysis relies heavily on the phenotypic screening of desirable traits and their identification. Fruit quality, in addition to yield, is crucial to the productivity and marketability of okra. The length, diameter, mucilage, greenness, and fiber content of okra fruit are among its many appealing characteristics (El-Sherbeny et al. 2018; Nagesh et al. 2014; Reddy 2010). Modern okra breeding programs are based on heterosis. A crucial step in the development of cultivars for modern okra production is the formation of F1 hybrids. Hybridization makes it easier to choose acceptable types for different conditions and recombine genes from various strains (Amiteye et al. 2019). To take advantage of heterosis in okra, homozygous inbred lines must be used for heterosis breeding. One of the most critical aspects of traditional breeding to produce the finest results is the combining ability between different parents, although rigorous cultivation techniques, as well as a time-consuming and arduous crossing and screening process, are necessary (Chowdhury and Kumar 2019; Dhankhar and Koundinya 2020). Heterosis offers the potential for greater yield. Hybrid seeds are typically more expensive than open-pollinated seeds, but the advantages, such as increased yield, uniformity, and stress tolerance, make the investment worthwhile (Maciel et al. 2018).

Both heterosis and combining ability can help to identify the best combiners that may be used in crosses to take advantage of heterosis or accumulate fixable genes and obtain desired segregates. The best crop combiners can be utilized to either take advantage of heterosis in F_1 or to accumulate fixable genes to evolve a variety. Combining ability analysis gives a guideline for evaluating the relative breeding potential of the parents or finding the best combiners in crops. The general and specific combining abilities will help analyze and interpret the genetic basis of significant traits. They offer guidelines on the type of gene action involved in expressing economic traits in distinct contexts (Abed et al. 2020; El-Gendy et al. 2012; El-Sherbeny et al. 2018; Nagesh et al. 2014). Improvement of okra yield could be achieved by natural gene action and the magnitude of heritability variation. In okra, it has been observed that both additive and nonadditive gene effects have a role in regulating yield and other factors (Abed et al. 2020; El-Sherbeny et al. 2018; Mudhalvan and Senthilkumar 2021).

The partial diallel approach maximizes human and financial resources while increasing the possibility of discovering novel combinations in breeding programs when crossings are conducted between genotypes with distinct genetic compositions. Even while thorough cultivation methods and a tedious and time-consuming screening and crossing process are required for traditional breeding to produce the finest results, one of the aspects that influences the aptitude among various parents to combine is one of them (Fasahat et al. 2016). There have been many studies in Egypt on okra's heterosis and combining ability. El-Gendy et al. (2012), El-Sherbeny et al. (2018), and Abed et al. (2020) examined the combining ability of the okra lines and elaborated on the cross combinations that performed much better in terms of both general and specific combining ability. The general combining ability is typically very beneficial for guiding future hybridization activities because it is predicted that high-valued individuals of this parameter will enable good hybrid combinations with the other parents. On the other hand, a hybrid combination that performs better or worse than anticipated based on the average performance of the involved parental inbred lines is referred to as having a specific combining ability (Lyngdoh et al. 2017; Silva et al. 2021). Diallel analysis obtains estimates of general and specific combining abilities, which are very helpful in choosing lines and hybrid combinations. In the okra hybrid, reciprocal differences in crossing direction have not demonstrated a pronounced variable effect. As a result, either of the parents may be used as male or female. Such cross-pollination will result in fruits that produce 100% hybrid seed (Dhankhar and Mishra 2004). However, studying the reciprocal effect is crucial for several traits that have a material impact (Mudhalvan and Senthilkumar 2021). Crop genetic improvement entails crossing many varieties, which results in considerable morpho-physiological and nutrient changes. The first and main task is to select superior parental lines before engaging in any desirable hybridization or breeding activity.

11.5 Opportunities and Challenges

11.5.1 Applications of Molecular Markers in Okra Breeding

The complementary nature of the parents' traits served as the foundation for traditional breeding techniques. However, it remained unclear how much each parent contributed to the genome. Molecular marker technologies have advanced and gained popularity because they enable monitoring the transfer of chromosomal segments in progeny (Lata et al. 2021). It is common practice to employ phenotypic descriptors based on morphological and agronomical features to define the genotypes of okra (Abed et al. 2020; Helmy and Ragheb 2019; Ibrahim et al. 2013, 2018; Mohamed 2018; Ragheb and Helmy 2022). Although the okra genetic basis is extensive in North Africa, the use of only phenotypic traits in genetic diversity studies is restrictive since phenotypes can be affected by environmental factors or plant development stages (Aaron et al. 2016; Akash et al. 2013; Mohammed et al. 2022; Saifullah et al. 2010). By contrast, the difficulty of plant breeding lies in simultaneously improving all desirable features, which is made more challenging by the genetic correlations between various traits. When one trait is selected, the correlated attributes will change, sometimes in a positive way and sometimes in a negative way. Thus, selection might lead to unforeseen changes (Alam et al. 2021; Vrunda et al. 2019). To overcome these constraints, marker-assisted selection can accelerate the improvement of breeding programs (Collard and Mackill 2008). Molecular markers are considered a crucial extra tool in studies of okra germplasm characterization since they are mostly unaffected by environmental conditions or other factors that affect plant development (Schafleitner et al. 2013). Using molecular markers supports morpho-agronomic description and evaluation instead of replacing them.

Using conventional approaches to utilize heterosis based on phenotypic screening is still a time-consuming, traditional strategy from the past. However, the use of marker technology in breeding programs enhances the screening process by advancing marker-assisted selection (Hasan et al. 2021). Further, improvements in okra varieties in North Africa will soon be made possible by molecular technology developments, including genome-wide sequencing, marker-assisted selection, and next-generation sequencing. By recognizing and choosing genotypes with desirable alleles in the early stages of plant development in segregating generations, marker-assisted selection can more quickly control huge populations (Collard and Mackill 2008).

Only a small number of research used molecular markers to assess genetic diversity in okra (Lata et al. 2021; Mishra et al. 2017). Several molecular markers have been applied to the okra plant to characterize its germplasm. Initially, genetic diversity investigations between various okra accessions used random amplified polymorphic DNA (RAPD) (Aladele et al. 2008; Prakash et al. 2011). For instance, 39 okra genotypes were effectively identified using RAPD-based molecular markers as having different genetic makeup. According to the study, morphological features that overlap because of environmental changes indicate variety (Haq et al. 2012).

Studies on diversity analysis in okra have also used sequence-related amplified polymorphism (SRAP) (Gulsen et al. 2007). Simple sequence repeats (SSR) and amplified fragment length polymorphism (AFLP) were employed. Because of their high repeatability, multiallelic nature, codominant inheritance, and good genome coverage, SSR markers are a vital tool in plant genetics and breeding (Kumar et al. 2017). AFLP markers are widely used and have several advantages, such as vast genome coverage, cost efficiency, repeatability, and independence from sequence information (Kyriakopoulou et al. 2014). The Inter-Simple Sequence Repeat (ISSR) markers were used to examine the genetic variation among 24 okra genotypes (Yuan et al. 2014) and detected low polymorphic variation among the okra accessions (Ahiakpa et al. 2017).

Molecular characterization represents the most effective method for utilizing genetic diversity (intra- and interspecific variations) in okra from various phytogeographical regions. This will open the door to improving the crop's nutritional value, growing conditions, yields, and compositional value. For six generations of selection among genotype cross pairings, okra was tested for its ability to tolerate water stress. It needed extensive phenotypic screening of numerous desired plant progenies in each generation, entirely dependent on the breeder's abilities. However, phenological trait-based selection in one generation could not provide the desired outcomes in subsequent generations (Naveed et al. 2009). Utilizing marker-assisted selection in this situation improves the laborious process of screening vast progeny populations, enabling greater precision in the choice of stress-tolerant segregants (Naveed et al. 2012). In North Africa, little information is available on the molecular breeding methods of okra. Helaly et al. (2017) used the RAPD-PCR method and six primers to assess the molecular variability of nine Egyptian okra cultivars. Additionally, 29 Egyptian okra accessions were analyzed using 5 AFLP primers and 42 ISSR primers (Younis et al. 2015).

Generally, through marker-assisted selection, molecular markers can be effectively employed to enhance the okra breeding program. The early selection of the desired trait by marker-assisted breeding speeds up the production of improved cultivars. For DNA fingerprinting, gene mapping and tagging, seed quality testing, and understanding the molecular basis of heterosis in various crops, molecular markers have been widely used. However, their usage in okra is still restricted, making it imperative to use these strategies to speed up the breeding program. Genomics and bioinformatics should also be thoroughly integrated into the program to best use markers in okra breeding. When genome information is lacking, okra breeding projects should take a comparative genomics approach to other crops (Lata et al. 2021). Plant breeding will continue to rely heavily on traditional methods; however, with the addition of marker technology, which will be employed in conjunction with the more traditional ones, it will indeed become more efficient and successful. As a result, in the future, the variety of techniques will diversify to maximize their benefits through integrated exploitation.

11.5.2 Abiotic Obstacles to Okra Production

Recently, plant breeders needed to deal hard not only in developing genotypes for increasing production but also in improving genotypes that are resistant to the current and anticipated abiotic effects, like moisture and salt stress, that recent climate change will bring (Brummer et al. 2011; Cairns et al. 2013). A hotspot for climate change has been identified as the North African region. Under a climate change scenario, it will be challenging to guarantee food and nutrition security for a growing population due to the restricted resources. The major factors that negatively affect the production of okra in dry and semi-arid areas include abiotic stresses like drought and salinity and the usage of genetically subpar and unimproved cultivars that have poor tolerance to those circumstances (Alake 2020; Mukhopadhyay et al. 2021). Salinity and drought are major abiotic factors impacting okra production in North Africa. Research on okra's genetic improvement in salty and water-scarce environments is scarce in North Africa. As a result, only a small number of improved varieties have been developed in this region. Generally, okra breeding programs must develop a wide range of varieties and hybrids to satisfy the varying needs of farmers and consumers and adapt to abiotic challenges, especially drought and salinity. There are two main strategies for enhancing okra tolerance to salinity and drought that have the potential to result in beneficial improvements in tolerance in a relatively short amount of time. First, it is possible to study okra varieties or landraces and choose promising lines. Second, okra's tolerance may be increased if genes from a wild relative are inserted into the genome of the cultivated cultivars, either naturally or by genetic engineering. Combining traditional and molecular breeding methods could accomplish this through germplasm screening, purification, evaluation, and promotion of the best lines. High abiotic stress tolerance, market demands, high yield, short maturation, longer harvesting durations, and adaptation to regional agro-climatic conditions are critical concerns during breeding programs (Cairns et al. 2013).

11.5.2.1 Genetic Diversity for Breeding Drought and Salt Tolerance

Some improved okra cultivars in North Africa and other regions can tolerate abiotic stress (Abd El-Fattah et al. 2020; Abed et al. 2020). In order to enhance okra programs, it is essential to search the genetic variation observed in okra germplasm for traits connected to abiotic stress resistance (Mkhabela et al. 2021). Moreover, locally accessible germplasm for stress tolerance must first be screened to find tolerant parental genotypes. Using landrace resources in okra breeding is crucial for creating plans for adapting to climate change. Okra species growing in North Africa exhibit a wide range of phenotypic and genetic diversity, which will speed up breeding for abiotic stress tolerance and higher yields. Characterizing the okra genotypes already present in North Africa would help breeders develop cultivars that could survive in harsh climates and expand how they improve food security (Abd El-Fattah et al. 2020; Abed et al. 2020; Ibrahim and Abed 2020). Okra pre-breeding and breeding are crucial for identifying the proper genotypes and can develop genotypes with high yield potential and abiotic stress tolerance.

11.5.2.2 Screening for Drought Tolerance

Numerous morphological features, including pod length, fresh and dried pod weight, the number of pods per plant, and the number of seeds per pod, can be used to evaluate a plant's tolerance to drought (Aaron et al. 2016; Abd El-Fattah et al. 2020; Abed et al. 2020; Alake 2020). For physical features, genotypic variation offers the possibility for genotype selection, population growth, and genetic development (Helmy and Ragheb 2019; Ibrahim et al. 2013). Simple phenotypic selection can fix traits that are highly heritable. Okra genotypes with substantial phenotypic heritability concerning fruit quality should be sought out and chosen to improve genetic gains for desirable characteristics and yield responses under water-scarce conditions. The plant's ability to withstand drought is also influenced by physiological traits such as chlorophyll content, photosynthetic rate, stomatal conductance, transpiration rate, and chlorophyll fluorescence factors (Düzyaman 2005; Mkhabela et al. 2021; Premadasa et al. 2019). By finding a balance between avoidance, escape, and tolerance while maintaining a suitable yield, crop adaptations to drought can be established (Prasanth et al. 2015). Most drought adaptations have drawbacks because genotypes with short durations are less productive than those with long durations (Prasanth et al. 2015). The drought-tolerant variety ought to be short-lived (drought escape), have a robust root system, have high stomatal tolerance, have a high and consistent yield under water stress (drought tolerance), and have improved water usage efficiency (drought avoidance) (Abd El-Fattah et al. 2020; Kumar et al. 2012; Prasanth et al. 2015).

In order to create drought-tolerant cultivars utilizing marker-assisted or genomic-assisted breeding, it is vital to comprehend the genetic basis and gene activities associated with drought tolerance (Shi et al. 2020). To develop a drought-tolerant variety, the breeding approach is the same as for previous trait improvement programs, viz., recurrent selection. The selection and breeding of genotypes that grow and produce an economic yield in drought conditions is a more durable and complementary method of reducing the consequences of the drought (Berger et al. 2016; Hussain et al. 2016).

Selection is one of the crucial elements of plant breeding to increase crop output when there is water stress. Breeding by selection can change the frequencies of specific genes, and new genes can be introduced into the population. The availability of a method that could accurately and quickly measure drought tolerance is a requirement for mass selection (Ashraf 2010). On the other hand, the backcross approach is used if adding a few drought tolerance characteristics to a high-yielding genotype is the goal. Contrarily, bi-parental mating (half-sibling and full-sibling) preserves a wide genetic foundation and allows for the development of the desired drought genotype. Plant breeders cannot use a single trait to improve a crop's productivity under drought stress. Therefore, an alternative potential systematic method that can be useful for enhancing a genotype's drought tolerance is to pyramid several features inside that genotype (Prasanth et al. 2015).

Trait correlations under drought stress conditions are required to create an appropriate selection criterion for drought tolerance breeding in okra. An integrated selection criterion of heritable agronomic traits strongly connected with okra yield under drought stress conditions has been presented for drought resistance breeding in Egypt (Abed et al. 2020). Abd El-Fattah et al. (2020) observed positive associations between the number of pods per plant, the weight of pods per plant, and overall yield. This suggests that combined selection for diverse qualities may increase Egyptian okra production's response to drought conditions. They also discovered that early maturing, high-yielding okra genotypes might be produced for dry, drought-prone areas that experience mid- and late-season drought stress. Additionally, comparable outcomes were observed by Abed et al. (2020). Despite the fact that these studies suggested that some traits are highly heritable and might be used in the selection process during development programs, further research is necessary to determine how okra trait correlations work.

In dry conditions, the number of pods per plant, fresh pod weight, and length were all positively correlated with pod output per plant, demonstrating the value of secondary features in enhancing okra. The okra genotypes discovered as being phenotypically desirable and complementary serve as valuable genetic resources for upcoming crossings and for selecting desirable traits under water-restricted settings utilizing a combination of ability analysis and heredity (Mkhabela et al. 2021). Leaf water potential was found to be a more important signal for selection criteria due to increased heredity under drought stress in okra, despite the number of pods per plant having a limited sense of heritability and genetic advancement during drought (Naveed et al. 2009; Prasanth et al. 2015). Thus, the chance of finding stress-tolerant genotypes in segregating populations with high heritability and genetic advancement may be higher. Some okra cultivars with drought tolerance have been identified in Egypt and can be used as valuable germplasm for developing drought tolerance (Abd El-Fattah et al. 2020; Abed et al. 2020).

11.5.2.3 Screening for Salt Tolerance

Thresholds for salt tolerance are often established based on the relative crop output under specified salt stress levels. Additionally, the biological characteristics of the plant play an essential part in the selection process because they represent the summation of genetic and environmental influences on plant growth brought about by physiological mechanisms that confer salinity resistance. As a result, the following two critical selection criteria for genotype selection can be established: the primary criteria in the selection process for salt tolerance should be the capacity of seeds to germinate and the survival of seedlings at this phase of their life cycle. Highly salinity-tolerant varieties provide the least amount of relative economic yield drop with each unit rise in stress (Cordea and Borsai 2021). The level of salt tolerance at various growth phases of a crop's life cycle must be known for a plant breeder to choose salt-tolerant lines. Without this knowledge, selection at a certain point could result in plants whose tolerance is lost in the following generation.

Research on how plants react to salt stress is extensive, but little is known about how okra tolerates it. Several growth parameters were used in certain research studies to examine the salt tolerance of okra genotypes (Haq et al. 2012; Tahira et al. 2014). Other research looked at the physiology of salt tolerance in several okra genotypes (Mahougnon et al. 2021; Shahid et al. 2011). Salinity tolerance testing was done on the okra germplasm during plant ontogeny and at the seedling stage in previous studies. Genotypes selected at the seedling stage maintained their resistance to NaCl throughout plant development, demonstrating that screening germplasm entries for salt tolerance at the seedling stage successfully creates salt-tolerant and high-yield okra cultivars (Haq et al. 2012). The best tolerant genotypes suited for breeding programs in Egypt are further determined based on an evaluation of the morpho-physiological responses of okra cultivars to salinity stress at the seedling stage (Ibrahim and Abed 2020). The utilization of physiological parameters is strongly recommended to get more trustworthy information and select suitable candidates for future saline agriculture (Cordea and Borsai 2021). High amounts of proline, a good K+/Na+ selectivity ratio, a good buildup of potassium ions, and the exclusion of sodium ions from leaves are all indicators of okra's relative salinity tolerance (Mahougnon et al. 2021).

11.5.2.4 Challenges of Okra Breeding for Drought and Salt Tolerance

At least three major issues impede the progress of okra breeding for salinity and drought tolerance. First, the plant's ability to withstand drought and salinity is a complex attribute. It is complicated to understand how plants can tolerate salt and drought. Numerous traits contribute to this stress tolerance mechanism, and several genes control each. As a result, increasing stress tolerance by changing just one gene appears challenging. Typically, a change in one gene influences hundreds of other genes' expression and metabolism, which hinders plant growth or productivity (Blum 2017; Zafar et al. 2017). According to this information, producing stress-tolerant and high-yielding crop cultivars will be facilitated by effectively integrating trait-based traditional breeding with molecular techniques (Fig. 11.4). Thus, the fundamental goal will be to identify traits that significantly contribute to the stress tolerance mechanism, which may be achieved by validation studies carried out in the target environment (Zafar et al. 2017).

Second, it is unclear how specific mechanisms operating at the tissue or organ level contribute to yield under these abiotic stresses. It is crucial to include all three categories of characteristics (physiological, chemical, and morphological traits) and analyze plant responses when assessing plants' salt and drought stress tolerance. Due to the enormous genetic variation of the plants, in some circumstances, it is not sufficient to evaluate only the physiological, chemical, or morphological profile, as they are interconnected (Cordea and Borsai 2021).

Third, the lack of data on the genetic regions for drought and salt tolerance makes breeding for drought or salt stress challenging. As a result, it might be difficult to create a genetic marker for okra's response to drought or salt stress. Furthermore, little is known about the correlation between phenotypic features in okra under drought or salt stress, which limits the development of effective selection criteria for breeding drought and salinity tolerance. Therefore, examining genotypic variance for phenotypic features in genetic resources of okra grown in dry or saline environments may help select prospective genotypes for use in breeding programs for drought and salinity tolerance (Gerrano 2018).

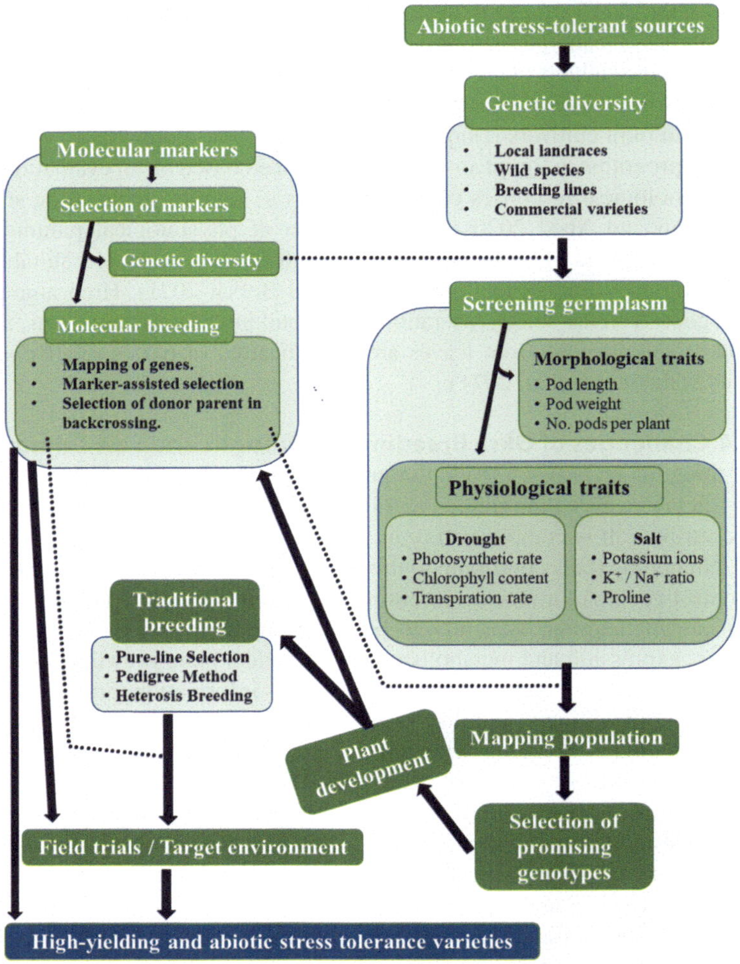

Fig. 11.4 An overview of the steps that are necessary for the development of new abiotic stress-tolerant varieties of okra via traditional and molecular breeding approaches

Breeders face unique challenges since drought and salinity tolerance are quantitatively inherited characteristics, making strategies for developing drought- and salt-tolerant okra genotypes time-consuming. Morphological characterization is the most dominant in okra germplasm. Growing the plants until they are fully mature before identification is necessary since the diversity of phenotypic and morphological attributes typically varies with surroundings (Aladele et al. 2008). Most okra improvement programs are hampered by the almost total dearth of available data on the morphological traits and characteristics of okra genotypes for breeders, particularly in the North African region. Breeding for abiotic stress tolerance in okra may greatly benefit from the knowledge of genetic diversity and interactions among okra germplasms. To increase yield and other desirable qualities, it is essential to understand the degree of genetic diversity among the available genotypes, the association between traits and yield, how much environmental factors affect these, and the heritability of the traits (Saifullah and Rabbani 2009).

Okra genotypes must be crossed and selected over several generations to achieve drought and salt tolerance. Each generation must include extensive phenotypic screening of many desired plant progenies. However, selection based on phenological traits in one generation might not provide the intended results in the following generations (Naveed et al. 2009). In this case, marker-assisted selection increases the accuracy of selecting stress-tolerant segregants by speeding up the laborious process of screening large progeny populations (Naveed et al. 2012). For instance, Yuan et al. (2014) investigated the genetic diversity across numerous genotypes and concluded that field screening in conjunction with the use of molecular markers increases the accuracy of selection.

The development of okra plants by altering traits through a traditional plant breeding program is labor-intensive and time-consuming. By integrating biotechnological approaches with conventional breeding, okra cultivars with increased tolerance to climate change issues are likely to develop and accelerate (Lata et al. 2021; Mishra et al. 2017; Zheng et al. 2016). It is simpler to map quantitative trait loci (QTL) affecting yield and yield components using molecular marker approaches, such as genomic selection and marker-assisted selection techniques. QTL are unique chromosomal loci containing many functional genes that affect yield and yield components. Numerous QTLs linked to adaptation to drought or salinity have been detected (Fan et al. 2015). The best-performing genotypes for drought- or salinity-stressed locations can be produced more speedily and successfully using QTL-assisted breeding. The molecular processes driving abiotic tolerance are clarified, and novel cultivars with improved salinity and drought resistance are developed by mapping QTL for both morphological and physiological traits (Zheng et al. 2016). Finding drought- and salt-resistant QTLs is the initial stage in creating crop breeding candidates beneficial in developing drought- and salt-resistant cultivars (Cordea and Borsai 2021). Some potential sub-features for drought tolerance have been identified using molecular markers to find QTLs for physiological variables related to stress tolerance (Athar and Ashraf 2009). Therefore, if quantitative genetics can identify QTL associated with drought or salinity tolerance, breeders would have better guidance to swiftly generate better genotypes having improved drought or salinity-tolerance traits, for example, using MAS selection. Therefore, plants that tolerate drought or salinity typically employ a drought or salinity escape strategy.

11.6 Conclusion and Future Perspectives

Due to the limited efforts made to improve okra by creating core collections for breeding to boost productivity or quality, farmers in Egypt and Sudan have few improved varieties to choose from. The problem of drought and salinity is expected to spread to semi-arid and arid locations worldwide due to global warming, which would decrease the productivity of okra. Further, significant output losses are also a result of the use of okra genotypes with low yield potential and poor adaptation. To increase okra producers' interest in large-scale farming and for adoption by the local and export markets, it is crucial to improve the productivity of the local landrace types of okra through breeding. There are many okra genotypes, with a wide variety of phenotypic and genetic diversity dispersed throughout North Africa. Although the native okra in North Africa has various genetic resources, not enough has been done to improve its genetic makeup. The vast diversity available in North Africa should be explored for better designing genotypes suitable for different conditions with abiotic stress tolerance, especially drought and salinity. To increase okra productivity, efforts must be made to gather, conserve, and analyze genetic resources, support okra breeders with various genetic materials, broaden the genetic base, and develop new varieties. Additionally, there is a need to guard against the genetic diversity of okra being lost. To improve the yield and other desired characteristics, a thorough study of those genotypes' morphological and chemical characteristics is required. Understanding the extent of genetic variation, how features relate to yield, and how the environment affects these materials is crucial. Additionally, molecular markers must be used to enhance the okra breeding effort by marker-assisted selection. Marker-assisted breeding enables early selection of the desired characteristic, hastening the production of superior cultivars.

References

Aaron TA, Elvis A-B, Faustina A, Kingsley T, Edmund OO (2016) Phenotypic traits detect genetic variability in okra (*Abelmoschus esculentus*. L. Moench). Afr J Agric Res 11:3169–3177

Abd El-Fattah BE, Haridy AG, Abbas HS (2020) Response to planting date, stress tolerance and genetic diversity analysis among okra (*Abelmoschus esculentus* (L.) Moench.) varieties. Genet Resour Crop Evol 67:831–851

Abed M, Ibrahim E, El-Shoura A (2020) Development of okra (*Abelmoschus esculentus* L. Moench) hybrids derived from selected inbreds under drought stress. J Plant Prod 11:61–69

Ahiakpa J, Magdy M, Werner O, Amoatey H, Yeboah M, Appiah A, Quartey E, Ros R (2017) Intra-specific variation in West African and Asian germplasm of okra (*Abelmoschus spp* L.). Ann Agric Sci 62:131–138

Ahmar S, Gill RA, Jung KH, Faheem A, Qasim MU, Mubeen M, Zhou W (2020) Conventional and molecular techniques from simple breeding to speed breeding in crop plants: recent advances and future outlook. Int J Mol Sci 21:2590. https://doi.org/10.3390/ijms21072590

Akash MW, Shiyab SM, Saleh MI (2013) Yield and AFLP analyses of inter-landrace variability in okra (*Abelmoschus esculentus* L.). Life Sci J 10:2771–2779

Aladele SE, Ariyo O, De Lapena R (2008) Genetic relationships among West African okra (*Abelmoschus caillei*) and Asian genotypes (*Abelmoschus esculentus*) using RAPD. Afr J Biotechnol 7:1426–1431

Alake CO (2020) Genetic variability and diversity in okra landraces using agromorphological traits and seed elemental minerals. Int J Veg Sci 26:127–149

Alam MA, Hafizuddin AAB, Gobilik J, Mijin SB, Khandaker MM (2021) Evaluation of best quality okra parental lines based on morpho-physiological and nutritional attributes. Bulgarian J Agr Sci 27:1127–1135

Alemu F, Mohammed W (2022) Genetic diversity assessment of okra (*Abelmoschus esculentus* (L.) Moench) collections in Ethiopia using multivariate analysis. Middle East J Agric Res 11:757–769

Ali A, Shah H, Gul R, Ahmad H, Nangyal H, Sherwani K (2014) Morpho-agronomic characterization of okra (*Abelmuscus esculentus* L.). World Appl Sci J 31:336–340

Amiteye S, Amitaaba T, Akama C, Amoatey HM (2019) Hybridization studies of okra (*Abelmoschus spp.* (L.) Moench) accessions. Int J Biotech Trends Technol 9:42–47

Ashraf M (2010) Inducing drought tolerance in plants: recent advances. Biotechnol Adv 28:169–183

Athar H, Ashraf M (2009) Strategies for crop improvement against salinity and drought stress: an overview. In: Ashraf M, Ozturk M, Athar HR (eds) Salinity and water stress. Springer, Cham, pp 1–16

Berger J, Palta J, Vadez V (2016) An integrated framework for crop adaptation to dry environments: responses to transient and terminal drought. Plant Sci 253:58–67

Blum A (2017) Osmotic adjustment is a prime drought stress adaptive engine in support of plant production. Plant Cell Environ 40:4–10

Brummer EC, Barber WT, Collier SM, Cox TS, Johnson R, Murray SC, Olsen RT, Pratt RC, Thro AM (2011) Plant breeding for harmony between agriculture and the environment. Front Ecol Environ 9:561–568

Cairns JE, Hellin J, Sonder K, Araus JL, MacRobert JF, Thierfelder C, Prasanna BM (2013) Adapting maize production to climate change in sub-Saharan Africa. Food Secur 5:345–360

Chowdhury S, Kumar S (2019) Okra breeding: recent approaches and constraints. Ann Biol 35:55–60

Collard BC, Mackill DJ (2008) Marker-assisted selection: an approach for precision plant breeding in the twenty-first century. Philos Trans R Soc B Biol Sci 363:557–572

Cordea MI, Borsai O (2021) Salt and water stress responses in plants. In: Hasanuzzaman M, Nahar K, Brzozowski T (eds) Plant stress physiology-perspectives in agriculture. IntechOpen

Dhankhar S, Koundinya A (2020) Accelerated breeding in okra. In: Accelerated plant breeding. Springer, pp 337–353

Dhankhar B, Mishra J (2004) Objectives of okra breeding. J New Seeds 6:195–209

Durazzo A, Lucarini M, Novellino E, Souto EB, Daliu P, Santini A (2018) *Abelmoschus esculentus* (L.): bioactive components' beneficial properties—focused on antidiabetic role—for sustainable health applications. Molecules 24:38

Düzyaman E (2005) Phenotypic diversity within a collection of distinct okra (*Abelmoschus esculentus*) cultivars derived from Turkish land races. Genet Resour Crop Evol 52:1019–1030

Elgendy RSS, Shaker GS, Ahmed OA (2012) Effect of foliar spraying with gibberellic acid and/or sitofex on bud behavior, vegetative growth, yield and cluster quality of Thompson Seedless grapevines. J Am Sci 8:21–34

El-Gendy SE, Obiadalla-Ali H, Ibrahim E, Eldekashy MJ (2012) Combining ability and nature of gene action in okra (*Abelmoschus esculentus* [L.] Moench). J Agric Chem Biotechnol Equip 3:195–205

Elkhalifa AEO, Alshammari E, Adnan M, Alcantara JC, Awadelkareem AM, Eltoum NE, Mehmood K, Panda BP, Ashraf SA (2021) Okra (*Abelmoschus esculentus*) as a potential dietary medicine with nutraceutical importance for sustainable health applications. Molecules 26:696

El-Sherbeny G, Khaled A, Obiadalla-Ali H, Ahmed A (2018) Estimates of heterosis and combining ability in okra under different environments. J Sohag Agrisci 3:50–64

Eltigani A, Müller A, Ngwene B, George E (2021) Physiological and morphological responses of okra (*Abelmoschus esculentus* L.) to Rhizoglomus irregulare inoculation under ample water and drought stress conditions are cultivar dependent. Plan Theory 11:89

Falconer DS, Mackay TFC (1996) Introduction to quantitative genetics. Pearson Education India

Fan Y, Shabala S, Ma Y, Xu R, Zhou M (2015) Using QTL mapping to investigate the relationships between abiotic stress tolerance (drought and salinity) and agronomic and physiological traits. BMC Genomics 16:1–11

FAOSTAT (2022) Food and Agriculture Organization of the United Nations, Production/Crops, Rome, Italy

Fasahat P, Rajabi A, Rad JM, Derera J (2016) Principles and utilization of combining ability in plant breeding. Biom Biostat Int J 4:1–24

Gemede HF, Ratta N, Haki GD, Woldegiorgis AZ, Beyene F (2015) Nutritional quality and health benefits of okra (*Abelmoschus esculentus*): a review. Food Process Technol 6:2

Gerrano AS (2018) Agronomic performance, nutritional phenotyping and trait associations of okra (*Abelmoschus esculentus*) genotypes in South Africa. In: Rediscovery of landraces as a resource for the future. IntechOpen, p 69

Govindaraj M, Vetriventhan M, Srinivasan M (2015) Importance of genetic diversity assessment in crop plants and its recent advances: an overview of its analytical perspectives. Genet Res Int 2015:1–14

Gulsen O, Karagul S, Abak K (2007) Diversity and relationships among Turkish okra germplasm by SRAP and phenotypic marker polymorphism. Biol Plant 62:41–45

Haq I-u, Khan AA, Khan IA, Azmat MA (2012) Comprehensive screening and selection of okra (*Abelmoschus esculentus*) germplasm for salinity tolerance at the seedling stage and during plant ontogeny. J Zhejiang Univ Sci B 13:533–544

Hasan N, Choudhary S, Naaz N, Sharma N, Laskar RA (2021) Recent advancements in molecular marker-assisted selection and applications in plant breeding programmes. J Genet Eng Biotechnol 19:128. https://doi.org/10.1186/s43141-021-00231-1

Helaly AA, Shalan MA, Abdou AA, Alabd MT (2017) Molecular identification of nine okra landraces by SDS-PAGE and RAPD markers technique. Nat Sci 15:36–44

Helmy EM, Ragheb EI (2019) Assessment of genetic behavior of some selected families from local okra cultivar (*Abelmoschus esculentus* L. Moench). Middle East J Agric Res 8:1384–1391

Hussain M, Kamran M, Singh K, Zouhar M, Ryšánek P, Anwar SA (2016) Response of selected okra cultivars to *Meloidogyne incognita*. Crop Prot 82:1–6

Ibitoye DO, Kolawole AO (2022) Farmers' appraisal on okra [*Abelmoschus esculentus* (L.)] production and phenotypic characterization: a synergistic approach for improvement. Front Plant Sci 13:787577–787577

Ibrahim EA, Abed MY (2020) Screening of Egyptian okra genotypes for salinity tolerance at the seedling stage. Singap J Sci Res 10:79–87

Ibrahim EA-A, Abed MY, Moghazy AM (2013) Genetic behavior of families selected from some local okra (*Abelmoschus esculentus* L. Moench) populations in Egypt. Plant Breed Biotechnol 1:396–405

Ibrahim E, Abed M, Moghazy A (2018) Improving some economical traits of local okra (*Abelmoschus esculentus* L. Moench) through selection and inbreeding. J Plant Prod 9:743–748

Joshi A, Hardas M (1956) Alloploid nature of okra, *Abelmoschus esculentus* (L.) Monech. Nature 178:1190–1190

Kumar S, Sehgal SK, Kumar U, Prasad P, Joshi AK, Gill BS (2012) Genomic characterization of drought tolerance-related traits in spring wheat. Euphytica 186:265–276

Kumar A, Kumar P, Nadendla R (2013) A review on: *Abelmoschus esculentus* (okra). Int Res J Pharm Appl Sci 3:129–132

Kumar S, Parekh MJ, Fougat RS, Patel SK, Patel CB, Kumar M, Patel BR (2017) Assessment of genetic diversity among okra genotypes using SSR markers. J Plant Biochem Biotechnol 26:172–178

Kyriakopoulou OG, Arens P, Pelgrom KT, Karapanos I, Bebeli P, Passam HC (2014) Genetic and morphological diversity of okra (*Abelmoschus esculentus* [L.] Moench.) genotypes and their possible relationships, with particular reference to Greek landraces. Sci Hortic 171:58–70

Lamont WJ (1999) Okra—a versatile vegetable crop. HortTechnology 9:179–184

Lata S, Yadav RK, Tomar B (2021) Genomic tools to accelerate improvement in okra (*Abelmoschus esculentus*). In: Landraces-traditional variety and natural breed. IntechOpen, p 240

Liu Y, Qi J, Luo J, Qin W, Luo Q, Zhang Q, Wu D, Lin D, Li S, Dong H (2021) Okra in food field: nutritional value, health benefits and effects of processing methods on quality. Food Rev Intl 37:67–90

Lyngdoh Y, Mulge R, Shadap A, Singh J, Sangwan S (2017) Combining ability analysis in near homozygous lines of okra [*Abelmoschus esculentus* (L.) Moench] for yield and yield attributing parameters. J Appl Nat Sci 9:324–331

Maciel A, Cullors A, Lukowiak AA, Garces J (2018) Estimating cost savings of pharmacogenetic testing for depression in real-world clinical settings. Neuropsychiatr Dis Treat 14:225

Mahougnon BGG, Julien KK, Armel CGM, Christophe BG (2021) Salinity resistance strategy of okra (*Abelmoschus esculentus* L. Moench) cultivars produced in Benin Republic. Int J Plant Physiol Biochem 13:19–29

Masoud A, El-Waraky Y, Shalaby T, Kasem M (2007) Developing new strains of okra. J Plant Prod 32:583–590

Massawe F, Mayes S, Cheng A, Chai H, Cleasby P, Symonds R, Ho W, Siise A, Wong Q, Kendabie P (2015) The potential for underutilised crops to improve food security in the face of climate change. J Procedia Environ Sci 29:140–141

Medagam TR, Kadiyala H, Mutyala G, Hameedunnisa B (2012) Heterosis for yield and yield components in okra (*Abelmoschus esculentus* (L.) Moench). Chil J Agric Res 72:316

Mir RA, Sharma A, Mahajan R (2020) Crop landraces: present threats and opportunities for conservation. In: Rediscovery of genetic and genomic resources for future food security. Springer, pp 335–349

Mishra GP, Singh B, Seth T, Singh AK, Halder J, Krishnan N, Tiwari SK, Singh PM (2017) Biotechnological advancements and begomovirus management in okra (*Abelmoschus esculentus* L.): status and perspectives. Front Plant Sci 8:360

Mkhabela SS, Shimelis H, Gerrano AS, Mashilo J (2021) Phenotypic and genotypic divergence in okra [*Abelmoschus esculentus* (L.) Moench] and implications for drought tolerance breeding: a review. S Afr J Bot 145:56

Mohamed RAO (2018) Molecular characterization and qualitative evaluation of some okra (*Abelmoschus esculentus*) L. (Moench) genotypes, Sudan. Horticultural Sciences, University of Gezira, p 53

Mohammed J, Mohammed W, Shiferaw E (2022) Phenotypic diversity assessment of okra (*Abelmoschus esculentus* (L.) Moench) genotypes in Ethiopia using multivariate analysis. Scientifica 2022:3306793

Mudhalvan S, Senthilkumar N (2021) Combining ability, reciprocal effects and heterosis for fruit yield characters in okra (*Abelmoschus esculentus* (L.) Moench). J Pharmacogn Phytother 10:96–101

Mukhopadhyay K, De S, Kundu S, Ghosh P, Chatterjee S, Chatterjee M (2021) Evaluation of levels of oxidative stress as a potential biomarker in patients with rheumatoid arthritis. J Family Med Prim Care 10:1981

Nagesh G, Mulge R, Rathod V, Basavaraj L, Mahaveer S (2014) Heterosis and combining ability studies in okra [*Abelmoschus esculentus* (L.) Moench] for yield and quality parameters. Bioscan 9:1717–1723

Naveed A, Khan AA, Khan IA (2009) Generation mean analysis of water stress tolerance in okra (*Abelmoschus esculentus* L.). Pak J Bot 41:195–205

Naveed A, Khan AA, Rauf S (2012) The potential of breeding okra (*Abelmoschus esculentus* L.) for water stress tolerance. In: Crop production for agricultural improvement. Springer, pp 217–235

Nwangburuka C, Denton O, Kehinde O, Ojo D, Popoola A (2012) Genetic variability and heritability in cultivated okra ["*Abelmoschus esculentus*" (L.) Moench]. Span J Agric Res 10:123–129

Patel B, Patel A (2016) Heterosis studies in okra (Abelmoschus esculentus (L.) Moench). Ann Agric Environ Sci 1:15–20

Petropoulos S, Fernandes Â, Barros L, Ferreira IC (2018) Chemical composition, nutritional value and antioxidant properties of Mediterranean okra genotypes in relation to harvest stage. Food Chem 242:466–474

Prakash K, Pitchaimuthu M, Ravishankar K (2011) Assessment of genetic relatedness among okra genotypes [*Abelmoschus esculentus* (L.) Moench] using RAPD markers. Electron J Plant Breed 2(1):80–86

Prasanth K, Sebastian A, Jayasheela DS (2015) Drought resilient breeding in vegetable crops; a review. Int J Adv Res 3:1126–1130

Premadasa S, Mahendran S, Hariharan G (2019) Physiological responses and yield of three okra (*Abelmoschus esculentus* L.) cultivars as affected by soil moisture stress during the flowering stage. In: Proceedings of the 2nd international symposium on agriculture. Eastern University, Sri Lanka, p 186

Priyanka V, Reddy T, Begum H, Sunil N, Jayaprada M (2018) Studies on genetic variability, heritability and genetic advance in genotypes of okra [Abelmoschus esculentus (L.) Moench]. Int J Curr Microbiol App Sci 7:401–411

Ragheb E, Helmy E (2022) Improving earliness-related traits of Egyptian local okra cultivar (*Abelmoschus esculentus* L. Moench). Sci J Agric Sci 4:1–11

Rambabu B, Waskar DP, Khandare VS (2019) Correlation and path coefficient analysis of fruits yield and yield attributes in okra [*Abelmoschus esculentus* (L.) Moench]. Int J Curr Microbiol App Sci 8:764–774. https://doi.org/10.20546/ijcmas.2019.804.084

Reddy M (2010) Genetic diversity, heterosis, combining ability and stability in okra (*Abelmoschus esculentus* (L.) Moench). Acharya N. G. Ranga Agricultural University, Hyderabad, p 313

Saifullah M, Rabbani M (2009) Evaluation and characterization of okra (*Abelmoschus esculentus* L. Moench.) genotypes. SAARC J Agric 7:92–99

Saifullah M, Rabbani M, Garvey E (2010) Estimation of genetic diversity of okra (*Abelmoschus esculentus* L. Moench) using RAPD markers. SAARC J Agric 8:19–28

Sandeep N, Dushyanthakumar BM, Sridhara S, Dasaiah L, Mahadevappa Satish K, El-Shehawi AM, Althaqafi M, Aloufi S, Sharma H, Alaklabi A (2022) Characterization of okra species, their hybrids and crossability relationships among Abelmoschus species of the Western Ghats region. Horticulturae 8:587

Schafleitner R, Kumar S, Lin C-Y, Hegde SG, Ebert AJG (2013) The okra (*Abelmoschus esculentus*) transcriptome as a source for gene sequence information and molecular markers for diversity analysis. Gene 517:27–36

Shahid MA, Pervez MA, Balal RM, Ahmad R, Ayyub CM, Abbas T, Akhtar N (2011) Salt stress effects on some morphological and physiological characteristics of okra (*Abelmoschus esculentus* L.). Soil Environ 30:66–73

Shalan M, Shanan S, Abdou A, Amer S, Alabd M, Alaaeldin A (2011) Comparative studies on nine Egyptian okra (*Abelmoschus esculentus* L.) genotypes via physical and chemical characteristics. Al-Azhar J Agric Res 11:81–95

Shi D, Wang J, Bai Y, Liu Y (2020) Transcriptome sequencing of okra (*Abelmoschus esculentus* L. Moench) uncovers differently expressed genes responding to drought stress. J Plant Biochem Biotechnol 29:155–170

Silva EHC, Candido WdS, Braz LT (2021) Obtaining okra hybrids through partial diallel analysis. Crop Breed Appl Biotechnol 21:e37192141

Singh A, Kumar PP, Kumar BA, Bahadur V (2017) Studies on genetic variability, heritability and character association in okra. Int J Bioresour Stress Manag 8:457–462

Tahira A, Pervez M, Ayyub C, Shaheen M, Sana T, Shahid M, Bilal R, Abdul M (2014) Evaluation of different okra genotypes for salt tolerance. Int J Plant Anim Environ Sci 4:23–30

Verma A, Sood S (2015) Genetic expression of heterosis for fruit yield and yield components in intraspecific hybrids of okra (*Abelmoschus esculentus* (L.) Moench). SABRAO J Breed Genet 47:221–230

Vrunda R, Patel A, Vashi JM, Chaudhari BN (2019) Correlation and path analysis studies in okra (*Abelmoschus esculentus* (L.) Moench). J Acta Sci Agric 3:65–70

Walsh B, Lynch M (1998) Selection under inbreeding. In: Evolution and selection of quantitative traits. Oxford University Press, p 1496

Yadav Y, Maurya PK, Devi AP, Jamir I, Bhattacharjee T, Banerjee S, Dutta S, Debnath D, Mandal AK, Dutta S (2018) Enation leaf curl virus (ELCV): a real threat in major okra production belts of India: a review. J Pharmacogn Phytochem 7:3795–3802

Younis RA, Hassan S, Itriby H (2015) Genetic diversity as assessed by molecular markers and morphological traits in Egyptian okra germplasm. Glob J Biol Agric Health Sci 4:117–128

Yuan C, Zhang C, Wang P, Hu S, Chang H, Xiao W, Lu X, Jiang S, Ye J, Guo X (2014) Genetic diversity analysis of okra (*Abelmoschus esculentus* L.) by inter-simple sequence repeat (ISSR) markers. Genet Mol Res 13:3165–3175

Zafar ZU, Manzoor H, Rasul S, Noreen S, Ali Q, Iqbal M, Javed M, Gul HS, Ahmad Z, Shahzad F (2017) Strategies to improve crop salt and drought tolerance: success and limitations. J Agrobios 11:265–298

Zheng J, Oluoch G, Riaz Khan MK, Wang X, Cai X, Zhou Z, Wang C, Wang Y, Li X, Liu F (2016) Mapping QTLs for drought tolerance in an $F_{2:3}$ population from an inter-specific cross between *Gossypium tomentosum* and *Gossypium hirsutum*. Genet Mol Res 15:15038477

Okra Improvement: Tales from West Africa

12

Issaka Housseini, Aichatou Salifou, and Sanjeet Kumar

12.1 Introduction

Okra (*Abelmoschus* spp.) is a versatile, nutritious, and staple crop in West Africa. Irrespective of Anglophone or Francophone, population of 16 countries in the region prefer and consume okra pods in chopped, dried, or fresh powdery forms, often cooked with meat or meat stew. Thus, okra holds cultural, nutritional, climate change adaptation, gender, and livelihood significance. In the region, okra is called Gombo (French), Miyan-gro (Hausa), Lafay (Djerma), Layre (Fulani), Gan (Bambara), Kandia (Manding), Nkruma (Akan), and Fetri (Ewe).

West Africa continues to produce almost 75% of the okra produced in Africa (Kumar et al. 2010). The total area and production are significantly higher in Nigeria, with Africa's highest population. Okra has been considered "a perfect villager's vegetable" because of its robust nature, dietary fibers, and distinct seed protein (NAP 2006). However, small-scale okra producers in the region are trapped in a vicious cycle of subsistence farming and face multiple socio-political obstacles and policy fatigue aggravated by climate change (Kumar 2022).

The first-hand anecdotes on okra improvement research for development (R4D) are shared with the intention of focusing on long-term research goals to strengthen the okra value chain. Hence, we describe some well-researched potential traits for validation research for enhanced okra demand creation. The traits for okra

I. Housseini
Horticulture Training Dov Center, Niamey, Niger

A. Salifou
Ainoma Seed Farm, Niamey, Niger

S. Kumar (✉)
Genetics and Plant Breeding, Acharya Narendra Deva University of Agriculture & Technology, Kumarganj, Ayodhya, India
e-mail: sanjeet.kumar@nduat.org

S. K. Tikoo et al. (eds.), *Okra: Status, Challenges and Opportunities*,
https://doi.org/10.1007/978-981-97-9963-3_12

improvement research, with articulated heftier outcomes and positive impact paths for small-scale producers, are described.

12.2 Domesticated Species and Major Stresses

West Africa is typified by diverse climates ranging from the dry-hot Sahel to wet-hot humid-coastal zones. The amphitetraploid common okra (*A. esculentus*) is cultivated in the Sahel, Sudano, and Savanna zones. The cultivation of newly identified amphipolyploid West African okra (*A. caillei*, also known as Guinean type) is limited to the coastal humid-tropical zones (Kumar et al. 2010). The size and number of epicalyx are distinguishable features of these two domesticated species. *A. esculentus* has more and thin epicalyx and *A. caillei* has up to six wider epicalyx (Fig. 12.1). The irrigated and rainfed crops in the dry and rainy seasons are grown under home gardens, mixed-, and mono-cropping commercial production systems. The crop management practices blend agroecological by default due to poor access to inputs and intensive agriculture due to developmental interventions such as introducing high-yielding seeds, chemical fertilizers, and pesticides.

The major biotic stress of okra in the region is leaf curl disease caused by a begomovirus (*Okra leaf curl virus*), followed by root-knot nematodes (*Meloidogyne* spp.). In Niger, fruit worms [the tomato fruit worm (TFW) *Helicoverpa armigera* (Hübner) and the spiny bollworm *Earias biplaga* (Walker)] are the most destructive pests (Ratnadass et al. 2014). The complete nucleotide sequences of DNA-A and associated DNA-β satellites from Niger were analyzed. The begomovirus and

Fig. 12.1 West African okra population in the Savanna zone; flower buds and epicalyx of common (green) and West African (red) okra

DNA-β were *Cotton leaf curl Gezira virus* and *Cotton leaf curl Gezira beta-satellite*, respectively (Shih et al. 2009).

12.3 West African Okra (*A. caillei*) Potential Donor

The West African okra (*A. caillei*), although accounting for only 5% of the total world production of okra, is a vital crop in the tropical-humid areas of Cote d'Ivoire, Benin, Cameroon, Nigeria, Ghana, and Togo. This amphipolyploid species (Siemonsma 1982) is known for possessing a gene pool that may be useful for okra improvement (Kumar et al. 2010; Table 12.1).

In India, *A. caillei* was used as a resistant source to develop the first *yellow vein mosaic virus*-resistant common okra "Prabahni Kranti" from a cross with Pusa Sawani (Nerkar and Jambhale 1986). Thus, inter-specific hybridization and gene introgression between *A. caillei* and *A. esculentus* are successful after overcoming partial hybrid breakdown barriers.

It is relevant to mention that *A. caillei* was referred to as *A. manihot* or Ghanian okra with about 200 chromosomes by Jambhale and Nerkar (1986) in their publications. Based on chromosome counts, *A. manihot* subsp. *manihot* from West Africa was named as the new species *A. caillei* ($2n = 196–200$, Siemonsma 1982). *A. caillei* has been located only in West Africa, and chromosome analysis suggests that *A. manihot* could be one of its parents along with common okra (Table 12.1).

12.4 Germplasm Exploration

Between 1982 and 1986, Bioversity International, collaborating with the Institut de Recherche pour le Development (IRD), explored okra germplasm in several West African countries. These accessions, however, underrepresent collections from countries like Niger (3) and Chad (5) (Fig. 12.2). The World Vegetable Center (WorldVeg) and its partners explored unexplored regions. Between 2008 and 2009, 102 new accessions from Mali, Senegal, Niger, and Guinea were collected and regenerated. The varietal data collected and analyzed on landraces (traditional

Table 12.1 Comparison between common and West African okra

Species	Cytogenetics	Contrasting traits
A. esculentus (Common okra) 95% cultivated area	Amphidiploid ($2n = 130–140$): *A. tuberculatus* ($2n = 58–60$) × unknown??	Poor adaptation in humid zones, more susceptible to biotic stresses, less vigorous, short life cycle (suitable as rainfed), usually day neutral, cultivated in both (dry and rainy) seasons.
A. caillei (West African okra) 5% cultivated area in humid regions	Amphipolyploid ($2n = 196–200$): *A. esculentus* ($2n = 130–140$) × *A. manihot* ($2n = 60–68$)	Better adaptation in humid zones, tolerant/resistant to biotic stresses, more vigorous, longer life cycle, mostly photoperiod sensitive and cross compatible with common okra.

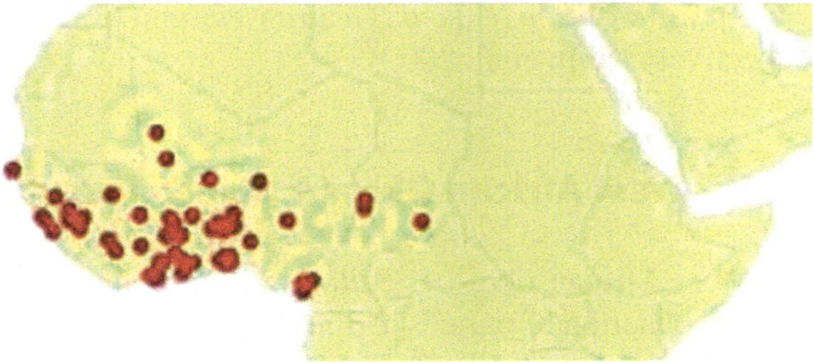

Fig. 12.2 Okra accessions from West Africa regenerated and available through the WorldVeg genebank, Taiwan

varieties) and improved cultivars used by farmers from Burkina Faso have revealed that considerable genetic diversity in the form of on-farm richness and community evenness is maintained in landraces (Jarvis et al. 2008). The World Vegetable Center genebank (https://genebank.worldveg.org/) with about 3000 cultivated and wild okra accessions and USDA-ARS Germplasm Resources Information Network (GRIN https://www.ars-grin.gov/) okra accessions are available for research and breeding use. In India, the National Bureau of Plant Genetic Resources (NBPGR), New Delhi, and its Regional Stations have more than 4000 okra accessions (http://www.nbpgr.ernet.in/Research_Projects/Base_Collection_in_NGB.aspx). Additionally, they also have excellent collections of pre-breeding materials derived from distant hybridization.

12.5 Germplasm Regeneration and Characterization

Between 2007 and 2011, the WorldVeg-ICRISAT Vegetable Project was operational at the ICRISAT Sahelian Center in Niamey, Niger. We employed a range of strategies to improve vegetable cultivars and production systems in the Sahel aligned with available market insights. With the passionate leadership of the Late Prof. Dov Pasternak, project largely focused on the introduction, evaluation, and promotion of locally adapted heat-tolerant cultivars with the aim of delivering short-term developmental outcomes (Pasternak et al. 2017). The consultation, socioeconomic significance, the okra price range in the markets, and its nonperishable nature in the West African context (dried forms are commonly consumed) were criteria for selecting okra for breeding research. A total of 250 okra accessions were regenerated, and collection gaps in West Africa were mapped (Fig. 12.2). We conducted combined leaf and pod transcriptome analyses produced by RNA sequencing and short-read assembly. More than 55% of the unigenes were annotated through sequence comparison with databases, and transcriptome sequences were mined for simple sequence repeat (SSR) markers. From 935 nonredundant SSR motifs, 199 were chosen for testing in a germplasm set, resulting in 161 polymorphic SSR markers. From this set, 19 markers were selected for a diversity analysis on 65 okra accessions comprising three different species, revealing 58 different genotypes and resulting in the clustering of the accessions according to species and geographic origin. The okra gene sequence information and the marker resource were made available to the research community for functional genomics and breeding research (Schafleitner et al. 2013).

12.6 Varietal Purification and Promotion

In the process of okra germplasm regeneration, we located and purified a local landrace and named it "Konni," as the original population was collected from the town of Birnie N'Konni, 375 km east of Niger's capital, Niamey. The agronomic performance of Konni was assessed through on-station and on-farm trials (Fig. 12.3), and we found it highly suitable for immediate promotion because (i) it is a day-length neutral genotype, thus can be planted and produce fruits year-round; (ii) the fruit has a regular shape acceptable to the local market; (iii) it is early fruiting and maturing, therefore highly successful under rainfed cultivation; (iv) the plant is short and therefore suitable for production under moringa plantations that reduce radiation load during the hot season; and (v) harvesting of shorter plants is also comfortable, i.e., no itching from stem spines and (vi) under rainfed high density planting can secure higher pod yield (Fig. 12.3, Pasternak et al. 2017). The Konni cultivar became remunerative under the Bio-reclamation of Degraded Land (BDL) (Pasternak et al. 2009) and African Market Garden systems adopted by the Sahelian Women's Associations. Both of these systems are holistically rainfed and irrigated horticultural production systems (Pasternak et al. 2017).

Other okra cultivars popular in West Africa are Clemson Spineless, Volta, Emerald, Gombo Paysan, Gombo Cafeier, Hire, Perkins Long Pod, Koto, Tomi (*A. caillei*), Indiana, Saloni (F1), Asontem, Torkor, Yellen, Sabalibougou, Keleya, V-35, V-2, NHAR-47-42, White Velvet, Ex-Borno, Lolli, Lima (F_1), etc. West African okra seed requirements are about 225 t per year, estimated from an area under okra of 1.8 m ha in 2021 (https://www.fao.org/faostat/en/#data). Several seed companies operational in the region are breeding locally adapted hybrid okra cultivars, including *A. caillei* hybrids for humid zones.

Fig. 12.3 Konni-okra under different production systems, rainfed (**a**), sprinkler (**b**), and drip (**c**) planted on same date, Rainy Season 2010, ICARISAT, Sadore, Niger

12.7 Capacity Building and Knowledge Transfer

For five consecutive years, annual training courses on vegetable seed production (Fig. 12.4) were conducted to empower technical officers and farmers in Niger, Burkina Faso, Mali, and Senegal. A couple of nascent seed producers from Niger came forward to undertake seed production of vegetables, including "Konni" okra, for commercialization.

The Ainoma Seed Farm in Niger, for the first time, started marketing vegetable seeds and had success with the Konni okra cultivar. As expected, women producers liked this improved landrace because the crop was early, spineless, highly productive, and harvesting was comfortable (Fig. 12.5). In Tillabery of Niger, Konni is renamed by local communities as Baadara, the ethnicity of the late Mr. M. Salifou (the founder of Ainoma Seed Farm), who introduced the Konni seeds. He also

Fig. 12.4 Annual training course on seed production of open-pollinated cultivars was a regular feature

Fig. 12.5 Rainfed Konni-okra crop stand in Niger (left) and Togo (right); note that due to the shorter plant height, harvesting is comfortable, i.e., no itching from stem spines

introduced the concept of fresh okra pods for marketing that fetch immediate cash. This opened new market opportunities for remotely located vulnerable communities who otherwise were consuming chopped dried okra and marketing surplus dried okra pods.

12.8 Seeds Sustain

Locally adapted improved seeds are the simplest technology, thus highly attractive to agricultural R4D. Seeds are the most critical inputs, containing many desirable biotic and abiotic resistance and quality traits. Breeding short-duration crops is always a dream of any plant breeder. The short-duration Konni-Okra is a perfect fit for rainfed cultivation prevalent in the Sudano-Sahel of West Africa. We probed the status of Konni cultivation in the region 12 years after project interventions.

Currently, the Konni okra is grown in about 200 BDL (Fig. 12.5) and 150 AMG, benefiting approximately 8500 women-led households in Niger. The average fresh pod yield of Konni ranged from 3 t/ha under rainfed to 12 t/ha under irrigated conditions, with an average annual farm gate price of 0.85 USD/kg (Housseini et al. 2022).

Despite the Pandemic, between 2019 and 2022, Ainoma Seed Farm, Niger, sold 7.8 t seed of Konni in Niger, Nigeria, and Togo (Fig. 12.5). Likewise, FERME BBK, Kollo, Niger, trained by the WorldVeg-ICRISAT project specifically on okra seed production (Fig. 12.6), has also commercialized Konni seeds in the region after getting it certified by the national agency.

Fig. 12.6 Mentored by us, Mr. Amadou Hassane, the founder of FERME BKK, in his okra seed production field, Kollo, Niger

12.9 Innovative Research Provocation

Okra, as a global crop, is triggered by the commercial interest of the seed industry in the tropics of Africa and Asia. Okra breeding programs are operational under big, medium, and large seed companies, as well as publicly funded national vegetable breeding programs. These breeding efforts are focused on the immediate needs of solving the problems of okra growers. Recently, the okra genome has been sequenced (1.19 Gb), which will facilitate the functional studies and genetic improvement of okra (Wang et al. 2023). Like other perishable commodities, fresh okra production scaling would be limited because perishability of fresh pods does not provide production price safety nets for small-scale producers.

Therefore, research gaps that strengthen okra production and the value chain warrant R&D investments. Okra seed flour has traditionally been used to supplement corn flour in Egypt to make better-quality dough (Taha El-Katib 1947). Supplementing maize ogi (a maize flour dish) with okra meal increases protein, ash, oil, and fiber contents (Akingbala et al. 2003; Adelakun et al. 2008). Okra seed protein is rich in tryptophan (94 mg/g N) and contains adequate amounts of a sulfur-containing amino acid (189 mg/g N)—a rare combination that makes okra seeds exceptionally useful in reducing human malnutrition (NAP 2006). Okra seed protein, with good protein efficiency ratio (PER) and net protein utilization (NPU) values, is comparable to cereals (except wheat), and its oil yield is comparable to most oilseed crops (Rao 1985).

Additionally, the ethnobotanical uses of West African okra (*A. caillei*) could be viewed as prior art for further research. In a comprehensive survey in Nigeria, 559 respondents from nine ethnic groups revealed that West African okra was also popular for nonfood uses, including medicine (27%), myth/religion (32%), soil fertility indicator (19%), rainy season indicator (8%), dry season/harvest time indicator (100%), fuel (15%), and sponge (11%) (Osawaru and Dania-Ogbe 2010).

These findings deserve validation to diversify innovative uses of okra for demand creation. For example, rodent/animal feeding trials of okra seed flour to eliminate the possibility of gossypol would be pertinent to expanding the okra value chain, helping millions of small-farm families in the tropics.

12.10 Conclusion

The improved okra seeds, resistant to economically important diseases and insects, are benefiting okra producers, the seed industry, and other actors involved in seed and crop value chains. However, there are limitations to scaling seeds in the socio-economic context of small-farm families in less-privileged regions. This is because increased okra seed sales, production, or productivity does not always guarantee increased profit for okra farmers. Therefore, besides improving market linkages, renewed attention on validating research on specific traits, such as okra seed flour and oil, innovative uses of biomass, etc., with heftier impact potential is imperative. This can ensure increased okra demand, leading to increased okra seed demand and assured income and profit for the growers—key factors for okra production sustainability. Besides R&D investment, committed leaders at various levels are required to foster team science and partnerships that have a holistic, larger vision, take calculative risks, and actions to generate proof of concept based on prior art and data to conduct piloting research. The provocative research interventions we highlighted can be considered potential entry points for expanding okra production, benefiting okra producers at scale.

References

Adelakun OE, Oyelade OJ, Ade-Omowaye BIO, Adeyemi IA, Van de Venter M (2008) Influence of pre-treatment on yield, chemical and antioxidant properties of Nigerian okra seed (*Abelmoschus esculentus* M.) flour. Food Chem Toxicol. https://doi.org/10.1016/j.fct.2008.12.023

Akingbala JO, Akinwande BA, Uzo-Peters PI (2003) Effects of color and flavor changes on acceptability of ogi supplemented with okra seed meals. Plant Foods Hum Nutr 58:1–9

Housseini I, Salifou A, Kumar S (2022) Okra improvement: tales from West Africa. In: Invited presentation, abstract, first global okra roundtable (GORT-1), 10–13 October, New Delhi

Jarvis DI et al (2008) A global perspective of the richness and evenness of traditional crop-variety diversity maintained by farming communities. Proc Natl Acad Sci USA 105:5327–5331

Kumar S (2022) Innovation and advocacy for climate-resilient small-scale producers. In: Abstract cum Souvenir book, international conference on biotechnological initiatives for climate resilient agriculture, 7–9 January, pp 21–23

Kumar S, Dagnoko S, Haougui A, Ratnadass A, Pasternak D, Kouame C (2010) Okra (*Abelmoschus* spp.) in West and Central Africa: potential and progress on its improvement. Afr J Agric Res 5(25):3590–3598

National Academies Press (2006) Lost crops of Africa: volume II: vegetables, pp 287–301. www.nap.edu/catalog/11763.html

Nerkar YS, Jambhale ND (1986) 'Prabhani Kranti' a yellow vein mosaic-resistant okra. HortScience 21:1470–1471

Osawaru M, Dania-Ogbe FM (2010) Ethnobotanical studies of West African okra [*Abelmoschus caillei* (a. Chev) stevels] from some tribes of southwestern Nigeria. Sci World J 5(1). https://doi.org/10.4314/swj.v5i1.61484

Pasternak D, Senbeto D, Nikiema A, Kumar S, Fatondji D, Woltering L, Ndjeunga J, Alein R (2009) Bioreclamation of degraded African lands with women empowerment. Chron Hortic 49(2):24–27

Pasternak D, Isska H, Kumar S (2017) Selection and dissemination of vegetable cultivars in the Sahel. Chron Hortic 57:23–30

Rao PU (1985) Chemical composition and biological evaluation of okra (*Hibiscus esculentus*) seeds and their kernels. Qual Plant Food Hum Nutr 35:389–396

Ratnadass A, Zakari-Moussa O, Kadi-Kadi HA, Kumar S, Grechi I, Ryckewaert P, Salha H, Akourki M, Maâzou A, Siaka S, Salami I (2014) Potential of pigeon pea as a trap crop for control of fruit worm infestation and damage to okra. Agric For Entomol 16:426–433

Schafleitner R, Kumar S, Chen-yu L, Hegde SG, Ebert A (2013) The okra (*Abelmoschus esculentus*) transcriptome as a source for gene sequence information and molecular markers for diversity analysis. Gene 517:27–36

Shih SL, Kumar S, Tsai WS, Lee LM, Green SK (2009) Complete nucleotide sequences of okra isolates of *Cotton leaf curl Gezira virus* and their associated DNA-β from Niger. Arch Virol 154:369–372

Siemonsma JS (1982) West African okra – morphological and cytogenetical indicators for the existence of a natural amphidiploid of *Abelmoschus esculentus* (L.) and *A. manihot* (L.) Mediks. Euphytica 31:241–252

Taha El-Katib MM (1947) Value of adding cotton okra, fenugreek seed to maize flour. Nature 156:716

Wang R, Li W, He Q, Zhang H, Wang M, Zheng X, Liu Z, Wang Y, Du C, Du H, Xing L (2023) The genome of okra (*Abelmoschus esculentus*) provides insights into its genome evolution and high nutrient content. Hortic Res 10. https://doi.org/10.1093/hr/uhad120

Part IV

Genomics and Genetic Engineering Research & Future in Okra

13 Genomics Studies in Okra and Marker Development: An Overview

M. Gayathri, M. Pitchaimuthu, and K. V. Ravishankar

13.1 Introduction

Okra polyploidy levels assessed through the chloroplast DNA (cpDNA) intronic spacer revealed that *A. esculentus* is the closest relative of two wild species, that is *A. ficulneus* and *A. moschatus* (Ramya and Bhat 2012).

Though *Abelmoschus* is a prominent genus, its phylogeny and evolution remain a mystery. In Asia and Africa, wild and cultivated forms coexist in nature, and it appears that *A. esculentus* is a relatively new species. The genus *Abelmoschus* has 11 species, four subspecies, and five variations (Li et al. 2020), the majority of which are economically valuable. Furthermore, chromosomal numbers vary greatly, ranging from $2n = 62$ to $2n = 144$, with $2n = 130$ being the most common chromosome number (Benchasri 2012). These findings have led to a new hypothesis about the origins of *A. esculentus*, including the possibility that a $2n = 58$ *A. tuberculatus*, native to Northern India, and a $2n = 72$ *A. ficulneus*, native to East Africa, hybridised, resulting in an alloploid *Abelmoschus* hybrid with $2n = 130$ (Joshi and Hardas 1956; Siemonsma 1982; Benchasri 2012). It has been reported that there are two kinds of *A. esculentus* L.: diploids ($2n = 60–70$) and tetraploids ($2n = 120–130$). This may be due to irregularities in chromosomal movement during the mitotic phase of cell division (Nwangburuka et al. 2011). Flow cytometry analysis revealed the nuclear DNA size, with 2C values ranging from 3.98 to 17.67 pg, corresponding to genome sizes of 3.8–17.3 Gbp (Salameh 2014). However, recent sequencing studies indicate that the genome size is around 1.2 Gbp (Wang et al. 2023).

M. Gayathri · K. V. Ravishankar (✉)
Division of Basic Sciences, ICAR-Indian Institute of Horticultural Research, Bangalore, India

M. Pitchaimuthu
Division of Vegetable Crops, ICAR-Indian Institute of Horticultural Research, Bangalore, India

S. K. Tikoo et al. (eds.), *Okra: Status, Challenges and Opportunities*,
https://doi.org/10.1007/978-981-97-9963-3_13

At the genome level, okra is regarded as an orphan crop, and up until recently, little effort has been made for its genetic advancement. Millions of sequences are produced by advanced sequencing technologies, especially next-generation sequencing (NGS). NGS has made it possible to find genetic markers on a massive scale (Ravishankar et al. 2015). Recently, the first report on genomic SSR markers in okra was developed using NGS, which were used for the assessment of genetic relatedness and cross-species transferability (Ravishankar et al. 2018). SSR markers, due to their codominant inheritance, multi-allelic nature, high reproducibility, and very good genome coverage, play a key role in many applications of plant genetics and breeding (Ravishankar et al. 2015).

Hence, in this chapter, attempts have been made to highlight the marker studies, transcriptomic, and whole-genome sequencing studies of okra.

13.2 Genetic Diversity Studies Using Various DNA Markers

Okra has been largely unknown in terms of genomic resources, as the genome has not been completely sequenced. Because okra is a tropical crop with insufficient research funding compared to highly industrialised countries, little effort has been made to improve its genetics (Werner et al. 2016). Molecular markers have paved the way for the assessment of genetic variations and genetic relationships among and within the species (Chakravarthi and Naravaneni 2006; Yuan et al. 2014, 2015). Molecular marker techniques like rapid fragment length polymorphism (RFLP), random amplified polymorphic DNA (RAPD), amplified fragment length polymorphism (AFLP), and simple sequence repeats (SSR) are widely used for genetic characterization and crop improvement programmes (Sawadogo et al. 2009) (Table 13.1).

From a scientific and breeding standpoint, molecular markers hold substantial promise in augmenting the okra breeding initiative via marker-assisted selection (MAS). This methodology facilitates the identification of traits and genes, thereby expediting the advancement of enhanced cultivars. Although molecular markers have been widely utilized for applications like DNA profiling, gene charting, gene labelling, seed purity assessment, and investigating the molecular mechanisms of hybrid vigour in diverse crops, their application in okra breeding has been relatively constrained. As a result, employing these methodologies becomes crucial to hasten the progress of the okra breeding programme.

To achieve the best possible results, it is crucial to integrate genomics and bioinformatics into the breeding process, ensuring the effective utilization of markers in okra breeding. This cooperative approach will improve the accomplishments and efficiency of the breeding protocol, ultimately resulting in the development of outstanding okra varieties.

Table 13.1 List of molecular markers used for genetic diversity studies in okra

	Markers used	Species	Application	Genotypes used	No. of primers	PIC (%)	Reference
1	ISSR	*A. esculentus*	Genetic diversity	48	22	54.55	Yuan et al. (2015)
2	ISSR	*A. esculentus*	Genetic diversity	24	22	0.53	Yuan et al. (2014)
3	AFLP		Genetic diversity	30	5 combinations	–	Massucato et al. (2019)
4	RAPD	*A. esculentus* and *A. caillei*	Genetic diversity	50	13	–	Aladele et al. (2008)
5	RAPD	*A. esculentus*	Genetic diversity	39	31	–	Leal et al. (2000)
6	AFLP	*A. esculentus*	Genetic diversity	48	8 primer pairs	–	Akash et al. (2013)
7	AFLP	*A. esculentus*	Genetic diversity		33	0.20	Kyriakopoulou et al. (2014)
8	SRAP	*A. esculentus*	Genetic diversity	23	39	–	Gulsen et al. (2007)
9	RAPD	*A. esculentus*	Genetic diversity	39	20	–	Ikram-ul-Haq et al. (2013)

13.3 Molecular Marker Studies in Okra Based on NGS Approach

Next-Generation Sequencing technology has brought about a revolution in the field of molecular biology, particularly in the development of microsatellite or SSR markers. Its cost-effectiveness, time-saving benefits, and ability to identify a large number of SSRs have made it a preferred choice in genetics and breeding applications. The technologies predominantly used for sequencing include Roche/454, Solexa, Illumina, and SOLID, as reported by Ravishankar et al. (2015).

In a recent development, NGS technology was employed to develop the first report on genomic SSR markers in okra, which were subsequently used to assess genetic relatedness and cross-species transferability. A total of 402 SSRs were designed from 3735 contigs, with a mean PIC of 0.519, as reported by Ravishankar et al. (2018). The multi-allelic nature, high reproducibility, and excellent genome coverage of SSR markers make them a valuable tool in various applications of plant genetics and breeding, owing to their codominant inheritance, as highlighted by Bertini et al. (2006). There have been a few more studies in okra based on SSR markers.

13.4 Molecular Marker Studies in Okra Based on Transcriptome Sequencing Approach and Transcriptomic Studies in Okra

A transcriptome is a set of all the transcripts in one cell or one population of cells at a certain stage. Transcriptome analysis helps to study the identification of genes that are differentially expressed in distinct cell populations or tissues. The list of transcriptome studies in okra is tabulated (Table 13.2).

The first study on the combined leaf and pod transcriptome in okra was conducted by Schafleitner et al. (2013). In their study, they identified 150,000 unigenes and 935 non-redundant SSR motifs, out of which 199 were selected for testing the germplasm set. The average polymorphism information content (PIC) was

Table 13.2 List of transcriptome studies in okra

	Species	Plant part/ organ	Sequencing platform	Reads obtained	Application	Reference
1	*Abelmoschus esculentus* (L.) cv. Mahnco *Arka Abhhay*	Leaf and pod	Illumina HiSeq™ 2000	150,000 unigenes	Diversity analysis	Schafleitner et al. (2013)
2	*Abelmoschus esculentus* (L.)	Roots, stem, leaves, flowers, and fruits	Illumina HiSeq X Ten platform	293,971 unigenes	Bioactive active constituents	Zhang et al. (2018)
3	*Abelmoschus esculentus* cv. Xianzhi	Roots, stems, and leaves	Illumina HiSeq X Ten platform	293,971 unigenes	Drought stress	Shi et al. (2020)
4	*Abelmoschus esculentus* cv. xian zhi	Leaf	Illumina Hiseq X Ten platform	121,360 unigenes	Molecular mechanism of melatonin-mediated salt tolerance	Zhan et al. (2021)
5	*Abelmoschus esculentus* (L.)	Root, stem, leaf, bud, flower, different parts of developing pods, drought stress, heat stress, and salt stress plantlet	Illumina Hiseq	112,597 transcripts	Genetic diversity	Gayathri et al. (2021)
6	*Abelmoschus esculentus* (L.) cv. Pusa Sawani and Pusa Bhindi-5	Leaf	Illumina Hiseq	–	Genetic diversity	Yadav et al. (2022)

determined to be 0.49, with a range spanning from 0.1 to 0.84. The identified SSRs were found to be valuable for understanding the germplasm classification of okra.

In a study by Shi et al. (2020), transcriptomic analysis was performed to investigate drought stress in okra under water-limited conditions during the seedling stage. A total of 94,769 unigenes were identified, out of which 2276, 5626, 11,719, 14,892, 855, 4593, and 5170 differentially expressed genes (DEGs) were identified in seven pairwise comparisons (A5 vs. A0; A7 vs. A0; A15 vs. A0; A20 vs. A0; A7 vs. A5; A15 vs. A7; and A20 vs. A15). Additionally, the study identified a total of 17,456 simple sequence repeat motifs, 479,831 single nucleotide polymorphisms, and 3746 transcription factors. The study provides insight into the molecular mechanisms involved in the response to drought stress in okra.

A first comprehensive transcriptome was studied in okra, where a total of 10,492 SSRs were identified, of which 18 SSRs were found to be polymorphic, which were screened over 36 okra genotypes and accessions. The comprehensive analysis of the dataset revealed that the expected heterozygosity values ranged from 0.000 to 0.971, with an average of 0.593. For observed heterozygosity, the values spanned from 0.000 to 0.794, with a mean of 0.254. The number of alleles per locus varied from 1 to 30. Regarding the Polymorphic Information Content (PIC), the values ranged from 0.000 to 0.940, with an average of 0.5546. The identified genic SSR markers were found to be helpful in germplasm characterization and genetic diversity studies in okra (Gayathri et al. 2021). In addition to this, RNA-seq from leaf samples of okra was performed, where a total of 106,224 EST-SSRs were identified, and it was found that 163 primer pairs produced the expected amplicon with an amplification efficiency of 89%. The identified markers were found to be a robust molecular tool for germplasm identification, genetic diversity studies, and also for cross-transferability.

13.5 Identification of lncRNA and TFs

Recent advancements in sequencing technology have enabled the decoding of genomes for lesser-studied crops, including okra (Ravishankar et al. 2018; Gayathri et al. 2021). However, there is a lack of reports on long non-coding RNA (lncRNA) and limited research on abiotic stress in okra. Nevertheless, the emergence of new technologies such as Next-Generation Sequencing (NGS) and RNA sequencing (RNA-seq) has made it possible to predict hundreds of lncRNA in various plant species, including *O. sativa*, *Arabidopsis thaliana*, *Populus trichocarpa*, and *Zea mays*, for both biotic and abiotic stress responses. RNA sequencing has proven to be particularly useful for obtaining gene sequence data from non-model crops (Strickler et al. 2012).

In an effort to predict lncRNA and identify transcription factors (TFs) implicated in abiotic stress response in okra, analysis was done using comprehensive pooled transcriptome data. The predicted lncRNAs were classified, and a few were confirmed using PCR. The study also identified a total of 57,675 lncRNAs and 1281

TFs, which were categorized into 51 transcription factor families (Muthaiah et al. 2023). Out of 15 lncRNA primers tested, only Ae_Lnc9, Ae_Lnc12, Ae_Lnc13, and Ae_Lnc14 showed amplification in all tissues, while the other lncRNAs exhibited tissue-specific expression patterns. These results suggest that lncRNA expression patterns vary across genotypes, tissues, developmental stages, and environmental conditions. This study provides insights into the structure of the okra genome and the molecular changes that occur during plant growth, development, and response to abiotic stresses.

In another study on cotton, a total of 8414 lncRNAs were identified, with 6516 intergenic lncRNAs (77%), 759 anti-sense lncRNAs (7%), 334 intronic lncRNAs (4%), and 805 sense lncRNAs (10%) (Zhang et al. 2019). Additionally, cotton was found to have 10,820 lncRNAs under three different environmental conditions (control, drought stress, and re-watering) (Lu et al. 2016).

13.6 Chloroplast Genome of Okra

Sequencing of the okra chloroplast genome through Illumina Hiseq technology revealed a total of 132 genes, including 88 protein-coding genes, 36 tRNA genes, and rRNA genes (Liu et al. 2023). Phylogenetic analysis revealed that *Abelmoschus* species are monophyletic.

13.7 The Okra (*Abelmoschus esculentus*) Genome

13.7.1 Genome Analysis and Annotation

The first published okra genome was obtained through a comprehensive sequencing approach using different technologies. Specifically, 58.66 Gb of HiFi long reads, 55.18 Gb of Hi-C paired reads, and 94.59 Gb of Illumina sequencing reads were utilized. The initial assembly resulted in a genome size of 1.19 Gb, comprising 128 contigs with a contig N50 size of 16.98 Mb (Wang et al. 2023). Notably, the assembled genome size closely approached the estimated genome size derived from flow cytometry (1.22 Gb) and k-mer analysis (1.23 Gb).

Aligning DNA and RNA short reads against the assembled genome showed an average mapping rate of 95.73% and 97.10%, respectively, indicating a reliable alignment.

An analysis of Benchmarking Universal Single-Copy Orthologs (BUSCO) revealed that the assembled genome contained 99.0% of complete homologous genes (Wang et al. 2023). Moreover, during the analysis, 117 telomeric repeat regions (TTTAGGG) were identified across the 65 assembled chromosomes. Among these chromosomes, 53 of them harboured two telomeric regions situated at the beginning and end of the chromosomes.

13.7.2 Sub-genome Division

Through gene family analysis, it was observed that okra sub-genomes A and B exhibited expansions in 3083 and 2427 gene families, respectively. The primary type of duplication responsible for these expansions in both sub-genomes was Whole Genome Duplication (WGD) or segmental duplication. Within these expanded gene families, there were 105 and 66 rapidly expanded gene families detected in sub-genomes A and B, respectively, with a significance threshold of P-value ≤ 0.01. Notably, the okra genome harboured a total of 2339 unique gene families, which were distinct from those involved in the expansions observed in the sub-genomes.

13.7.3 Whole Genome Duplication/Karyotype Analysis

In the investigation of Whole Genome Duplication (WGD) events that took place in Malvaceae species during their evolution, a comparative analysis of syntenic depth ratios among four different groups was studied (Wang et al. 2023). The results provided insights into the WGD history of okra in relation to three other species. First, the syntenic depth ratio between okra and *Gossypium raimondii* showed an 8:1 pattern. This observation supported the suggestion that okra had undergone three additional rounds of WGD events compared to *G. raimondii*. Next, the syntenic depth ratio between okra and *Hibiscus cannabinus* was found to be 4:1. This finding was consistent with the previous hypothesis that okra experienced two additional WGD events relative to *H. cannabinus*. Thirdly, the syntenic depth ratio between okra and *Hibiscus mutabilis* observed an overall 4:3 pattern. It is worth noting that three WGD events had been previously reported in *H. mutabilis*.

Additionally, the syntenic depth ratio between *Hibiscus cannabinus* and *Hibiscus mutabilis* resulted in an overall 1:3 pattern. This finding provides further evidence supporting the occurrence of an additional Whole Genome Triplication (WGT) event in the genome of *H. mutabilis* compared to *H. cannabinus* (Wang et al. 2023). The comparative analysis of syntenic depth ratios provided valuable insights into the WGD events that occurred during the evolutionary history of the Malvaceae species, particularly highlighting the multiple WGD events in the genome of okra.

13.8 Biosynthesis Pathway of Sinapic Acid/S-lignin in Okra

Metabolomics analysis of the okra samples revealed the identification of 589 compounds. Among these compounds, flavonoids were the most abundant, comprising 33.11% of all metabolites, followed by phenolic acids (25.81%) and alkaloids (15.45%) (Wang et al. 2023).

In parallel, transcriptome analysis across the samples led to the identification of 37,493 differentially expressed genes (DEGs), which were further classified into eight clusters based on their expression patterns. As the mature pod of okra is the primary part consumed as food, special attention was given to genes highly expressed only in mature pods. Functional enrichment analysis revealed that these genes were mainly associated with the biosynthesis of various plant secondary metabolites, including phenylpropanoid biosynthesis (K00430, $P = 6.50 \times 10–4$) and phenylalanine metabolism (K13064, $P = 9.38 \times 10–5$) (Wang et al. 2023). These findings suggest a highly active status of material synthesis in mature pods, which could contribute to their nutritional value and overall quality as a food source. As this is the first report on the whole genome sequence, we expect improvements in assembly and annotation in the years to come.

13.9 Conclusion

Genome sequencing and transcriptome sequencing can contribute to the identification of genes and transcripts responsible for agronomic traits, such as disease resistance and pod quality, in okra. Okra is considered a non-model crop due to its complex genome. A recent study indicated a genome size of approximately 1.2 Gbp with 65 chromosomes. Understanding the entire genome structure and assembly would aid in designing a crop improvement programme. The identification of genes and their functions would be valuable for both physiological and breeding studies. Additionally, this knowledge could facilitate the development of genetically modified crops. Identifying genes and markers associated with these genes/traits could streamline the breeding of resistant varieties. To address multiple diseases across diverse genetic backgrounds, improving whole genome sequencing is essential for gene pyramiding and genetic modification. Furthermore, developing varieties or hybrids that can tolerate abiotic stresses, such as cold, moisture, and salt stress, is crucial given the changing climatic conditions.

References

Akash MW, Shiyab SM, Saleh MI (2013) Yield and AFLP analyses of inter-landrace variability in okra (Abelmoschus esculentus L.). Life Sci J 10(2):2771–2779

Aladele SE, Ariyo OJ, De Lapena R (2008) Genetic relationships among West African okra (Abelmoschus caillei) and Asian genotypes (Abelmoschus esculentus) using RAPD. Afr J Biotechnol 7(10):1426–1431

Benchasri S (2012) Okra (Abelmoschus esculentus (L.) Moench) as a valuable vegetable of the world. Ratar Povrt 49(1):105–112

Bertini CH, Schuster I, Sediyama T, Barros EGD, Moreira MA (2006) Characterization and genetic diversity analysis of cotton cultivars using microsatellites. Genet Mol Biol 29:321–329

Chakravarthi BK, Naravaneni R (2006) SSR marker-based DNA fingerprinting and diversity study in rice (Oryza sativa. L). Afr J Biotechnol 5(9):684–688

Gayathri M, Pitchaimuthu M, Ravishankar K (2021) SSR marker development in Abelmoschus esculentus (L.) Moench using transcriptome sequencing and genetic diversity studies. J Hortic Sci 16(2):206–214

Gulsen O, Karagul S, Abak K (2007) Diversity and relationships among Turkish okra germplasm by SRAP and phenotypic marker polymorphism. Biologia 62(1):41–45

Ikram-ul-Haq, Khan AA, Azmat MA (2013) Assessment of genetic diversity in okra (Abelmoschus esculentus L.) using RAPD markers. Pak J Agric Sci 50(4):655–662

Joshi AB, Hardas MW (1956) Alloploid nature of okra, Abelmoschus esculentus (L.) Monech. Nature 178(4543):1190–1190

Kyriakopoulou OG, Arens P, Pelgrom KT, Karapanos I, Bebeli P, Passam HC (2014) Genetic and morphological diversity of okra (Abelmoschus esculentus [L.] Moench.) genotypes and their possible relationships, with particular reference to Greek landraces. Sci Hortic 171:58–70

Leal NR, Amaral AT Jr, Pereira MG, Daher RF, Martinello GE (2000) Comparison of morphological characteristics and RAPD for estimating genetic diversity in Abelmoschus spp. In: *International symposium on molecular markers for characterizing genotypes and identifying cultivars in horticulture 546*. International Society for Horticultural Science, Leuven, pp 101–104

Li J, Ye GY, Liu HL, Wang ZH (2020) Complete chloroplast genomes of three important species, Abelmoschus moschatus, A. manihot and A. sagittifolius: genome structures, mutational hotspots, comparative and phylogenetic analysis in Malvaceae. PLoS One 15(11):e0242591

Liu Y, Wang J, Bai Y, Zhang T, Shi D, Liu Z, Jiang L, Ye L (2023) The whole chloroplast genome in Abelmoschus esculentus L. Moench. N Z J Crop Hortic Sci 51(1):123–135

Lu X, Chen X, Mu M, Wang J, Wang X, Wang D, Yin Z, Fan W, Wang S, Guo L, Ye W (2016) Genome-wide analysis of long noncoding RNAs and their responses to drought stress in cotton (Gossypium hirsutum L.). PLoS One 11(6):e0156723

Massucato LR, Nakamura KK, Ruas PM, Zeffa DM, Silva DJHD, Gonçalves LSA (2019) Genetic diversity among Brazilian okra landraces detected by morphoagronomic and molecular descriptors. Acta Sci Agron 42:e43426

Muthaiah G, Elangovan D, Mottaiyan P, Ravishankar KV (2023) Comprehensive transcriptome analysis in okra (Abelmoschus esculentus L. Moench): analysis of LncRNA and transcription factors involved in abiotic stress. Russ J Plant Physiol 70(3):47

Nwangburuka CC, Kehinde OB, Ojo DK, Denton OA, Popoola AR (2011) Morphological classification of genetic diversity in cultivated okra, Abelmoschus esculentus (L) Moench using principal component analysis (PCA) and single linkage cluster analysis (SLCA). Afr J Biotechnol 10(54):11165–11172

Ramya P, Bhat KV (2012) Analysis of phylogenetic relationships in Abelmoschus species (Malvaceae) using ribosomal and chloroplast intergenic spacers. Indian J Genet Plant Breed 72(4):445–453

Ravishankar KV, Dinesh MR, Nischita P, Sandya BS (2015) Development and characterization of microsatellite markers in mango (Mangifera indica) using next-generation sequencing technology and their transferability across species. Mol Breed 35:1–13

Ravishankar KV, Muthaiah G, Mottaiyan P, Gundale SK (2018) Identification of novel microsatellite markers in okra (Abelmoschus esculentus (L.) Moench) through next-generation sequencing and their utilization in analysis of genetic relatedness studies and cross-species transferability. J Genet 97:39–47

Salameh NM (2014) Flow cytometric analysis of nuclear DNA between okra landraces (Abelmoschus esculentus L.). Am J Agric Biol Sci 9(2):245–250

Sawadogo M, Ouedraogo JT, Balma D, Ouedraogo M, Gowda BS, Botanga C, Timko MP (2009) The use of cross species SSR primers to study genetic diversity of okra from Burkina Faso. Afr J Biotechnol 8(11):2476–2482

Schafleitner R, Kumar S, Lin CY, Hegde SG, Ebert A (2013) The okra (Abelmoschus esculentus) transcriptome as a source for gene sequence information and molecular markers for diversity analysis. Gene 517(1):27–36

Shi D, Wang J, Bai Y, Liu Y (2020) Transcriptome sequencing of okra (Abelmoschus esculentus L. Moench) uncovers differently expressed genes responding to drought stress. J Plant Biochem Biotechnol 29(2):155–170

Siemonsma JS (1982) La culture du Gombo (Abelmoschus spp.), legume-fruit tropical (avec reference speciale a la Cote d'Ivoire). Wageningen University and Research

Strickler SR, Bombarely A, Mueller LA (2012) Designing a transcriptome next-generation sequencing project for a nonmodel plant species. Am J Bot 99(2):257–266

Wang R, Li W, He Q, Zhang H, Wang M, Zheng X, Liu Z, Wang Y, Du C, Du H, Xing L (2023) The genome of okra (Abelmoschus esculentus) provides insights into its genome evolution and high nutrient content. Hortic Res 10:uhad120

Werner O, Magdy M, Ros RM (2016) Molecular systematics of Abelmoschus (Malvaceae) and genetic diversity within the cultivated species of this genus based on nuclear ITS and chloroplast rpL16 sequence data. Genet Resour Crop Evol 63:429–445

Yadav RK, Lata S, Iquebal MA, Tomar BS, Choudhary H, Tomer A, Sharma PK, Jaiswal S (2022) Development and validation of novel SSR markers using transcriptome data in okra (*Abelmoschus esculentus* L.)-a genetically orphan crop. Res Sq:1–13. https://doi.org/10.21203/rs.3.rs-1493839/v1

Yuan CY, Zhang C, Wang P, Hu S, Chang HP, Xiao WJ, Lu XT, Jiang SB, Ye JZ, Guo XH (2014) Genetic diversity analysis of okra (Abelmoschus esculentus L.) by inter-simple sequence repeat (ISSR) markers. Genet Mol Res 13(2):3165–3175

Yuan CY, Wang P, Chen PP, Xiao WJ, Zhang C, Hu S, Zhou P, Chang HP, He Z, Hu R, Lu XT (2015) Genetic diversity revealed by morphological traits and ISSR markers in 48 okras (Abelmoschus escullentus L.). Physiol Mol Biol Plants 21:359–364

Zhan Y, Wu T, Zhao X, Wang Z, Chen Y (2021) Comparative physiological and full-length transcriptome analyses reveal the molecular mechanism of melatonin-mediated salt tolerance in okra (Abelmoschus esculentus L.). BMC Plant Biol 21:1–16

Zhang C, Dong W, Gen W, Xu B, Shen C, Yu C (2018) De novo transcriptome assembly and characterization of the synthesis genes of bioactive constituents in Abelmoschus esculentus (L.) moench. Genes 9(3):130

Zhang J, Yang Z, Feng P, Zhong X, Ma Q, Su Q, Wang X, Li C, Yang Y (2019) Identification and the potential roles of long non-coding RNAs in cotton leaves damaged by Aphis gossypii. Plant Growth Regul 88:215–225

Utilization of Wild Species in Okra (*Abelmoschus esculentus*) Improvement 14

K. V. Bhat, K. Joseph John, A. Suma, C. N. Murali, M. Latha, C. J. Lakshmi, M. Pitchaimuthu, and V. A. Muhammed Nissar

14.1 Introduction

India annually produces over 63 lakh metric tons of okra from an area of 5.24 lakh hectares, valued at Rs. 534,037 lakhs at current market prices. It is cultivated in all the states of India, with West Bengal leading in both area and production, followed by Andhra Pradesh, Gujarat, and Odisha. However, the widespread occurrence of Okra Yellow Vein Mosaic Virus (OYVMV) and Okra Enation Leaf Curl Virus (OELCV) has been the major factor limiting the yield. OYVMV has been reported to cause yield losses ranging from 20% to 90%, depending on the disease's severity. Furthermore, the productivity of conventional okra varieties has remained static over the years due to a lack of sufficient diversity in cultivars. Traditionally, *A. manihot* and *A. tetraphyllus* have been used as sources of BYVMV tolerance genes in backcross breeding programs to manage the virus. However, over the years, this single source has proven insufficient to control the disease due to the frequent breakdown of tolerance conferred by a single dominant gene.

Intensive studies of the natural habitats of wild species of the genus *Abelmoschus*, followed by taxonomic studies conducted by a consortium of three organizations led by ICAR-National Bureau of Plant Genetic Resources, have resulted in the

K. V. Bhat (✉) · C. N. Murali
ICAR-National Bureau of Plant Genetic Resources, Pusa Campus, New Delhi, India

K. J. John · A. Suma · M. Latha · C. J. Lakshmi
ICAR-National Bureau of Plant Genetic Resources, Regional Station, Thrissur, Kerala, India

M. Pitchaimuthu
ICAR-Indian Institute of Horticultural Research, Bengaluru, Karnataka, India

V. A. M. Nissar
ICAR-Indian Institute of Spices Research, Kozhikode, Kerala, India

S. K. Tikoo et al. (eds.), *Okra: Status, Challenges and Opportunities*,
https://doi.org/10.1007/978-981-97-9963-3_14

systematic collection of diverse populations of over 17 *Abelmoschus* species from different ecological zones. These species were used for field screening under epiphytotic conditions for OYVMV and OELCV at locations in Delhi and Bangalore, leading to the identification of sources of tolerance (John et al. 2013). *A. enbeepeegearense*, a newly described species endemic to the Western Ghats, was observed to possess a very high degree of tolerance to both OYVMV and OELCV. Additionally, other sources in *A. angulosus* var. *grandiflorus*, *A. pungens* var. *mizoramensis*, *A. moschatus*, and alternate populations of *A. tetraphyllus* were also identified and used to generate F_1 hybrids with okra. Sterility in these hybrids was overcome by diploidizing the F_1s through colchicine treatment. Tolerant amphidiploids were identified after testing in three diverse locations in the country, namely Delhi, Bengaluru, and Guntur, and were used in this pre-breeding program. Despite improved fertility in the amphidiploids, the lack of segregants in the progenies indicated the need to follow alternative strategies to transfer the genes for virus tolerance from these valuable amphidiploids to the cultivated okra genetic background. In addition, the study of plant characteristics in the F_1s, amphidiploids, and other segregating generations of the crosses between *A. esculentus*, *A. caillei*, *A. angulosus* var. *grandiflorus*, and *A. pungens* var. *mizoramensis* indicated a high potential for incorporating prolificacy and prolonged fruit bearing in these populations within the cultivated gene pool to improve the productivity of this important vegetable crop. However, success in wide hybridizations to transfer useful genes from wild species has largely hindered by the presence of several crossability barriers, pronounced linkage drag, and higher-order polyploidy in these donor species. Considering the immense value of the genetic resources generated and the progress made, this work continued for another 3 years with the aim of achieving the goals through the use of appropriate tools, such as bridging crosses and backcrossing amphidiploids to the cultigen, to overcome the barriers to gene flow between the wild and cultivated gene pools in okra.

14.2 Diversity in *Abelmoschus* Species

The genus *Abelmoschus* (Malvaceae) is native to South and Southeast Asia and Africa (Borssum Waalkes 1966). Except for *A. ficulneus*, the natural habitat of all other wild species is restricted to Asia and Australia, while *A. ficulneus* is found in both Asia and Africa. In Africa, five species are reported: *A. esculentus*, *A. caillei*, *A. manihot*, *A. moschatus*, and *A. ficulneus* (Siemonsma 1982a, 1982b). The species *A. hispidissimus* Chev. is considered a synonym of *A. moschatus*. However, confusion exists over the identity of *A. manihot* specimens from Africa. Several recent germplasm collection missions in West and Central Africa have failed to find true *A. manihot* (Hamon 1988; Hamon and Hamon 1992; Hamon and Yapo 1986; Siemonsma 1982b). The reported occurrence of *A. manihot* in West Africa should ideally be considered as incorrect designations or misidentifications of *A. caillei*.

The genus was originally established by Medikus (1787) and was accepted by Gaertner (1791) and Moench (1794). Later authors like Masters (1874), Prain (1903), and Dunn (1915) did not agree with its elevation as a distinct genus and instead followed De Candolle (1824) in treating it under the broad generic concept of *Hibiscus*. However, following Hochreutiner (1924), who clearly defined the taxonomic traits of *Abelmoschus*, there is a general consensus for its existence as a separate genus distinct from *Hibiscus*. The genus *Abelmoschus* is characterized by an asymmetrical, spathaceous, deciduous calyx that falls off as a unit, in contrast to the campanulate or cupular regular, accrescent calyx in *Hibiscus*. The genus derives its name from the Arabic word "Kaab-el-misk," meaning musk seed (Don 1831), or "Abu-al-misk," meaning father of musk (Sivarajan and Pradeep 1996). The type species, *A. moschatus* Medik, is the source of Ambrette seed oil characterized by its musk scent. The number of species described under *Abelmoschus* exceeds 50. The *Index Kewensis* lists more than 30 species in Asia and Africa, with four each in the Americas and Australia. The taxonomy of the genus is complex, especially that of the infraspecific categories of *A. manihot* and *A. moschatus*.

In the revision of the genus in 1966, van Borssum Waalkes retained only six species: *A. manihot*, *A. moschatus*, *A. angulosus*, *A. crinitus*, *A. ficulneus*, and *A. esculentus*. He simplified Hochreutiner's infraspecific subdivision of *A. manihot* and regrouped them under two subspecies: subsp. *manihot* (L.) Medikus, which includes all the cultivated forms, and subsp. *tetraphyllus*, comprising all the wild forms. Within subsp. *tetraphyllus*, he recognized two varieties: var. *tetraphyllus* and var. *pungens*, both based on morphological distinctness and geographical discontinuity. Var. *tetraphyllus* is predominantly distributed in lower elevations from sea level to 400 m that are subjected to an annual dry season, whereas var. *pungens* grows at higher elevations (400–1600 m). There are no reports of the occurrence of *A. manihot* subsp. *manihot* in India, either in cultivation or in the wild. A summary of *Abelmoschus* species reported (Yadav et al. 2014) from India is provided in Table 14.1.

Bates (1968), in a revision of the Malvaceae, generally agreed with van Borssum Waalkes' species descriptions but considered *A. manihot* as a single polymorphic species. He did not support the subspecific classification, especially the placement of cultivated *A. manihot* under one subspecies, which he viewed as a highly artificial assemblage. According to him, it is a highly illogical division because the subspecies *tetraphyllus* represents a natural biological entity, whereas the cultivars are only selections from it. Hamon (1988) also expressed doubts about the discrete boundary between subspecies *manihot* and subspecies *tetraphyllus*. Vredebregt (1991) recognized a third botanical variety within subspecies *tetraphyllus* based on seed characters, var. *megaspermus*, while expressing doubts about the distinctness of var. *pungens* and var. *tetraphyllus*. However, Paul and Nair (1988), in their revision of the Malvaceae for India, recognized all three taxonomic varieties within subspecies *tetraphyllus*. Nevertheless, var. *megaspermoides* clearly requires further study with live samples to ascertain its distinctness.

Table 14.1 Distribution of the genus in the World and India

Species	Distribution based on literature	Distribution based on field and herbarium study by the proponent and team
A. moschatus Medikus	Southeast Asia, Africa, Australia	
subsp. *moschatus* var. *moschatus*	In India, occasional in Andhra Pradesh, Tamil Nadu, Karnataka (Deccan), Maharashtra, Rajasthan, Gujarat, Kerala, Assam, and Meghalaya	Mizoram, Tripura, Assam, Meghalaya, Arunachal Pradesh (all wild state), peninsular India (naturalized escape)
subsp. *moschatus* var. *betulifolius* (Mast.) Hochr.	Indochinese peninsula, Myanmar, North Australia	–
subsp. *biakensis* (Hochr.) Borss.	New Guinea	–
subsp. *tuberosus* (Span.) Borss.	Indochinese peninsula, Myanmar, North Australia; in India, Meghalaya (Khasi & Jaintia Hills)	Rare in Assam, Meghalaya, and elsewhere in North East India (Myanmar border)
A. manihot (L.) Medikus	Presently under cultivation in Papua New Guinea, Indonesian islands, Malesia, and elsewhere in Southeast Asia as a leafy vegetable and vegetatively propagated through cuttings	–
A. tetraphyllus (Roxb. ex Hornem.) R. Graham	Southeast Asia, Africa, Australia; occasional in Andhra Pradesh, Tamil Nadu, Karnataka (Deccan), Maharashtra, Rajasthan, Gujarat, and Goa in India	West coast, Konkan (Goa, Dakshin Kannada, Uttar Kanada, Uduppi, Malabar districts of Kerala), Sindhdurg, and Ratnagiri districts of Maharastra
var. *pungens* (Roxb.) *Hochr.*	Nepal, Sikkim, Uttarakhand, and Meghalaya (rare) in India	Arunachal Pradesh, Mizoram, Uttaranchal, Sikkim
A. esculentus (L.) Moench	Cultivated throughout Southeast Asia, Africa, Australia, and India	All over India under cultivation
A. tuberculatus Pal & Singh	Occasional in Andhra Pradesh, Maharashtra, Rajasthan, Uttarakhand, Madhya Pradesh, and Gujarat in India	Andhra Pradesh, Gujarat, Rajasthan, Madhya Pradesh, Tharai (Uttaranchal), Maharastra
A. ficulneus (L.) Wight & Arn. ex Wight	Southeast Asia, Africa, Australia; occasional in Andhra Pradesh, Tamil Nadu, Karnataka (Deccan), Maharashtra, Uttar Pradesh, Rajasthan, and Gujarat in India	Kerala (rare-Attapady), common in Tamil Nadu, Andhra Pradesh, Karnataka (Deccan), Uttar Pradesh, Rajasthan, Gujarat
A. crinitus Wall.	Southeast Asia, Africa, Australia; occasional in Andhra Pradesh, Bihar, Chathisgarh, Jharkhand, Tamil Nadu, Karnataka (Deccan), Maharashtra, Rajasthan, Uttarakhand, and Gujarat in India	Occasional in Andhra Pradesh, Bihar, Chathisgarh, Jharkhand, Orissa, Uttarakhand, and Gujarat in India

(continued)

Table 14.1 (continued)

Species	Distribution based on literature	Distribution based on field and herbarium study by the proponent and team
A. angulosus Wall. ex Wight & Arn.		
var. *angulosus* Sivarajan et al.	Sri Lanka, India; in India very rare in Nilgiri and Annamalai hills of Southern Western Ghats	Kerala, Munnar, Marayur, and Shanthanpara (Idukki district)
var. *grandiflorus* Thwaites	Western Ghats of Kerala, Tamil Nadu, Karnataka, and Goa (occasional); Maharastra and Gujarat (rare)	Lower Western Ghats (Kerala, Karnataka, T N, and Goa)
var. *purpureus* Thwaites	Rare in the high ranges of Southern Western Ghats; Bababudan Hills, Nilgiris, and Silent Valley National Park	Rare in the high ranges of Southern Western Ghats; Bababudan Hills, Gudallur, Nilgiris, Idukki district, and Silent Valley National Park
A. caillei (A. Chev.) Stevels	South West Africa	Introduced and naturalized under cultivation in Andhra Pradesh, Tamil Nadu, Karnataka (Deccan), Maharashtra, Kerala, and Mizoram in India
Abelmoschus enbeepeegearense J John, Scariah, Nissar, KV Bhat *et* Yadav	New species	Low elevation Western Ghats of India, including Kerala, Tamil Nadu, and Karnataka states
A. pungens var. mizoramensis K.J. John., Krishnaraj & K. Pradheep	New species	Collected from Mizoram, full distribution pattern yet to be studied

14.3 Utilization of Wild Species as Source of Novel Genetic Variations

Intensive collections and taxonomic studies conducted at ICAR-NBPGR, along with Shivaji University Kolhapur, North Eastern Hill University Shillong, and ICAR-IIHR, have resulted in the assembly of over 17 species from India. These species were crossed in different combinations to study their relationships and group them based on their genetic affinities. Studies of floral biology, pollen–pistil interaction, crossability barriers, etc., were useful in resolving species identities (John et al. 2013). Field screening of the accessions under epiphytotic conditions for Okra Yellow Vein Mosaic Virus (OYVMV) and Okra Enation Leaf Curl Virus (OELCV) at Guntur, Delhi, and Bangalore locations resulted in the identification of sources of tolerance. *A. enbeepeegearense*, a newly described species endemic to the Western Ghats, was observed to possess a very high degree of tolerance to both OYVMV and OELCV. In addition, other sources in *A. tetraphyllus*, *A. angulosus* var. *grandiflorus*, and *A. pungens* var. *mizoramensis* were also identified and used to generate F_1 hybrids with okra. Sterility in these hybrids was overcome by diploidizing the F_1 through colchicine treatment. These amphidiploids served as useful bridging genotypes for backcross breeding for the transfer of virus tolerance genes.

The utility of wild relatives of cultivated okra as sources of genes for virus tolerance had been demonstrated by earlier studies, resulting in the release of several varieties with OYVMV tolerance, such as Arka Anamika, Punjab Padmini, and Arka Abhay. Studies on the inheritance pattern have suggested varied genetics, which include the role of two complementary dominant genes in susceptible X susceptible (S X S) and susceptible X resistant (S X R) crosses, and two duplicate dominant genes in resistant X resistant (R X R) crosses (Pullaiah et al. 1998). However, in interspecific crosses between *A. manihot* and *A. tetraphyllus*, a single dominant gene controlling virus tolerance was reported (Jambhale and Nerkar 1981a, b; Dutta 1984), while tolerance to OYVMV in *A. manihot* ssp. *manihot* was reported to be controlled by two dominant genes (Sharma and Dhillon 1983). These results indicate the need for more elaborate studies in each interspecific cross combination developed, as these might be due to the presence of variations in somatic chromosome numbers of the wild species, as well as variations existing within *A. esculentus* and the virus. Differences in inheritance patterns in different combinations might also be due to ploidy variations in cultivated okra.

14.4 Wide Hybridization

Distant hybridization involving different species/genera helps to combine diverse genomes into one individual. It is often a man-made attempt to break the species barrier to gene flow that exists in nature, thereby effecting new genomic combinations. Many cultivated plants have an allopolyploid origin resulting from interspecific hybridization that occurred in nature. The gene flow facilitating the incorporation of genes from one species to another is called introgression, which can be either natural or man-mediated. Introgressive hybridization, the incorporation of genetic material from one species to another through wide hybridization and repeated backcrossing, plays an important role in the evolution of plant species, genetic modification, and enriching the gene pool for breeding. Breeders are increasingly interested in introgressing genes conferring desirable traits from the wild to crop species in breeding programs.

Different ploidy levels and often genotype-specific variations in chromosome numbers have been noticed in okra as well as some of its wild relatives (Merita et al. 2013). Even though there is a general consensus regarding the distinctness of species, the lack of consensus on infraspecific taxonomy leads to misreporting of wide crosses within species as interspecific hybridization. In general, okra wild species have many desirable traits, such as dark green color, high mucilage, extended bearing, perennation tendency, high branching, reduced fruit length, drought resistance, high-temperature tolerance, and YVMV resistance, which can be incorporated into the cultivars. Kuwada (1966) was the first to develop an amphidiploid of okra with *A. tuberculatus* and described it as *Abelmoschus* × *tuberculatus*. Further, F_1 hybrids of *A. tuberculatus* with *A. manihot* were also attempted (Kuwada 1974). Pal et al. (1952) had attempted wide hybridization between cultivated okra and its closest progenitor, *A. tuberculatus*. Akhond et al. (2000) and Samarajeewa (2003) described the success of wide hybridization attempts involving okra with *A. moschatus* and *A. angulosus*, respectively.

Wide hybridization was attempted between various *Abelmoschus* species, including okra, primarily to establish genetic relationships based on F_1 fertility and to derive useful new combinations. The material could be useful for OYVMV and enation leaf curl virus resistance, besides cytoplasmic male sterility in cultivated okra. Screening of wild *Abelmoschus* germplasm at hot-spot localities (New Delhi and Varanasi) showed field resistance of *A. enbeepeegearense* and *A. angulosus* var. *grandiflorus* to BYVMV and OELCV, whereas all other species and okra genotypes were susceptible to the virus. *A. caillei* has very high mucilage in its fruits, besides an extended bearing trend beyond 12 months.

Elaborate field and controlled evaluations of wild *Abelmoschus* germplasm to identify alternative sources of tolerance to BYVMV and OELCV are lacking. This is essential to overcome the problems of the frequent breakdown of tolerance introduced to cultivated okra from the limited wild sources. There is a need to manage the disease more effectively through the deployment of alternative genes conferring tolerance. Systematic wide hybridization programs aimed at the genetic enhancement of cultivated okra are much needed. The introgression of useful genes from the wild is required to overcome domestication bottlenecks in genetic diversity. The presence of high-order polyploidy and gene duplications needs to be tackled through scientific pre-breeding programs.

14.5 Materials and Methods

Generation Advancement and Introgression of Useful Genes

The accessions of wild species identified as tolerant to OYVMV and OELCV were planted along with the amphidiploid in four locations, namely, ICAR-NBPGR Thrissur and Delhi, ICAR-IIHR Bengaluru, and at Guntur in farmers' fields, to identify the tolerant hybrid combinations. Wild species, crosses, and amphidiploids exhibiting field tolerance to OYVMV were used for backcrossing to cultivated okra in order to ensure the recovery of the genetic background of cultivated okra, with the genes for tolerance incorporated. Backcrosses were carried out between F_2s, tolerant colchiploids, and released okra varieties for the introgression of genes from wild species into the cultivated genomic background. New crosses were also attempted with okra and tolerant accessions of *A. enbeepeegearense*, *A. angulosus* var. *grandiflorus*, *A. pungens* var. *mizoramensis*, *A. moschatus*, and alternative collections of *A. tetraphyllus* to generate sufficient numbers of segregants for backcrossing and genetic enhancement.

14.6 Status of the Work

The F_1 and backcross generations of 55 different combinations and parental species were planted in both net house and field conditions, alongside susceptible okra varieties serving as controls. These crosses involved okra and the five tolerant species, namely, *A. tetraphyllus*, *A. enbeepeegearense*, *A. angulosus* var. *grandiflorus*, *A. moschatus*, and *A. pungens* var. *mizoramensis*. The F_1s also included crosses among amphidiploids of the above species with *A. esculentus*.

The fieldwork was conducted under net house conditions in Delhi for 3 years. During one season in 2018, a field trial was also conducted at GKVK, Bengaluru. The performance of the crosses under Bengaluru conditions was excellent, while the poor soil conditions in Delhi resulted in suboptimal trait expressions. Nevertheless, the conditions in Delhi, a hotspot for the diseases, were helpful in carrying forward only those segregants with a higher degree of tolerance to viruses.

Fresh crosses were also attempted between three okra cultivars and the tolerant species, *A. enbeepeegearense*, *A. angulosus* var. *grandiflorus*, *A. moschatus*, and *A. pungens* var. *mizoramensis*. The observations indicated successful fruit set in crosses involving okra and *A. enbeepeegearense*, *A. angulosus* var. *grandiflorus*, and *A. pungens* var. *mizoramensis*. However, fruit drop was frequent in the cross combination of *A. esculentus* and *A. moschatus* (Table 14.2).

Furthermore, previously produced amphidiploids were also planted for use in the crossing program since they were selected based on their response to virus infestation. The selection process was based on the comprehensive field screening conducted earlier under hotspot conditions in Guntur, Bangalore, and Delhi. Screening of over 155 amphidiploids resulted in the identification of 11 amphidiploids with significant tolerance to BYVMV and OELCV (Table 14.3).

These crosses were planted both in pots and under field conditions. Due to constraints related to the availability of high-quality field facilities with proper irrigation and protection, the initial results indicated varying seed viability and seedling survival. These outcomes seemed to be influenced by soil moisture and temperature. Consequently, out of the 174 cross combinations sown, only 96 combinations successfully established plants. Observations were made regarding plant morphology, floral characteristics, and response to virus infection (see Figs. 14.1 and 14.2).

14.7 Backcrossing of F_1s for Recovery of the Okra Genetic Background

Backcrossing attempts were made with amphidiploids and interspecific hybrids in all combinations. In all cross combinations, the female plant was *A. esculentus*, so that cytoplasmic incompatibilities, if any, can be avoided. Hence, in the crossing program, both amphidiploids and F_1s, as well as cultivated okra, were used as female parents. The fruit set percentage varied among the cross combinations. Initial observations indicated a good frequency of seed set in amphidiploids, whereas the success rate was lower in backcrosses involving cultivated okra as the

Table 14.2 List of interspecific cross combinations with seed set

A. Crosses involving *esculentus* × amphidiploids
(A. esculentus IC433641 × A. tetraphyllus IC141029) × A. esculentus Arka Anamika
(A. esculentus IC433641 × A. tetraphyllus IC141029) × A. palianus IC0624218
(A. esculentus ICA433641 × A. pungens var. mizoramensis IC0624222) × A. palianus IC0624218
[(A. esculentus IC433641 × A. angulosus var. grandiflorus IC613527) × (A. esculentus IC433641 × A. pungens var. mizoramensis IC0624222)]
[(A. esculentus IC433641 × A. angulosus var. grandiflorus IC613527) × (A. esculentus IC433641 × A. tetraphyllus IC253122)]
[(A. esculentus IC433641 × A. pungens var. mizoramensis IC0624222) × (A. esculentus IC433641 × A. tetraphyllus IC253122)]
B. Crosses involving amphidiploids and species
AM23 × A. pungens var. mizoramensis IC0624222
AM23 × A. tetraphyllus IC253122
AM23 × A. esculentus Ruchi
AM23 × SP37 (A. manihot IC566867)
AM23 × SP5 (A. ficulneus IC550640)
AM23 × WAS16 (A. enbeepeegearense IC582757)
AM24 × A. pungens var. mizoramensis IC0624222
AM25 × A. pungens var. mizoramensis IC0624222
AM6 × A. esculentus var. 2288
AM6 × Ruchi *(A. esculentus)*
AM17 × Arka Anamika *(A. esculentus)*
AM6 × A. pungens var. mizoramensis IC0624222
AM12 [A. esculentus IC433641 × A. pungens var. mizoramensis IC0624222] × AM25 [A. caillei IC556513]
C. New crosses attempted
A. esculentus Arka Anamika × [A. esculentus IC433641 × A. enbeepeegearense IC582757]
A. esculentus Arka Anamika × A. angulosus var. grandiflorus IC613527
A. esculentus Arka Anamika × A. pungens var. mizoramensis IC0624222
A. esculentus Arka Anamika × A. moschatus IC0624232
A. esculentus Arka Anamika × A. tetraphyllus IC253122
A. esculentus Arka Anamika × A. tetraphyllus ex. Brahmavara IC253122
A. esculentus Arka Anamika × CR4 [A. esculentus IC260106 × A. sp. *nova Mizoram IC0624236]*
A. esculentus Arka Anamika × MS (A. moschatus var. tuberosus IC470750 × A. moschatus IC0624232)
A. esculentus Arka Anamika × PN28 (A. esculentus IC433639 × A. manihot IC566867)
A. esculentus Arka Anamika × PN30 (A. esculentus IC433641 × A. manihot IC566867)
A. esculentus Arka Anamika × PN32 [A. esculentus IC433641 × A. manihot IC141029]
A. esculentus Arka Anamika × PN36 (A. caillei IC566817 × A. manihot IC141020)
A. esculentus Arka Anamika × SP47 (A. moschatus var. tuberosus IC470750 × A. moschatus IC0624232)

(continued)

Table 14.2 (continued)

A. esculentus Arka Anamika × SP49 (A. moschatus var. tuberosus IC470750 × A. moschatus IC0624232)
D. Crosses involving two amphidiploids
AM12 [A. esculentus IC265657 × A. pungens var. mizoramensis IC0624222] × SP49 (A. moschatus var. tuberosus IC470750 × A. moschatus IC0624232)
AM15 [A. esculentus IC265657 × A. pungens var. mizoramensis IC0624222] × (A. esculentus Arka Anamika × A. tetraphyllus IC253122) amphidiploid
AM15 × AM6 [A. esculentus IC265657 × A. pungens var. mizoramensis IC0624222]
AM23 [A. esculentus IC265657 × A. angulosus var. grandiflorus IC613527] × (A. esculentus IC265657 × A. tetraphyllus IC253122)
AM23 [A. esculentus Arka Anamika × A. angulosus var. grandiflorus IC613527] × AM6 [A. esculentus Arka Anamika × A. pungens var. mizoramensis IC0624222]

The crosses attempted included new F1s with seed set (with varying degrees of viability), combinations involving amphidiploids and five target species, and crosses between two amphidiploid combinations

Table 14.3 List of OYVMV and OELCV tolerant amphidiploids between *A. esculentus* (female) and the tolerance donors, *A. pungens* var. *mizoramensis*, *A. angulosus* var. *grandiflorus*, and *A. moschatus*

ID	Cross combination
AM2	*c/50mizo30 [A. esculentus IC265657 × A. pungens* var. *mizoramensis IC0624222]*
AM3	*c2/106mizo30 [A. esculentus IC265657 × A. pungens* var. *mizoramensis IC0624222]*
AM4	*c2/50mizo30 (bulk) [A. esculentus IC265657 × A. pungens* var. *mizoramensis IC0624222]*
AM6	*c3/50mizo24; [A. esculentus IC265657 × A. pungens* var. *mizoramensis IC0624222]*
AM7	*A. sagittifolius IC470750 × A. moschatus IC0624232 F6 generation selections*
AM8	*c/50mizo30 [A. esculentus IC265657 × A. pungens* var. *mizoramensis IC0624222]*
AM12	*c3/50mizo24; A. esculentus IC265657 × A. pungens* var. *mizoramensis IC0624222*
AM13	*c2/106/mizo30; A. esculentus IC265657 × A. pungens* var. *mizoramensis IC0624222*
AM22	*c2/741/gr12-1; A. esculentus Arka Anamika × A. angulosus* var. *grandiflorus IC613527*
AM23	*c2/87/gr4; A. esculentus Arka Anamika × A. angulosus* var. *grandiflorus IC613527*
AM24	*c2/50/mizo34 [A. esculentus IC265657 × A. pungens* var. *mizoramensis IC0624222]*

female parent. The successful backcross combinations, along with some seed recoveries, are summarized in Table 14.4.

Fresh crosses were also attempted with the five tolerant species, namely, *A. tetraphyllus*, *A. enbeepeegearense*, *A. angulosus* var. *grandiflorus*, *A. moschatus*, and *A. pungens* var. *mizoramensis* as males and the cultigen as the female parent. Subsequently, selected plants from these crosses were self-pollinated, and seeds were collected separately from each plant. The F_2, F_3, and F_4 generations of these crosses were cultivated for further studies in a controlled environment within a nethouse (see Fig. 14.3). Observations were made to document morphological variations and segregations within the populations.

The F_2 and BC2 generations resulting from crosses between three okra cultivars and the tolerant species, namely, *A. enbeepeegearense*, *A. angulosus* var.

Fig. 14.1 OYVMV and OELCV tolerant amphidiploids along with new undescribed wild species with potential

Fig. 14.2 (**a**) Infestation of whitefly, the vector for OYVMV and OELCV in the nethouse. The host plant is *A. enbeepeegearense*. However, no virus symptom was noticed. (**b**) Amphidiploid of cross between *A. esculentus* and *A. pungens* var. *mizoramensis* exhibiting tolerance to both OYVMV and OELCV under field conditions. (**c**) Leaf of *A. moschatus* showing enation symptoms in the nethouse

grandiflorus, *A. moschatus*, and *A. pungens* var. *mizoramensis*, were planted during the 2019–20 season under nethouse conditions. The susceptibility reactions of the segregating populations were recorded, along with variations in morphological traits. Notably, the populations exhibited enhanced seed fertility, as indicated by increased fruit set.

Table 14.4 Summary of backcrosses recovered with varied success

Cross ID	No seeds obtained	Remarks
[A. esculentus IC433641 × A. enbeepeegearense IC582757] × A. esculentus Ruchi	25	All shriveled
14 [A. esculentus IC433641 × A. tetraphyllus IC141029] × A. mizoramensis IC0624222	88	50% shriveled
2 = [(C2/50mizo30) [A. esculentus IC265657 × A. pungens var. mizoramensis IC0624222] × A. esculentus var. Arka Anamika] F1 self	140	Appear viable
A. esculentus var. 2288 okra × [A. esculentus IC433641 × A. tetraphyllus IC141029]	71	Appear viable
A. esculentus var. 2288 × 2[(C2/50miz30xAA) [A. esculentus IC265657 × A. pungens var. mizoramensis IC0624222] × A. esculentus var. Arka Anamika]	40	Appear viable
A. esculentus var. 2288 × A. mizoramensis IC0624222	87	Appear viable
A. esculentus var. 2288 × A. moschatus IC0624232	22	Appear viable
A. esculentus var. 2288 × B9 [A. esculentus Arka Anamika × A. angulosus var. grandiflorus IC613527] × [A. esculentus IC433641 × A. tetraphyllus IC141029]	15	Appear viable
A. esculentus var. 2288 × MS45 [A. sagittifolius IC470750 × A.moschatus IC0624232]	19	Appear viable
A. esculentus var. 2288 × MS-A [A. sagittifolius IC470750 × A. moschatus IC0624232]	1	Appear viable
A. esculentus var. 2288 × PN29 [A. esculentus IC433639 × A. manihot IC141020]	1	Appear viable
A. esculentus var. 2288 × SP47 [A. sagittifolius IC470750 × A. moschatus IC0624232]	13	Appear viable
A. esculentus var. 2288 × WAS42 (A. tetraphyllus IC253122)	36	Appear viable
4 [(C2/50mizo34) [A. esculentus IC265657 × A. pungens var. mizoramensis IC0624222] × A. esculentus var. Parbhani Kranti F1]	149	Appear viable
4 [A. esculentus IC265657 × A. pungens var. mizoramensis IC0624222] × A. esculentus var. Parbhani Kranti F1 × A. esculentus var. 2288	7	Appear viable
4 [A. esculentus IC265657 × A. pungens var. mizoramensis IC0624222] × A. esculentus var. Parbhani Kranti F1 × A. esculentus var. Parbhani Kranti	4	Appear viable
5 [A. esculentus IC265657 × A. pungens var. mizoramensis IC0624222] × A. esculentus var. Arka Anamika × A. esculentus var. Ruchi	4	Appear viable
7 [(C2/50mizo34) [A. esculentus IC265657 × A. pungens var. mizoramensis IC0624222] × A. esculentus var. Arka Anamika] × A. esculentus var. 2288	0	Aborted

(continued)

Table 14.4 (continued)

Cross ID	No seeds obtained	Remarks
7 [(C2/50mizo34) [A. esculentus IC265657 × A. pungens var. mizoramensis IC0624222] × A. esculentus var. Arka Anamika] × A. esculentus var. 2288	2	Appear viable
7 [(C2/50mizo34) [A. esculentus IC265657 × A. pungens var. mizoramensis IC0624222] × A. esculentus var. Arka Anamika] × A. esculentus var. Ruchi	2	Appear viable
A. moschatus KSG × MS39 [A. sagittifolius IC470750 × A. moschatus IC0624232]	50	Appear viable
A. esculentus var. Arka Anamika × [C2/miz50 [A. esculentus IC265657 × A. pungens var. mizoramensis IC0624222] × A. esculentus var. Arka Anamika]	39	Appear viable
A. esculentus var. Arka Anamika × 4 [C2/50miz34 [A. esculentus var. × Parbani Kranti]]	86	Appear viable
A. esculentus var. Arka Anamika × PN21 [A. esculentus IC433685 × A. manihot IC141020]	26	Appear viable

Backcrossing of F_1s to cultivated okra for the recovery of the okra genetic background was also carried out. Backcrosses of amphidiploids and interspecific hybrids from all previously mentioned combinations were planted for further studies under field conditions. These crosses involved *A. esculentus* as the female parent, and in the interspecific crosses, cultivated okra was also used as the female parent, thus ensuring compatibility between cytoplasmic and nuclear genes. Individual plant selections were made in the F_2 and BC_2 generations to minimize the accumulation of undesirable genes. The selected F_2 and BC_2 crosses that have been carried forward are listed in Table 14.5.

Field screening of the segregating populations aimed to study the genetic control of tolerance to OYVMV and OELCV in cross combinations:

The presence of a whitefly population on susceptible *Abelmoschus* plants with distinct virus symptoms inside the net house resulted in the expression of virus infection symptoms on the susceptible plants. The F_1s from crosses between okra and wild species in all combinations exhibited symptoms of OYVMV and OELCV, indicating that tolerance is possibly of a recessive nature. This reaction aligns with earlier reports indicating the recessive nature of gene expression in okra.

Amphidiploids and F_1s resulting from crosses between the amphidiploids displayed tolerance. These crosses involved the following species: *A. tetraphyllus*, *A. enbeepeegearense*, *A. angulosus* var. *grandiflorus*, *A. rugosus*, and *A. pungens* var. *mizoramensis*. All the amphidiploid and F_1 plants exhibited vigorous growth, a branching pattern, and a perennial habit. The fruits were smaller, with prickly hairs on the surface and a high fiber content. Some plants from susceptible cultivars and certain plants in the segregating generations displayed virus infection symptoms. Individual plant reactions were documented, and the tolerant plants were labeled and carried forward to the next generation.

Fig. 14.3 Segregating populations of the interspecific crosses between cultivated okra and *A. angulosus* var. *grandiflorus* and *A. pungens* var. *mizoramensis* in F_2 generation

The individual plant selections made earlier, following virus screening, were planted during the 2019–20 and 2020–21 seasons under field conditions (see Fig. 14.4a–c). These selected plants exhibited a significant recovery of cultivated okra characteristics. Favorable characteristics included longer fruits with significantly higher seed set, reduced prickliness of hairs on the fruits, decreased fiber content in the fruits, seed size, and desirable plant characteristics. However, the gigantism associated with the amphidiploids was still noticeable in the selections. Despite this, it was evident that several valuable characteristics, such as virus tolerance, prolonged flowering and fruiting, and a greater number of branches, had been successfully transferred to the derived plants (see Fig. 14.4a–c).

The pattern of virus infection among individual plants in the populations revealed that tolerance was governed by a recessive gene, segregating in the ratio of 3:1 for

Table 14.5 Performance of selected combinations of backcross populations planted

Cross ID	Remarks about recovery of fertility
[A. esculentus IC433641 × A. enbeepeegearense IC582757] × A. esculentus Ruchi	Variation in seed size, low seed viability, short spines on fruits
A. esculentus var. 2288 okra × [A. esculentus IC433641 × A. tetraphyllus IC141029]	Fruits intermediate, short, less prickly spines
A. esculentus var. 2288 × 2[(C2/50miz30xAA) [A. esculentus IC265657 × A. pungens var. mizoramensis IC0624222] × A. esculentus var. Arka Anamika]	Fruits intermediate, short, short prickly spines
A. esculentus var. 2288 × A. mizoramensis IC0624222	Fruits intermediate, short, short prickly spines
A. esculentus var. 2288 × A. moschatus IC0624232	Low plant establishment
A. esculentus var. 2288 × MS45 [A. sagittifolius IC470750 × A. moschatus IC0624232]	Fruits intermediate, short, medium-sized prickly spines
A. esculentus var. 2288 × MS-A [A. sagittifolius IC470750 × A. moschatus IC0624232]	Fruits intermediate, short, medium-sized prickly spines
A. esculentus var. 2288 × WAS42 (A. tetraphyllus IC253122)	Fruits intermediate, short, medium-sized prickly spines
4 [(C2/50mizo34) [A. esculentus IC265657 × A. pungens var. mizoramensis IC0624222] × A. esculentus var. Parbani Kranti F1]	Fruits intermediate, short, medium-sized prickly spines
5 [(A. esculentus IC265657 × A. pungens var. mizoramensis IC0624222) × A. esculentus var. Arka Anamika] × A. esculentus var. Ruchi	Fruits intermediate, short, medium-sized prickly spines
A. moschatus KSG × MS39 [A. sagittifolius IC470750 × A. moschatus IC0624232]	Fruits intermediate, short, medium-sized prickly spines
A. esculentus var. Arka Anamika × [C2/miz50 [A. esculentus IC265657 × A. pungens var. mizoramensis IC0624222] × A. esculentus var. Arka Anamika]	Fruits intermediate, short, medium-sized prickly spines
A. esculentus var. Arka Anamika × 4 [C2/50miz34 [A. esculentus var. × Parbani Kranti]]	Fruits intermediate, short, medium-sized prickly spines
A. esculentus var. Arka Anamika × PN21 [A. esculentus IC433685 × A. manihot IC141020]	Fruits intermediate, short, medium-sized prickly spines

susceptible to tolerant (see Fig. 14.5) in the cross between *A. esculentus* and *A. angulosus* var. *grandiflorus* (19–57 = ES33).

The results obtained and the genetic enhancements reported here provide alternative sources of tolerance genes to OYVMV and OELCV within the cultivated okra gene pool. However, it is important to note that further evaluation of tolerance genes against different virus strain variations is necessary. With the generous support from the Education Division of ICAR to the Emeritus Scientist, this effort has led to the generation of 127 virus-tolerant derivatives in the cultivated okra genetic background, which have been deposited with ICAR-NBPGR. This collection is of immense value, as it comprises genotypes with genes for virus tolerance transferred from five highly diverse wild species sources. It is likely to serve as a source of diverse genes for tolerance, as well as other important traits for okra breeding, such

Fig. 14.4 (**a**) Differences in fruit characteristics among derivatives from *A. angulosus* var. *grandifloras* (38–42, 44, 47, 51), *A. mizoramensis* (43), *A. tetraphyllus* (41), and *A. enbeepeegearense* (48) in *A. esculentus* and *A. caillei* (44, 45, 50) background. (**b**) Characteristics of the derivatives in F_3 and F_4 generations in the crosses of okra involving *A. ficulneus* (NB71); *A. pungens* var. *mizoramensis* (ES113); *A. tetraphyllus* (B13), and *A. angulosus* var. *grandiflorus* (ES111). (**c**) Fruits and seeds of selected derivatives of interspecific crosses. The selections involve *A. tetraphyllus*, *A. pungens* var. *mizoramensis*, *A. enbeepeegearense*, and *A. angulosus* var. *grandiflorus*. Improved seed set indicate restoration of fertility in the derivatives

Fig. 14.4 (continued)

Fig. 14.5 Segregation in the ratio 3:1 for susceptibility and tolerance to OYVMV in the cross *A. esculentus* × *A. angulosus* var. *grandiflorus* (19–57 = ES33) indicating recessive gene action for tolerance in this donor

as extended fruit bearing and profuse branching. This valuable material may be made available to national institutes, such as ICAR-IIVR and ICAR-IIHR, and leading state agricultural universities such as ICAR-IARI and PAU, for further screening and utilization in breeding programs. It is to be noted that much systematic research work remains to be done, which relates to characterization and naming the genes conferring tolerance to viruses and traits of importance to okra breeding.

Acknowledgments The authors extend their gratitude to ICAR-NBPGR, New Delhi, for providing the necessary facilities to carry out the research on okra genetic enhancement. They also appreciate the funding support from the ICAR-National Agriculture Innovation Project, the Extra-mural Project of ICAR-Horticultural Sciences Division, and the ICAR-Education Division Emeritus Scientist grant awarded to KVB.

References

Akhond AYM, Molla AHM, Islam OM, Mohammad A (2000) Cross compatibility between *Abelmoschus esculentus* and *A. moschatus Euphytica* 114:175–180. https://doi.org/10.1023/A:1003985728630

Bates DM (1968) Notes on the cultivated Malvaceae 2, *Abelmoschus*. Baileya 16:99–112

Borssum Waalkes T (1966) Malesian Malvaceae revised. Blumea 14:1–251

de Candolle AP (1824) Malvaceae. In: A.P. de Candolle (Ed.), Prodromus Systematis naturalis regni vegetabilis, 1. Paris. pp. 427–474

Don G (1831) Malvaceae. In: General history of the Dichlamydeous plants. Vol. 1. London. pp. 458505

Dunn ST (1915) Malvaceae. In: Gamble, J.S., Flora of Presidency of Madras. Vol. 1. London. pp. 86–102

Dutta OP (1984) Breeding in okra for resistance to yellow vein mosaic virus and enation leaf curl virus. Annual Report, IIHR, Bangalore.

Gaertner J (1791) De fructibus et seminibus plantarum. Vol. 2. Tubingen

Hamon S (1988) Organisation evolution of genus *Abelmoschus* (gombo): co adaptation, Travaux et documents microédités, vol 46. ORSTOM, Paris

Hamon S, Hamon P (1992) Future prospects of genetic integrity of two cultivated species of okra (*A. esculentus and A. caillei*) cultivated in West Africa. Euphytica 58:101–111

Hamon S, Yapo A (1986) Perturbation induced within the genus *Abelmoschus* by the discovery of a second edible okra species in West Africa. In Maesen, L.J.G. Van Der, First international symposium on taxonomy of cultivated plants. Acta Hortic 182:133–144

Hochreutimer BPG (1924) Centres of origin for family Malvaceae. Candolla 2:79

Jambhale ND, Nerkar YS (1981a) Inheritance of resistance to okra yellow vein mosaic disease in interspecific crosses of *Abelmoschus*. Theor Appl Genet 60:313–316. Online pdf: https://sci-hub.mksa.top/10.1007/bf00263725

Jambhale ND, Nerkar YS (1981b) Occurences of spontaneous amphiploidy in an interspecific cross between *Abelmoschus esculentus* and *A. tetraphyllus*. J MAU 6:167

John JK, Muhammed NVA, Latha M, Sutar S, Patil P, Malik SK, Negi KS, Merita K, Rao SR, Yadav SR, Bhat KV (2013) Genetic resources and crossability relationships among various species of *Abelmoschus*. Curr Hortic 1:35–46

Kuwada H (1966) The new amphidiploid plant named "*Abelmoschus tuberculatus-esculentus*", obtained from the progeny of the reciprocal crossing between *A. tuberculatus* and *A. esculentus*. Japanese Journal of Breeding 16(1): 21–30

Kuwada H (1974) F1 hybrids of *Abelmoschus tuberculatus x A.* manihot with reference to the genome relationship. Japanese Journal of Breeding 24:207–210

Masters MT (1874) Malvaceae. In: Hooker J.D. (ed.), Flora of British India, Vol. 1, London. pp. 317–353.
Medikus FK (1787) Ueber eininge Kunstliche Geschlechter aus der Malvenfamilie, den der Klasse der, Monadelphien. University Library Basel, pp 45–46
Merita K, Kattukunnel JJ, Yadav SR, Bhat KV, Rao SR (2013) Chromosome counts in wild and cultivated species of *Abelmoschus* Medikus. from the Indian sub-continent. J Hortic Sci Biotechnol 87(6):593–599
Moench C (1794) Methodus plantas horti botanici et agri Marburgeniss, e staminum situ describendi. Marburg.
Pal BP, Singh HB, Swarup V (1952) Taxonomic relationships and breeding possibilities of species of *Abelmoschus* related to okra (*A. esculentus*). Bot Gaz 113:455–464
Paul T, Nair MP (1988) Malvaceae. Flora of India, Fascicle. 19: 61–73, Botanical Survey of India, Kolkata
Prain D (1903) Bengal Plants. Vol. 1. Calcutta, Botanical Survey of India
Pullaiah N, Reddy TB, Moses GJ, Reddy BM, Reddy DR (1998) Inheritance of resistance to yellow vein mosaic virus in okra (*Abelmoschus esculentus* L. Moench). J Appl Genet 58: 349–52
Samarajeewa PK (2003) Wild *Abelmoschus* species in the improvement of okra. In: Conservation and use of crop wild relatives. In: Proceedings of the joint department of agriculture, Sri Lanka and National Institute of Agribiological sciences, Japan, workshop held on 3rd February 2003, pp. 97–108
Sharma BR, Dhillon TS (1983) Genetics of resistance to yellow vein mosaic virus in inter-specific crosses of okra (*Abelmoschus species*). Genetica Agraria 37:267–76
Siemonsma JS (1982a) La culture du gombo (*Abelmoschus* spp.), legume-fruit tropical (avec reference speciale a la Cote d'Ivoire). Thesis, Wageningen University, The Netherlands. https://edepot.wur.nl/201749
Siemonsma JS (1982b) West African okra – morphological and cytogenetical indications for the existence of a natural amphidiploid of *Abelmoschus esculentus* (L.) Moench and *A. manihot* (L.) Medikus. Euphytica 31(1):241–252. https://doi.org/10.1007/BF00028327
Sivarajan VV, Pradeep AK (1996) Malvaceae of Southern Peninsular India: A taxonomic monograph. Daya Pub. House, Delhi, India. pp. 1–73
Vredebregt JH (1991) Taxonomic and ecological observations on species of *Abelmoschus Medik.* In Report of an International workshop on okra genetic resources held at NBPGR, New Delhi, India, 8–12 October 1990. pp. 60–76
Yadav SR, Bhat KV, Rao SR (2014) Genus *Abelmoschus* Medik in India: an illustrated guide for species identification. ICAR-Nationa Bureau of Plant Genetic Resources, New Delhi, p 44. http://pgrinformatics.nbpgr.ernet.in/cwr/Publication/AbelmoschusinIndia

Genome Editing: A Potential Tool for Improving Okra

15

Hukum Singh Panwar, Suhas G. Karkute, Ram Krishna, and Achuit K. Singh

15.1 Introduction

Even though conventional breeding is presently the most commonly employed method in crop improvement, it has limitations of being labour-intensive, resource-consuming, and takes a long time, often requiring several years to progress from early phenotype and genotype screening to the initial crossbreeding for commercial varieties (Tester and Langridge 2010). Conventional breeding faces difficulties in achieving simultaneous improvement of all the desired traits while avoiding undesirable ones. This challenge arises from factors such as genetic correlations between traits, population structure, linkage disequilibrium, and the possibility of linkage drag (Hartl and Clark 1997).

Modern molecular breeding methods have addressed several problems and significantly reduced the time required for crop genetic improvement (Ribaut et al. 2010). These methods include Marker-Assisted Selection (MAS), Marker-Assisted Backcrossing (MAB), and Genome-Wide Selection (GWS). MAB has significantly sped up the development of new cultivars by enabling breeders to detect and choose desirable features precisely with less time and effort than the conventional breeding techniques. However, one of the drawbacks of modern breeding techniques, particularly MAB, is the availability of genetic resources. In order to introduce novel features and enhance crop varieties, breeding programmes must have access to a broad range of genetic material that has been well-characterized. When dealing with species that have crossing barriers, conventional and molecular breeding approaches sometimes have their limitations. Embryo rescue techniques can help to some extent, although they are not always effective, need specialized knowledge, and take a lot of time. On the contrary, transgenic technology offers a way to overcome these

H. S. Panwar · S. G. Karkute · R. Krishna · A. K. Singh (✉)
Division of Crop Improvement, ICAR-Indian Institute of Vegetable Research, Varanasi, Uttar Pradesh, India

S. K. Tikoo et al. (eds.), *Okra: Status, Challenges and Opportunities*,
https://doi.org/10.1007/978-981-97-9963-3_15

limitations by incorporating desired DNA traits from different sources directly into the plant genome.

In recent years, there has been a notable advancement in modern breeding practices due to various biotechnological tools and techniques, particularly transgenics and genome editing. These innovations have brought about a revolution in crop improvement programmes. These techniques have complemented and enhanced MAB by providing new avenues for precision and control in modifying crop traits. Transgenics have led to the creation of Genetically Modified (GM) crops with enhanced traits, although they come with concerns about safety and environmental impact. GM crops with valuable traits are developed by adding specific genes to well-performing crop varieties. While GM crops offer potential benefits for global food security, concerns about their health and environmental safety have restricted their use. Government regulations, meant to ensure safety, have raised costs and limited the adoption of new GM traits. Also, the commercialization of a genetically engineered (GE) crop involves a multifaceted, time-consuming, and resource-intensive journey (Prado et al. 2014).

The approval of genetically modified (GM) crops is subject to regulations, and different countries have distinct authorization processes. As a result, new GM crops do not receive approval simultaneously. Consequently, only a few crops have the advantages of GM traits (Stein and Rodríguez-Cerezo 2010). Therefore, the acceptance of GM crops is limited to only a few countries, and their commercial cultivation faces challenges. At present, in India, only insect-resistant Bt cotton is cultivated, whereas the commercial cultivation of Bt brinjal is not possible because of persistent opposition. So, this emphasizes the necessity for additional strategies. A possible approach that offers precision and transgene-free crop improvement is genome editing technology. While there are many obstacles to the broad acceptance of GM crops, genome editing offers the possibility of developing crops with desired features without incorporating foreign genes. This invention opens up a new field for agricultural improvement while lowering regulatory barriers and addressing transgenic crop issues. Its adoption could open the way for more sustainable and effective agriculture while getting around restrictions on conventional GM crops.

15.2 Genome Editing Techniques

Genome editing utilizes a variety of biotechnological techniques that make it possible to precise, effective, and targeted modification of an organism's DNA (Chen and Gao 2014). It has become a powerful tool in agriculture, especially for crop improvement, with the potential to develop plants with enhanced characteristics, higher resilience, and improved adaptability (Zhang et al. 2018). Several genome editing techniques are utilized in genome editing, but CRISPR-Cas9 is the most widely used. It uses guide RNA to accurately target particular DNA sequences and the Cas9 enzyme to perform precise cuts or alterations at the targeted site. It was quickly adopted in plant research because of its simplicity, efficiency, low cost, and potential to target numerous genes (Shan et al. 2013). An additional method known as TALENs (Transcription Activator-Like Effector Nucleases) provides a means of customization by manipulating proteins to selectively attach to particular DNA sequences and trigger double-strand breaks, which facilitate genome editing (Mahfouz et al. 2011). Zinc Finger Nucleases (ZFNs), which began the era of genome editing, were developed by combining a non-specific DNA cleavage domain derived from the FoK1 endonuclease with a DNA-binding zinc finger protein, thus imparting specificity to the process (Carroll 2011).

15.3 CRISPR-Cas9 System

CRISPR/Cas systems, particularly the type II CRISPR/SpCas9 system derived from Streptococcus pyogenes, have emerged as versatile genome-editing tools with a wide range of potential applications. A wide range of crops, such as rice, maize, wheat, soybean, barley, sorghum, potato, tomato, flax, rapeseed, cotton, cucumber, lettuce, grapes, grapefruit, apple, oranges, and watermelon, have undergone successful genome editing using the CRISPR/Cas9 technology (Zhang et al. 2016; Ricroch et al. 2017; Karkute et al. 2017; Bao et al. 2019; Gupta and Karkute 2021). The most common application has been the creation of null alleles or gene knockouts, typically achieved by introducing small insertions or deletions (indels) that cause frame-shift mutations or introduce premature stop codons. This versatile technology has revolutionized crop improvement by allowing precise modifications to their genetic makeup. These edited crops exhibit enhanced traits, improved resistance to pests and diseases, and increased tolerance to environmental stresses. The widespread adoption of CRISPR/Cas9 underscores its pivotal role in advancing agriculture, offering the potential to develop more resilient and productive crop varieties to meet the challenges of global food security and sustainability.

15.4 Genome Editing for Improvement of Okra

Researchers have enhanced various economic traits such as yield, quality, resistance to biotic and abiotic stresses, and improved nutritional content of crops with greater precision and accuracy by employing advanced biotechnological tools such as clustered regularly interspaced short palindromic repeats (CRISPR) and CRISPR-associated protein-9 (Cas9). However, the utilization of CRISPR/Cas9 in vegetable crops is currently constrained by several factors, including the limited availability of regeneration protocols and insufficient sequence information in many species. Only a few vegetable crops, namely potato, tomato, cucumber, lettuce, and watermelon, have taken the benefits of CRISPR/Cas9 technology for successful genome editing to achieve improvements (Pareek et al. 2021). The status of some important vegetables that have taken advantage of this technology for successful genome editing is given below (Table 15.1).

Table 15.1 Enhancements achieved in various vegetable crops through CRISPR/Cas9 techniques

Crop	Targeted genes	Enhanced traits	Reference
Tomatoes	SlMLO1, SlJAZ2	Disease resistance (powdery mildew, bacterial speck)	Nekrasov et al. (2017); Ortigosa et al. (2019)
Potatoes	Wx1	Creation of waxy potatoes suitable for industrial starch production	Andersson et al. (2017)
Watermelons	Genes related to herbicide resistance	Development of herbicide-resistant plants	Tian et al. (2018)
Cucumbers	eIF4E	Broad spectrum virus resistance	Chandrasekaran et al. (2016)

However, there are many other vegetable crops that have not yet utilized these tools; one such significant crop is okra, which holds great potential for enhancement through the application of CRISPR/Cas9 technology.

Despite its significant agricultural importance, okra faces various challenges during its cultivation, such as vulnerability to pests, diseases, and abiotic stresses like drought and high temperatures, etc. Okra is susceptible to a variety of begomoviruses, with at least 27 distinct types identified. Among these viruses, Yellow Vein Mosaic Virus (YVMV) and Okra Enation Leaf Curl Virus (OELCV) are the most devastating and have a significant impact on quality and yield (Swanson and Harrison 1993). The occurrence of these viruses, along with abiotic stresses, has led to a decline in overall okra production in India. Effectively managing and developing resistance to these particular viruses and abiotic stresses is crucial for enhancing okra yields. Biotechnological methods hold great promise for tackling these problems, but their application in okra breeding is limited by several factors. One major challenge is the scarcity of readily available polymorphic molecular markers and well-defined genetic maps. Furthermore, the complex allopolyploid nature of the okra genome, characterized by a variable number of chromosomes (ranging from $2n = 56$ to 196), adds further complexity to genetic study (Sastry and Zitter 2014). Presently, genomic resources for the *Abelmoschus* genus are limited, with only a limited number of mRNA sequences and a small collection of coding sequences accessible in the public domain (Schafleitner et al. 2013). This scarcity of genomic data poses significant challenges and resource demands when studying plants like okra. However, researchers have developed an alternative approach for investigating non-model organisms like okra, which involves focusing on transcriptome sequencing instead of whole-genome sequencing (Hirsch and Robin Buell 2013). Transcriptome sequencing, also known as RNA-seq, has emerged as a potent tool in molecular biology. It involves the sequencing of the RNA molecules present in a sample, offering insights into the genes actively expressed at a specific moment. RNA-seq can facilitate the discovery of new transcripts and gene sequences and assist in the development of molecular markers for non-model plants such as okra

(Lata et al. 2021). Sequencing the transcriptome can provide valuable insights into gene expression, alternative splicing, and regulatory elements, even when a complete genome sequence is lacking. However, the efforts for sequencing the okra genome resulted in significant genome information with the sequence of the chromosome-scale okra genome of 1.19 Gb (Wang et al. 2023). This whole genome sequence information will greatly accelerate the mining of the candidate genes associated with important agronomic traits, including disease resistance and stress tolerance. The sequence information of host susceptibility genes for various diseases available in other plant species can be used to identify homologs in okra and can be targeted through genome editing. Similarly, the homologs of negative regulators of stress tolerance can be manipulated for developing stress tolerance in okra. To address the various challenges and enhance okra production, a comprehensive approach that combines advanced biotechnological and genomic tools with conventional breeding methods is essential. Biotechnological tools such as RNA interference (RNAi) and genome editing techniques like CRISPR/Cas9 can be powerful tools for developing okra varieties with improved disease resistance and other desirable traits. RNAi can be used to suppress the expression of genes associated with diseases like YVMV and OELCV, while genome editing can precisely modify the okra genome to introduce resistance genes or enhance stress tolerance. Recently, the RNAi technology targeting viral genes such as AC1, AC2, and AC4 has been successfully implemented in okra for developing resistance against begomoviruses (Sharma et al. 2023).

15.5 Conclusion

While conventional breeding approaches have their constraints, molecular techniques such as marker-assisted selection and genome-wide selection have helped a lot in increasing crop improvement. However, these methods encounter obstacles such as limited genetic resources and species barriers. Transgenic technology has provided solutions to some of these challenges but has also given rise to safety and environmental concerns, resulting in regulatory hurdles and restricted adoption in specific regions. Genome editing technology, with the CRISPR-Cas9 system, is one of the important recent advancements. This robust tool enables precise, efficient, and targeted modifications to an organism's DNA, presenting substantial potential for crop improvement. Despite the potential advantages, genome editing in vegetable crops, including okra, encounters obstacles like the absence of efficient transformation and regeneration protocols and insufficient sequence data. Okra, being an important vegetable crop, could benefit immensely from the precision and control offered by genome editing. By harnessing the power of CRISPR-Cas9, we can aspire to develop improved okra varieties that are climate-resilient, disease-resistant, and address the challenges of the twenty-first century, contributing to a more secure and sustainable global food supply.

References

Andersson M, Turesson H, Nicolia A, Fält AS, Samuelsson M, Hofvander P (2017) Efficient targeted multiallelic mutagenesis in tetraploid potato (Solanum tuberosum) by transient CRISPR-Cas9 expression in protoplasts. Plant Cell Rep 36:117–128

Bao A, Burritt DJ, Chen H, Zhou X, Cao D, Tran LSP (2019) The CRISPR/Cas9 system and its applications in crop genome editing. Crit Rev Biotechnol 39(3):321–336

Carroll D (2011) Genome engineering with zinc-finger nucleases. Genetics 188(4):773–782

Chandrasekaran J, Brumin M, Wolf D, Leibman D, Klap C, Pearlsman M et al (2016) Development of broad virus resistance in non-transgenic cucumber using CRISPR/Cas9 technology. Mol Plant Pathol 17(7):1140–1153

Chen K, Gao C (2014) Targeted genome modification technologies and their applications in crop improvements. Plant Cell Rep 33:575–583

Gupta OP, Karkute SG (eds) (2021) Genome editing in plants: principles and applications. CRC Press

Hartl DL, Clark AG (1997) Principles of population genetics, vol 116. Sinauer Associates, Sunderland, p 542

Hirsch CN, Robin Buell C (2013) Tapping the promise of genomics in species with complex, non-model genomes. Annu Rev Plant Biol 64:89–110

Karkute SG, Singh AK, Gupta OP, Singh PM, Singh B (2017) CRISPR/Cas9 mediated genome engineering for improvement of horticultural crops. Front Plant Sci 8:1635

Lata S, Yadav RK, Tomar BS (2021) Genomic tools to accelerate improvement in okra (Abelmoschus esculentus). In: *Landraces-traditional variety and natural breed*. IntechOpen

Mahfouz MM, Li L, Shamimuzzaman M, Wibowo A, Fang X, Zhu JK (2011) De novo-engineered transcription activator-like effector (TALE) hybrid nuclease with novel DNA binding specificity creates double-strand breaks. Proc Natl Acad Sci 108(6):2623–2628

Nekrasov V, Wang C, Win J, Lanz C, Weigel D, Kamoun S (2017) Rapid generation of a transgene-free powdery mildew resistant tomato by genome deletion. Sci Rep 7(1):482

Ortigosa A, Gimenez-Ibanez S, Leonhardt N, Solano R (2019) Design of a bacterial speck resistant tomato by CRISPR/Cas9-mediated editing of Sl JAZ 2. Plant Biotechnol J 17(3):665–673

Pareek A, Solanke AU, Mangal V, Jiwani G, Tasleem M, Gupta OP, Karkute SG (2021) Towards "CRISPrY" vegetables: status of genome editing in vegetable crops. In: Genome editing in plants. CRC Press, pp 127–145

Prado JR, Segers G, Voelker T, Carson D, Dobert R, Phillips J et al (2014) Genetically engineered crops: from idea to product. Annu Rev Plant Biol 65:769–790

Ribaut JM, De Vicente MC, Delannay X (2010) Molecular breeding in developing countries: challenges and perspectives. Curr Opin Plant Biol 13(2):213–218

Ricroch A, Clairand P, Harwood W (2017) Use of CRISPR systems in plant genome editing: toward new opportunities in agriculture. Emerg Top Life Sci 1(2):169–182

Sastry KS, Zitter TA (2014) Plant virus and viroid diseases in the tropics, vol 2. Springer, Dordrecht, p 489

Schafleitner R, Kumar S, Lin CY, Hegde SG, Ebert A (2013) The okra (Abelmoschus esculentus) transcriptome as a source for gene sequence information and molecular markers for diversity analysis. Gene 517(1):27–36

Shan Q, Wang Y, Li J, Zhang Y, Chen K, Liang Z et al (2013) Targeted genome modification of crop plants using a CRISPR-Cas system. Nat Biotechnol 31(8):686–688

Sharma N, Sarao NK, Mohanpuria P, Sharma A (2023) Effective Agrobacterium-mediated genetic transformation of okra (*Abelmoschus esculentus L.*) and generation of RNAi plants resistant to Begomovirus infecting okra. J Hortic Sci Biotechnol 99:146. https://doi.org/10.1080/14620316.2023.2251978

Stein AJ, Rodríguez-Cerezo E (2010) Low-level presence of new GM crops: an issue on the rise for countries where they lack approval. AgBioForum 13(2):173–182

Swanson MM, Harrison BD (1993) Serological relationships and epitope profiles of isolates of okra leaf curl geminivirus from Africa and the Middle East. Biochimie 75(8):707–711
Tester M, Langridge P (2010) Breeding technologies to increase crop production in a changing world. Science 327(5967):818–822
Tian S, Jiang L, Cui X, Zhang J, Guo S, Li M et al (2018) Engineering herbicide-resistant watermelon variety through CRISPR/Cas9-mediated base-editing. Plant Cell Rep 37(9):1353–1356
Wang R, Li W, He Q, Zhang H, Wang M, Zheng X, Xing L (2023) The genome of okra (Abelmoschus esculentus) provides insights into its genome evolution and high nutrient content. Hortic Res 10:uhad120
Zhang D, Li Z, Li JF (2016) Targeted gene manipulation in plants using the CRISPR/Cas technology. J Genet Genomics 43(5):251–262
Zhang Y, Massel K, Godwin ID, Gao C (2018) Applications and potential of genome editing in crop improvement. Genome Biol 19:1–11

Genetic Improvement of Okra: Present Status and Bottlenecks

16

Suman Lata, P. V. Puneeth, S. B. Kiran, R. K. Yadav, P. B. Chaukhande, and B. S. Tomar

16.1 Introduction

Okra (*Abelmoschus esculentus* L. Moench) is an important vegetable crop grown for its pods. Okra pod is rich in dietary fibers, amino acids and vitamins but is free from fat and cholesterol, making it an essential element for a balanced diet. There is significant variation in the chromosome numbers and ploidy levels in *Abelmoschus*. It has an allopolyploid genome with a huge variation in the chromosome numbers ($2n = 56$–196). *A. esculentus* ($2n = 130$) is an amphidiploid believed to have originated from a cross between *A. tuberculatus* ($2n = 58$) and *A. ficulneus* ($2n = 72$) followed by chromosome doubling. India is the top global producer of okra, but its production is hampered by yellow vein mosaic virus (YVMV) disease. It has been the most devastating disease in okra transmitted by the white fly (*Bemisia tabaci*). The crop improvement program for okra has been slowed due to the limited and unstable nature of resistance in cultivated types. Although several wild relatives are identified as potential sources of resistance to biotic stresses in okra (Dhankhar et al. 2005), three YVMV-resistant wild species are *A. manihot* (L.) Medik, introduced from Japan, *A. manihot* (L.) Medik ssp. m*anihot* from Ghana, and *A. tetrahyllus* Wall, a collection from Maharashtra, India. Another important disease is okra enation leaf curl disease (OELCuD), caused by the okra enation leaf curl virus (OELCV) and transmitted by whitefly. However, newly identified wild species *A. enbeepeegearense* and *A. mizoremensis* (John et al. 2013, 2020) and *A. moschatus* (Santhiya et al. 2022) have been reported to be resistant to both YVMV and OELCV viruses. Further, okra is considered an orphan crop, where biotechnological

S. Lata (✉) · P. V. Puneeth · S. B. Kiran · R. K. Yadav · P. B. Chaukhande · B. S. Tomar
Division of Vegetable Science, ICAR-Indian Agricultural Research Institute, New Delhi, India
e-mail: sumanlata@iari.res.in

S. K. Tikoo et al. (eds.), *Okra: Status, Challenges and Opportunities*,
https://doi.org/10.1007/978-981-97-9963-3_16

techniques like tissue culture, recombinant DNA technology, genetic transformation, molecular markers, marker-assisted selection (MAS), genome-wide selection (GWS), and targeted mutagenesis have not been employed fully for its genetic improvement (Mishra et al. 2017). However, genomic resources are gradually developing in okra with the help of next-generation sequencing techniques like whole genome sequencing and transcriptome sequencing, also known as RNA-seq.

16.2 Screening Studies in Okra for YVMV and ELCV Resistance Using Wild Species

Wild relatives are a great source of genetic resistance to diverse biotic and abiotic stresses, but due to various pre- and post-zygotic crossing barriers, they have not been utilized to a great extent in breeding programs. *Abelmoschus manihot* var. *tetraphyllus* is widely used in breeding programs as a source of resistance/tolerance to YVMV (Gangopadhyay et al. 2016; Badiger and Yadav 2019). Another perennial species, *A. moschatus,* was found to be resistant to drought and low temperature (Akhond et al. 2000), along with YVMV, whitefly, Jassids, powdery mildew, and fruit and shoot borer (Badiger and Yadav 2019; Yadav et al. 2019). For developing multiple pest and disease resistance breeding genotypes, many resistant lines of *Abelmoschus manihot* var. *tetraphyllus* (IC90476-1) and *Abelmoschus moschatus* (IC141055) were crossed with Pusa A-4 (a commercially grown variety) of *A. esculentus* in a reciprocal manner (Badiger et al. 2023). To examine the crossability barriers Badiger et al. (2023) raised only direct cross (*Abelmoschus esculentus* as female parent) F_1s, amphidiploidized (C1), selfed, and backcrossed with *A. esculentus* variety Pusa A-4. No serious pre- and post-zygotic barriers were reported in *Abelmoschus manihot* var. *tetraphyllus* and found bi-directionally crossable with *A. esculentus.* But failure in producing F_1 seeds upon selfing and backcrossing might be due to differences in ploidy levels. Therefore, to introgress resistant traits from *A. manihot* var. *tetraphyllus* to *A. esculentus,* amphidiploidization of the F_1 hybrid followed by selfing and backcrossing (amphidiploid as seed parent) is suggested (Badiger et al. 2023). On the other hand, *A. moschatus* × *A. esculentus* crosses lead to the formation of aborted embryos with partially filled seeds due to prezygotic barriers (Badiger et al. 2023). Santhiya et al. (2022) showed less incidence of YVMD and no incidence of okra enation leaf curl disease (OELCuD) in *A. tetraphyllus* (IC 90476-1), *A. moschatus* (IC 141055), and *A. caillei* (Sikkim) genotypes compared to other okra species under natural epiphytotic screening. In a study conducted in Delhi, 64 distinct okra genotypes were screened for resistance to BYVMV; significant variations were found in disease incidence (DI), coefficient of infection (CI), vulnerability index (VI), and adult whitefly count (AWC) among the genotypes. Pusa Sawani exhibited the highest mean DI (86.5%), while genotypes Pusa Bhindi-5 (DOV-66), DOV-89, and DOV-92 showed 0% DI, indicating resistance. Pusa Bhindi-5 and DOV-92, identified as the most resistant genotypes, hold promise for transferring BYVMV resistance to other lines or varieties in future okra improvement programs, including hybrid development (Puneeth et al. 2022).

16.3 Improvement Through Wide Hybridization

To use wild species of the genus in interspecific combinations with cultivated species or between them is more difficult. In some cases, the F_1 generation can be obtained, but due to sterility issues, the process is blocked in the F_2 generation. Various researchers in the past have utilized wild species. Martin reported the development of a new variety, "Winter Bush," in 1982 by interspecific hybridization between Guinean okra and *A. esculentus* followed by backcross to Clemson Spineless. Research workers in India applied interspecific hybridization followed by backcross in the development of YVMV-resistant varieties, namely, P-7 (Thakur and Arora 1986), Parbhani Kranti (Jambhale and Nerkar 1982), Arka Anamika, and Akra Abhay (Dutta 1991). Another YVMV-resistant variety, Punjab Padmini, was also developed by interspecific hybridization but without backcrossing (Sharma 1982). These workers used *A. manihot* ssp. and *A. tetraphyllus* for interspecific hybridization and backcrossing with *A. esculentus*. The hybrid plants obtained through ovule and embryo culture techniques were sterile (Gadwal et al. 1968). Hybrids of the interspecific cross *A. esculentus* × *A. tuberculatus* readily obtained are sterile. Interspecific crosses of *A. esculentus*, readily obtained, are also sterile. Interspecific hybrid combining resistance could not become commercial, as they have a high degree of pollen sterility, resulting in low or no seed set. Seedless fruits being low in nutritive value have no acceptance by consumers. Presently, Regional Station of National Bureau of Plant Genetic Resources (NBPGR), Thrissur, has had success in interspecific crosses of almost all wild okra species except *A. crinitus* following colchicines treatment. Further improvement for the development of desirable traits is in progress.

16.4 Genetic Diversity Studies in Okra Using Molecular Markers

Genetic diversity analysis of 39 Turkish okra and 27 world genotypes, using peroxidase gene markers, identified 43 polymorphic bands out of 84 generated. Polymorphism information content ranged from 0.03 to 0.99 (mean: 0.48). Gene diversity values ranged from 0.03 to 0.42 (mean: 0.25). Shannon's information index varied from 0.07 to 0.62 (mean: 0.40). Divergent genotypes had a GS value of 0.23, with IN-7 and TR20-3 being the most distant. Okra germplasm was divided into three populations using different analysis methods. Shared genes significantly influence okra's genetic composition, as revealed for the first time by Yildiz et al. (2022).

Genetic analysis of 96 okra genotypes using 13 morphological parameters and 65 simple sequence repeat (SSR) markers determined high diversity among cultivated and wild okra germplasm. Okra genotypes were grouped into three clusters, with Cluster I comprising cultivated species and Clusters II and III comprising wild species. Population structure analysis revealed two distinct genetic groups: cultivated and wild-type okra. Genotype DOV-92 showed the highest performance for total fruit yield. Principal Component Analysis highlighted significant genetic variation. The findings suggest that the diverse germplasm could be valuable for improving okra genetic architecture (Das et al. 2022). Puneeth et al. (2023a) utilized 25 CAAT box-derived polymorphism (CBDP) and 36 start codon-targeted (SCoT) markers to analyze the genetic diversity of 89 okra accessions from cultivated and wild species. The CBDP markers displayed higher values for polymorphic information content (PIC), marker index (MI), and resolving power (Rp) compared to SCoT markers. Nei's genetic diversity (h) was higher for CBDP (1.7) than SCoT (0.13). Analysis of molecular variance (AMOVA) showed variation within populations was more significant for SCoT markers (76%) than CBDP markers (89%). Principal Coordinates Analysis (PCoA) indicated that CBDP markers explained 79.82% of the variation, while SCoT markers explained 82.72%. The study demonstrated that CBDP markers outperformed SCoT markers in estimating genetic diversity among cultivated and wild okra accessions (Puneeth et al. 2023a).

In another study by Puneeth et al. (2023b), RNA sequencing was performed on two okra varieties to identify simple sequence repeat (SSR) markers. Mononucleotide repeats were the most abundant (72.20%), followed by trinucleotide (15%) and dinucleotide (10.68%) repeats. From 106,224 identified EST-SSRs, 183 primer pairs (156 di- and 27 trinucleotides) were selected and validated in 12 okra accessions, resulting in 89% amplification efficiency. 14.1% of the primer pairs showed polymorphism, with an average polymorphic information content (PIC) of 0.32. The developed EST-SSR markers were useful for germplasm identification, genetic diversity analysis, and comparative mapping in okra and related species, as they exhibited high polymorphism and cross-transferability. A study by Gayathri et al. (2021) provides the first comprehensive transcriptome data for okra (*Abelmoschus esculentus*), an important vegetable crop in various regions. The researchers identified 10,492 SSRs (simple sequence repeats) from different parts of the okra plant, including those under heat, drought, and salt stress conditions. Among the identified SSRs, tri-repeats were the most abundant, followed by di-, tetra-, penta-, and hexa-repeats. Thirty-four SSRs were standardized and screened in 36 okra genotypes and accessions, with 18 SSR primers showing high polymorphism (PIC values >0.5). The genic SSR markers developed in this study have potential applications in germplasm characterization, mapping, genetic diversity studies, molecular-assisted breeding, and gene discovery in okra. Details of gene diversity studies carried out in okra are presented in Table 16.1.

16.5 Genome Sequencing Studies in Okra

The genome of okra, a valuable vegetable crop, has been analyzed, revealing its large chromosome number and genetic changes compared to other plants. The estimated chromosome-scale assembled genome size of okra is 1.19 Gb. Comparative genomic analysis showed the closest relative of okra is *Hibiscus cannabinus*, followed by *Hibiscus syriacus*. Phylogenetic analysis revealed whole-genome

Table 16.1 Gene diversity studies in okra

Species	No. of accessions/ genotypes	Types of markers	No. of primers	PIC	Reference
A. esculentus	48	ISSRS	–	1	Yuan et al. (2015)
A. esculentus	24	ISSRS	22	1	Yuan et al. (2014)
A. esculentus	66	(iPBS)-retrotransposons and SSRs	83 iPBS, 9 SSRs	0.66 iPBS, 0.62 SSRs	Yıldız et al. (2015)
A. caillei (50), *A. esculentus* (43)	93	RAPD	13	–	Aladele et al. (2008)
Abelmoschus spp.	39	RAPD	31	–	Martinello et al. (2001)
A. esculentus	50	AFLP	33	12	Kyriakopoulou et al. (2014)
A. esculentus	22	AFLP	8	0.26	Akash et al. (2013)
A. esculentus	23	SRAP	39	–	Gulsen et al. (2007)
A. esculentus, *A. moschatus*, *A. manihot*	65	SSR	19	0.49	Schafleitner et al. (2013)
A. esculentus (21), *A. tuberculatus* (1), *A. moschatus*(1), *A. manihot*(1)	24	SSR	18	0.53	Fougat et al. (2015)
A. esculentus (92), A. manihot(1) A. moschatus subspecies tuberosus (1), and A. moschatus(1), A. tuberculatus (1)	96	SSR	40	0.52	Kumar et al. (2017)
A. esculentus	39 Turkish okra genotypes	Peroxidase gene markers	84	0.48	Yildiz et al. (2022)
A. esculentus	36	SSR	34	0.56	Gayathri et al. (2021)
A. esculentus	96	SSR	65	0.38	Das et al. (2022)
A. esculentus, A. manihot, A. moschatus subspecies tuberosus, and A. moschatus, A. tuberculatus	89	CBDP, SCOT	25 CBDP, 36 SCOT	0.28 CBDP, 0.20 SCOT	Puneeth et al. (2023a)
A. esculentus	12	SSR	183	0.32	Puneeth et al. (2023b)

Adapted and modified from Lata et al. (2021)

duplication (WGD) events in *A. esculentus* and other Malvaceae species. Moreover, okra has undergone three additional WGDs, leading to the formation of a large number of chromosomes ($2n = 130$) compared with *Gossypium raimondii*. Okra has experienced large-scale genomic deletions following three WGD, resulting in unique gene families and differences in cellulose synthesis compared to *G. raimondii*. By studying gene expression and metabolite content, researchers found variations across different stages of okra development, shedding light on its nutrient content. The study provides insights into the evolutionary history of related plant species and the genetic factors influencing okra's nutritional properties, aiding future studies and improving okra varieties (Table 16.2; Wang et al. 2023).

Okra organelle (plastid and mitochondria) genome sequencing was attempted to understand the evolution of organelles and comparative analysis between the plastid and mitochondria genomes. Okra's plastid genome is structurally conserved, while its mitochondrial genome exhibits various configurations. The mitochondrial genome comprises two independent molecules that can further divide into four molecules through repeats. Integration of these molecules can occur at low frequency. Sequence transfer between the organelles reveals intact and nonfunctional genes in the mitochondrial genome. Genes from plastids showed sequence migration to the mitochondrial genome termed mitochondrial plastid sequences (MTPTs). Among these MTPTs (*psaA, rps7, and psbJ*) stayed as such in the mitochondrial genome. Moreover, MTPTs like *psbJ, psbF, psbE,* and *psbL* migrated to the mitochondrial genome and became non-functional or pseudo genes during the evolutionary process. RNA editing events in protein-coding genes of both organelles were also examined in this study (Table 16.2; Li et al. 2022).

16.6 Transcriptomic Studies in Okra

Transcriptome analysis, or RNA-seq, is a powerful technique to discover the transcript/gene sequences, transcription factors, and molecular markers like SSR and SNPs. The first report on leaf transcriptome assembly in okra was published in 2013 (Schafleitner et al. 2013). Gradually, there has been progress in transcriptomic studies for various biotic and abiotic traits (Table 16.2). Transcriptomic studies for abiotic stresses involve using high-throughput technologies like microarrays and next-generation sequencing (RNA-seq) to simultaneously analyze and quantify the expression levels of thousands of genes. By comparing the transcriptomes of organisms exposed to abiotic stresses with those grown under normal conditions, researchers can identify genes differentially expressed in response to stress. These differentially expressed genes are often associated with stress tolerance mechanisms and play crucial roles in the plant or organism's ability to adapt to adverse environmental conditions.

Table 16.2 Omics studies in okra

Okra species (plant organs)	Objective	Sequencing platform	Raw reads (M) and final assembly	Marker discovery	N50	NCBI accession	Reference
A. esculentus L. Moench cv. Arka Abhay (leaf and pod)	Transcriptome assembly	Illumina HiSeq™ 2000	26,32,45,57,263 (150,000 unigenes)	935 SSRs	321 bp	SRX206126	Schafleitner et al. (2013)
A. esculentus cv. Xianzhi (Roots, stems, and leaves)	Transcriptome assembly	Illumina HiSeq X Ten platform	7,16,33,02,52,716 (293,971 unigenes)	–	1885 bp	SRP130180	Zhang et al. (2018)
A. esculentus (Leaves)	Transcriptome assembly	Illumina NextSeq 500	206.3 million (66,382 unigenes)	9578 SSRs	1408 bp	SRX2995608, SRX2995609, SRX2995611, and SRX2995612	Priyavathi et al. (2018)
A. esculentus cv. Arka Anamika (Leaves)	Genome assembly	Roche 454 GS FLX Titanium	61,722 (3735 contigs)	402 SSRs	–	–	Ravishankar et al. (2018)
A. esculentus (Leaves)	miRNA identification	Illumina and Ion torrent	207,285,863 (845 novel miRNAs)	–	–	PRJNA352593	Kumar et al. (2017)
A. esculentus (Leaves)	Potential of exogenous melatonin pretreatment to alleviate salt stress	PacBio and Illumina	– (121,360 unigenes)	–	–	–	Zhan et al. (2021)

(continued)

Table 16.2 (continued)

Okra species (plant organs)	Objective	Sequencing platform	Raw reads (M) and final assembly	Marker discovery	N50	NCBI accession	Reference
A. esculentus (Leaves)	To provide high-quality organelle genomes for okra and understanding of the gene dialogue between organelle genomes	Illumina short reads and Nanopore long reads	1,454,069 (12 contigs)		17.40 kb.	OL348387.1 (mtDNA 1); OL348388.1 (mtDNA 2); OL348389.1 (cpDNA)	Li et al. (2022)
A. esculentus (Leaves)	Identification and validation of EST-SSRs	Illumina HiSeq™ 3000	– (106,224 EST SSRs)	–	–	PRJNA819417	Puneeth et al. (2023b)
A. esculentus (Leaves)	Genome assembly	PacBio Sequel 606 II	94.59 Gb Illum ina sequencing reads (128 contigs)	–	16.98 Mb	China-NCBIGWHBWBG00000000	Wang et al. (2023)
A. esculentus (Leaves)	Explore transcription factors (TFs) and long non-coding RNAs (lncRNAs) involved in abiotic stress	–	– (112,597 Unigenes, 1281 TFs and 57,675 lncRNA)	–	–	–	Muthaiah et al. (2023)

Adapted and modified from Lata et al. (2021)

16.6.1 Drought and Heat Stress

As cotton also belongs to malvaceae family. Genomic resources of cotton like EST are used in okra. Sixty-one expressed sequence tags (ESTs) known for heat and drought tolerance in cotton were identified and searched in the okra unigene database. Of this, 55 functional homologous unigenes were selected as markers and used for further analysis. Okra genotypes exhibited significant variations in physio-morphological parameters under heat and drought stress. Five genotypes performed better under drought stress, while four showed heat tolerance. 50 markers were successfully amplified in okra, and the genotypes were grouped into three distinct populations. Linkage disequilibrium analysis revealed a significant linkage between markers Unigene43786 and Unigene3662. Marker trait association analysis identified 23 markers significantly associated with different traits under control and stressed conditions, including relative water content and electrolyte leakage under drought and heat stress, respectively (Ahmad et al. 2022).

Shi et al. (2020) studied the impact of drought on okra and found that next-generation sequencing revealed 94,769 unigenes, with 38.18% annotated in various databases. The analysis identified 22,751 up-regulated and slightly fewer down-regulated differentially expressed genes (DEGs) in response to drought stress. DEGs were associated with categories such as "integral component of membrane," "oxidation-reduction process," and "metal ion binding." Pathway analysis indicated enrichment in "Biosynthesis of secondary metabolites," "Carbon metabolism," and "Glycolysis/Gluconeogenesis pathways." The study also identified simple sequence repeat motifs, single nucleotide polymorphisms, and transcription factors. RNA-Seq data were validated using RT-qPCR, contributing to a better understanding of the molecular mechanisms underlying the response to drought stress in okra. Similarly, Wang et al. (2022) explore the complex response of okra to drought stress, using transcriptomics and metabolomics to understand water-stress effects on the leaf. Drought negatively impacted photosynthesis and glycol metabolism, disrupting carbohydrate processes and secondary plant metabolism.

The tyrosine-derived pathway and arginine/proline metabolism were significantly affected under water-withholding conditions. Tyrosine metabolism plays a key role in drought tolerance by influencing carbon and nitrogen metabolism. This research provides valuable insights into the regulation and relationships of major transcripts, peptides, proteins, and metabolites in okra's response to drought stress, particularly in amino acid metabolism changes.

16.6.2 Salt Stress

Response of okra seedlings to soil salinization (300 mM NaCl) and the role of protein phosphorylation in detoxification and tolerance investigated (Yu et al. 2019). A large-scale phosphoproteome analysis was conducted using affinity enrichment and mass spectrometry, identifying 4341 phosphorylation sites on 2550 proteins. Gene ontology and pathway analysis revealed differentially expressed proteins associated with 'photosynthesis antenna proteins' and 'RNA degradation,' indicating their importance in salt stress responses in okra. These findings enhance our understanding of plant post-translational modifications and their molecular mechanisms in adapting to salt stress, crucial for improving land use efficiency and crop yields worldwide. Another study by Zhan et al. (2021), investigated the potential of exogenous melatonin pretreatment to alleviate salt stress (300 mM NaCl) in okra plants. Melatonin (50 μM) significantly improved plant tolerance to salt stress, as evidenced by enhanced photosynthetic rate, chlorophyll fluorescence, and chlorophyll content. Melatonin pretreatment reduced lipid peroxidation, H2O2 content, and O2•- scavenging, possibly due to increased enzyme activities such as POD and GR. The researchers used a combination of PacBio and Illumina sequencing to obtain 121,360 unigenes of okra, identifying differential expression genes (DEGs) in response to salt stress, melatonin, and their combination, enriched in various GO terms and KEGG pathways, including nitrogen metabolism, sulfur metabolism, and alanine, aspartate, and glutamate metabolism, as well as different transcription factors like MYB, WRKY, and NAC.

16.6.3 Transcription Factors (TFs) and Long Non-coding RNAs (lncRNAs) Identification

Previous studies mainly focused on the physiological and biochemical aspects of these stresses. However, this study aimed to explore transcription factors (TFs) and long non-coding RNAs (lncRNAs) on a genome-wide scale in okra. The researchers used comprehensive pooled transcriptome data from stressed 20-day-old seedlings and different plant parts to identify and validate 57,675 lncRNAs, 112,597 unigenes, and 1281 TFs. These findings provide valuable insights into the structure of the okra genome and its molecular changes during growth, development, and response to abiotic stresses (Muthaiah et al. 2023).

16.6.4 Reference Gene Selection

A total of 18 potential reference genes from okra were analyzed for their expression stability under different abiotic stress conditions (Zhu et al. 2023). *GAPC1, CYP, ACT7, HIS6,* and *60 s* were the most stable genes under cold stress. The most stable genes for salt stress were *EF-1α, UBQ, TUA4, Eif, HIS6*, and *Eif.* Under drought stress, *UBQ, Eif, HIS6*, and *TUA4* were most stably expressed in okra. According to sequencing results, *HIS6* and *ACT7* in the leaf, *HIS6* and *Eif* in the root, *UBC5B* and *60 s* in the stem, and *HIS6* and *Eif* in different tissues were the most suitable reference genes for okra. The study aimed to provide a molecular basis for functional gene analysis and regulatory mechanism research of okra.

16.7 Regeneration Studies in Okra

Micro-propagation is used to preserve, perpetuate, and achieve large-scale plant propagation through in vitro regeneration, resulting in healthy plants. Effective utilization of technologies such as transformation and soma clonal variation in any crop improvement, including okra, heavily relies on developing efficient in vitro regeneration techniques (Patel et al. 2016). However, the absence of an efficient regeneration protocol has hindered the progress in genetic engineering research for okra. Limited research has been conducted on in vitro plant multiplication of okra and successfully regenerating the plant by utilizing various explants such as hypocotyl, cotyledonary nodes, leaves, cotyledons, and apical shoots by direct and indirect organogenesis.

Direct shoot regeneration from cotyledon and cotyledonary node explants was observed by Mangat and Roy (1986), while the regeneration of okra plants from callus tissue derived from cotyledonary node was achieved by Roy and Mangat (1989). Direct organogenesis refers to the development of an explant directly by shooting and rooting through manipulation of plant growth regulators and culture conditions, whereas indirect organogenesis refers to the development of an explant into a complete plant undergoing the callus phase. Another approach by Ganesan et al. (2007) involved a plant-regeneration protocol through somatic embryogenesis from suspension culture, whereas Anisuzzaman et al. (2008a) developed a protocol for mass in vitro propagation using meristem culture to produce disease-free plants. Anisuzzaman et al. (2008b) successfully optimized a viable protocol for indirect shoot organogenesis of okra from leaf-disc and hypocotyl explants via the callus phase. Direct regeneration of plants by utilizing shoot apices and nodal cuttings as explants; successful regeneration has been observed, respectively (Dhande et al. 2012; Belkhodja et al. 2023). A comprehensive overview of different explants and media compositions employed by various researchers in successful okra regeneration has been mentioned in Table 16.3.

16.8 Transformation Studies in Okra

Okra cultivation faces significant challenges due to various biotic and abiotic stresses, viz., shoot and fruit borers, aphids, and the yellow vein mosaic virus. Unfortunately, none of the currently available edible okra cultivars possess natural resistance against these pests and diseases. Moreover, it's been challenging to modify and regenerate okra plants using *Agrobacterium*-mediated methods due to their recalcitrant nature. Therefore, an urgent requirement is to effectively enhance the genetic traits to combat these biotic stresses in okra using recombinant DNA techniques. Narendran et al. (2013) successfully devised a tissue culture protocol using zygotic embryo explants and integrated it with *Agrobacterium*-mediated transformation. The resulting transgenic *Bt* okra plants exhibited increased resistance against the okra shoot and fruit borer (*Earias vitella*).

Table 16.3 Description of regeneration studies in okra

Okra species/explant	Mode of regeneration	Organogenesis medium composition	Shooting medium	Rooting medium	Reference
A. esculentus cv. Parkin mammoth long pod (Shoot tip and Cotyledonary node)	Direct	MS medium: IAA (0.2 mg L^{-1}) and 2,4-D (0.2 mg L^{-1})	MS medium: NAA (1.0 rag L^{-1}) and BA (1.0 mg L^{-1})	MS medium NAA (1.0 mg L^{-1})	Mangat and Roy (1986)
A. esculentus cv. Parkin mammoth long pod (Cotyledonary axil)	Indirect	MS medium: Benzyladenine (BA) (1.0 mg L^{-1})	BA-enriched MS media containing various concentrations of silver nitrate (0–40 mg L^{-1})	MS medium containing BA (1.0 mg L^{-1}) and NAA (1.0 mg L^{-1})	Roy and Mangat (1989)
A. esculentus (Hypocotyl)	Indirect	Medium containing 0.1–0.3 mg L^{-1} BA and 1–3 mg L^{-1} NAA	MS medium supplemented with 1–3 mg L^{-1} benzyl adenine (BA) and 0.1–0.3 mg L^{-1} alpha-naphthaleneacetic acid (NAA)	Medium containing 1.0 mg L^{-1} BA	Haider et al. (1993)
A. esculentus cv. Surbhi-10 (Hypocotyl)	Indirect	MS salts, Gamborg (B5) vitamins, 2.0 mg dm-3 2,4 dichloro phenoxy acetic acid (2,4- D), 1.0 mg dm-3 naphthaleneacetic acid (NAA), 25 mg dm-3 polyvinylpyrrolidone, and 30 g dm-3 sucrose	Suspension culture containing MS salts, B5 vitamins, 2.0 mg dm-3 2,4-D, and 1.0 mg dm-3 kinetin	Half MS salts, B5 vitamins, 0.2 mg (BAP) and 0.2 mg (GA3)	Ganesan et al. (2007)
A. esculentus cv.Parbhanikranti and SL 44 (Meristem culture)	Direct	MS medium containing 1.0 mg L^{-1} of BAP	1.0 mg L^{-1} + 0.5 mg L^{-1} GA3	1.0 mg L^{-1} IBA	Anisuzzaman et al. (2008a)

A. esculentus cv. BARI Dherosh-1 (Leaf disc and Hypocotyl)	Indirect	Combination of 2.0 mg L^{-1} NAA + 0.5 mg L^{-1} TDZ	2.0 mg L^{-1} BAP + 0.1 mg L^{-1} IBA	MS medium: 1.5 mg L^{-1} NAA	Anisuzzaman et al. (2008b)
A. esculentus cv. BARI SL-44 (Hypocotyl and leaf)	Indirect	Combination of 0.5 mg L^{-1} BAP and 2.0 mg L^{-1} NAA	Combination of 2.0 mg L^{-1} BAP + 0.1 mg L^{-1} IAA and 2.0 mg L^{-1} BAP + 0.5 mg L^{-1} NAA	MS medium containing 2.0 mg L^{-1} IBA	Kabir et al. (2008)
A. esculentus N-550 Line (Shoot tips)	Direct	Combination of 1.0 mg L^{-1} IBA + 0.5 mg/L NAA	MS medium supplemented with kinetin 0.5 mg L^{-1}	MS medium containing 0.5 mg L^{-1} IAA and 1.0 g activated charcoal	Dhande et al. (2012)
A. esculentus cv. BARI Dherosh-1 (Cotyledonary node, hypocotyl)	Indirect	Combination of 2.0 mg L^{-1} BAP + 0.1 mg L^{-1} NAA	MS medium supplemented with 2.0 mg L^{-1} BAP and 0.5 mg L^{-1} NAA	MS medium containing NAA and 1.0 mg L^{-1}	Kabir et al. (2016)
A. esculentus cv. Xiang Fu (Cotyledonary node)	Indirect	Combination of 1 mg l^{-1} naphthalene acetic acid (NAA) + 0.5 mg l^{-1} benzyl adenine (BA)	MS medium containing 1.5 mgL^{-1} BAP + 1.5 mgL^{-1} KIN +3.0 mgL^{-1} AgNO3	MS medium +1.0 mgL^{-1} IBA (Indole 3-butyric acid) + 200 mgL^{-1} AC (Activated charcoal	Rizwan et al. (2018)
A. esculentus cv. CoBhH1 (cotyledonary leaves)	Indirect	Combination 2, 4-D 1.5 mg/l with Kinetin 1.0 mg/l	MS supplemented with 1 mg/l BAP and 0.5 mg/l GA3 along with 300 mg/l L-glutamine and 400 mg/l casein hydrolysate	MS medium supplemented with 300 mgL^{-1} glutamine and 400 mgL^{-1} casein hydrolysate	Daniel et al. (2018)
Farmer material from the Guelma region (Nodal and shoot explants)	Direct	MS	MS	Half MS devoid of growth regulator	Belkodhia et al. (2023)

Table 16.4 Different transformation methods and target traits in okra

Okra Species	Explant	Method	Targeted trait	Reference
A. esculentus cv. Parbhani Kranti	Hypocotyl, cotyledonary nodes, leaf segments, shoot tips	In vitro (Callusing)	GUS reporter gene assay	Mallela et al. (2009)
A. esculentus (Inbred line)	Embryo	In vitro	Insect resistance	Narendran et al. (2013)
A. esculentus cv. Arka Anamika	Seed	*In planta*	–	Manickavasagam et al. (2015)
A. esculentus cv. Pusa Sawani, Punjab Padmini, Parbhani Kranti	Seed	*In planta*	–	Rohit et al. (2018)
A. esculentus cv. Salkeerthi	Seed	*In planta*	YVMV resistance	Ganesh et al. (2022)
Unknown germplasm	Embryo	Biolistic method	–	Farooq et al. (2022)

Developing transgenic okra lines using coat protein (CP) genes can confer resistance against various viral diseases like YVMV and OELCV. Addressing this aspect holds great promise for enhancing the resilience of okra crops and ensuring sustainable agricultural practices. Recent advancements in genetic engineering and plant transformation techniques have enabled the identification and isolation of pest and disease-resistant genes from both bacterial and plant sources, facilitating their transfer into okra plants (Table 16.4).

Mallela et al. (2009) used a GUS reporter gene assay for confirming the transformation in the regenerated calli. Transformation efficiency through *In planta* was found to be very low, with 0.42%, 0.335%, and 0% in the varieties Pusa Sawani, Punjab Padmini, and Parbhani Kranti, respectively, as observed by Rohit et al. (2018). However, Manickavasagam et al. (2015) made significant progress by establishing a tissue culture-independent genetic transformation approach. In this method, they utilized seeds as explants and introduced *Agrobacterium tumefaciens* containing the binary vector pCAMBIA 1301–bar for transformation by achieving an impressive transformation efficiency of 18.3% in variety Arka Anamika. A highly YVMV-susceptible okra cv. Salkeerthi was transformed using the *In planta* transformation strategy with the ihpRNA-βC1 cassette by Ganesh et al. (2022). Transformation success was confirmed by detecting the amplification of the sense strand using *VLIC1* and *VLIC5* primers, with an efficiency of 11.42%.

Farooq et al. (2022) did a novel gene transformation technique by employing a combination of various Magnetofection factors. They found that the highest gene transformation efficiency occurred when the DNA was mixed with iron-nanoparticles at a ratio of 1:20, followed by rotation of the mixture (containing Plasmid DNA, iron-nanoparticles, and seed embryos) at 800 rpm for 5 h. They successfully transformed the green fluorescent protein (GFP) gene using this innovative approach. To validate the DNA transformation, they utilized a Laser Scanning Confocal Microscope (LSCM) and PCR to observe the expression of the transgene GFP. This technique offers several advantages, making it highly promising for practical application. It is cost-effective, adaptable to different genotypes, environmentally friendly, and significantly reduces the time required for the transformation process. Overall, this method presents a potential solution to address the challenges related to plant yield and immunity against pathogens, thus providing a new avenue for improving the performance and resilience of the okra crop.

16.9 Constraints/Bottlenecks in Okra Crop Improvement

Breeding and biotechnology approaches have been instrumental in improving crop traits, including yield, disease resistance, and stress tolerance in various crops, including okra. However, several constraints and bottlenecks exist when it comes to okra crop breeding and biotechnology. Limited genetic diversity within cultivated okra varieties is a significant constraint. In many regions, only a few traditional varieties are grown, leading to a narrow genetic base. This lack of diversity hinders the development of new and improved varieties with enhanced traits. Genetic transformation is a crucial biotechnological approach for introducing desirable traits into okra. However, okra's recalcitrant nature to tissue culture and low transformation efficiency pose challenges in the successful incorporation of new genes. Compared to some other crops, okra has limited genomic and genetic resources available. This includes a lack of comprehensive genetic maps, genomic sequences, and molecular markers, which makes trait mapping and marker-assisted breeding more difficult. Many of the important traits in okra, such as fruit yield, disease resistance, and stress tolerance, are controlled by multiple genes and are influenced by environmental factors. This complex inheritance makes it a very challenging task to select and breed for these traits. Okra is susceptible to various insect pests and diseases, which can significantly impact its productivity. Developing resistant varieties to YVMV and ELCV through conventional breeding or biotechnological approaches is essential. The acceptance of genetically modified (GM) okra by consumers and regulatory bodies can be a significant step for successfully deploying biotechnological approaches. Public perception, concerns about biosafety, and regulatory requirements can hinder the deployment of GM okra varieties. Intellectual property rights and regulations related to biotechnology can be complex and may restrict access to important genetic resources, technologies, and research tools for okra breeding. Gene flow between GM and non-GM okra varieties can occur, leading to the unintentional spread of transgenes into wild or traditional varieties. These issues will need establishing the best protocols to prevent such issues. Adequate infrastructure and funding are essential for carrying out successful breeding and biotechnology research programs but are currently not adequately met. Addressing these constraints and bottlenecks requires collaborative efforts from researchers, breeders, policymakers, and farmers. Due to okra's polyploidy nature, genetic improvement must be addressed through state-of-the-art technologies such as CRISPR/Cas-9.

With large investments in genetic resources, biotechnology research and infrastructure can help overcome these challenges and contribute to the development of improved okra varieties that are resilient, productive, and sustainable.

References

Ahmad I, Rana RM, Hassan MU, Khan MA, Sajjad M (2022) Association mapping for abiotic stress tolerance using heat-and drought-related syntenic markers in okra. Mol Biol Rep 49(12):11409–11419

Akash MW, Shiyab SM, Saleh MI (2013) Yield and AFLP analyses of inter-landrace variability in okra (*Abelmoschus esculentus* L.). Life Sci J 10(2):2771–2779

Akhond M, Molla M, Islam MO, Ali M (2000) Cross compatibility between *Abelmoschus esculentus* and *A. moschatus*. Euphytica 114(3):175–180

Aladele SE, Ariyo OJ, De Lapena R (2008) Genetic relationships among West African okra (*Abelmoschus caillei*) and Asian genotypes (*Abelmoschus esculentus*) using RAPD. Afr J Biotechnol 7(10)

Anisuzzaman M, Kabir AH, Sarker KK, Jarin S, Alam MF (2008a) Micropropagation of *Abelmoschus esculentus* L. (Moench.) for disease free plantlets 547 through meristem culture. Arch Phytopathol Plant Protect 43(5):460–466. https://doi.org/10.1080/03235400701875430

Anisuzzaman M, Jarin S, Naher K, Akhtar MM, Alam MJ, Khalekuzzaman M, Alam I, Alam MF (2008b) Callus induced organogenesis in okra (*Abelmoschus esculentus*). Asian J Plant Sci 7(7):667–681

Badiger M, Yadav RK (2019) Screening of germplasm of *Abelmoschus* against biotic stresses. Indian J Agric Sci 89(12):2085–2090

Badiger M, Yadav RK, Sharma BB, Bhat KV, Tomar BS, Lata S, Das A (2023) Pollen germination, pollen–pistil interaction and crossability studies in interspecific and induced colchiploid population of *Abelmoschus* species. Genet Resour Crop Evol 71:107. https://doi.org/10.1007/s10722-023-01610-y

Belkhodja L, Belkhodja M, Ghomari S (2023) *In vitro* regeneration of okra [*Abelmoschus esculentus* (L.) Moench] through nodal and shoot apex explants. Indian J Agric Res 57(2):218–223. https://doi.org/10.18805/IJARe.AF-714

Daniel MA, David RHA, Caesar SA, Ramakrishnan M, Duraipandiyan V, Ignacimuthu S, Al-Dhabi NA (2018) Effect of L-glutamine and casein hydrolysate in the development of somatic embryos from cotyledonary leaf explants in okra [*Abelmoschus esculentus* (L.) Monech]. S Afr J Bot 114:223–231

Das A, Yadav RK, Choudhary H, Lata S, Singh S, Kumar C, Talukdar A (2022) Population structure, gene flow and genetic diversity analyses based on agro-morphological traits and microsatellite markers within cultivated and wild germplasms of okra (*Abelmoschus esculentus* (L.) Moench.). Genet Resour Crop Evol 69(2):771–791

Dhande GA, Patil VM, Raut RV, Rajput JC, Ingle AG (2012) Regeneration of okra (*Abelmochus esculentus* L.) via apical shoot culture system. African J Biotechnol 11(86):15226–15230. https://doi.org/10.5897/AJB12.907

Dhankhar BS, Mishra JP, Narendra S, Thamburaj T (2005) Textbook of okra vegetables, tuber crops and spices. ICAR, New Delhi, pp 222–237

Dutta OP (1991) Okra germplasm utilization at IIHR [Indian Institute of Horticultural Research], Bangalore. International Crop Network Series (IBPGR)

Farooq N, Ather L, Shafiq M, Nawaz-ul-Rehman MS, Haseeb M, Anjum T, Abbas Q (2022) Magnetofection approach for the transformation of okra using green iron nanoparticles. Sci Rep 12(1):16568

Fougat RS, Purohit AR, Kumar S, Parekh MJ, Kumar M (2015) SSR based genetic diversity in *Abelmoschus* species. Indian J Agric Sci 85:1223–1228

Gadwal VR, Joshi AB, Iyer RD (1968) Interspecific hybrids in Abelmoschus through ovule and embryo culture. Indian J Genet Plant Breed 28(3):269–274

Ganesan M, Chandrasekar R, Kumari BDR, Jayabalan N (2007) Somatic embryogenesis and plant regeneration of *Abelmoschus esculentus* through suspension culture. Biol Plant 51(3):414–420

Ganesh KV, Mathew D, Augustine R, Soni KB, Alex S, Shylaja MR, Cherian A (2022) Development of transgenic okra (*Abelmoschus esculentus* L. Moench) lines having RNA mediated resistance to yellow vein mosaic virus (Geminiviridae). J Virol Methods 301:114457

Gangopadhyay KK, Singh A, Bag MK, Ranjan P, Prasad TV, Roy A, Dutta M (2016) Diversity analysis and evaluation of wild *Abelmoschus* species for agro-morphological traits and major biotic stresses under the north western agro-climatic condition of India. Genet Resour Crop Evol 64(4):775–790

Gayathri M, Pitchaimuthu M, Ravishankar K (2021) SSR marker development in *Abelmoschus esculentus* (L.) Moench using transcriptome sequencing and genetic diversity studies. J Hortic Sci 16(2):206–214

Gulsen O, Karagul S, Abak K (2007) Diversity and relationships among Turkish okra germplasm by SRAP and phenotypic marker polymorphism. Biologia 62(1):41–45

Haider SA, Islam R, Kamal AHM, Rahman SM, Joarder OI (1993) Direct and indirect organogenesis in cultured hypocotyl explants of *Abelmoschus esculentus* (L.) Moench. Plant Tissue Cult 3:85–89

Kumar S, Parekh MJ, Fougat RS, Patel SK, Patel CB, Kumar M, Patel BR (2017) Assessment of genetic diversity among okra genotypes using SSR markers J Plant Biochem. Biotechnol 26:172–178

Jambhale ND, Nerkar YS (1982) An induced Amphiploid of *Abelmoschus*. Indian J Genet Plant Breed 42(3):372–375

John KJ, Scariah S, Nissar VM, Bhat KV, Yadav SR (2013) *Abelmoschus enbeepeegearense* sp. nov. (Malvaceae), an endemic species of okra from Western Ghats, India. Nord J Bot 31(2):170–175. https://doi.org/10.1111/j.1756-1051.2012.01624.x

John JK, Krishnaraj MV, Pradheep K, Patil P, Harish GD, Bhat KV (2020) A new variety of *Abelmoschus pungens* (Malvaceae) from Indo-Burma Biodiversity Hotspot. J Indian Assoc Angiosperm Taxon 30(2):270–277

Kabir AH, Sarker KK, Sharmin SA, Islam MS, Alam MF (2008) Callus induction and plantlet regeneration in *Abelmoschus esculentus* (L.) Moench. J Agric Technol 4(1):193–204

Kabir M, Ahmed S, Akhond M (2016) Organogenesis in okra [*Abelmoschus esculentus* (L.) Moench]: a plant recalcitrant to tissue culture. Bangladesh J Agric Res 41(3):521–528. https://doi.org/10.3329/bjar.v41i3.29723

Kyriakopoulou OG, Arens P, Pelgrom KT, Karapanos I, Bebeli P, Passam HC (2014) Genetic and morphological diversity of okra (*Abelmoschus esculentus* [L.] Moench.) genotypes and their possible relationships, with particular reference to Greek landraces. Sci Hortic 171:58–70

Lata S, Kumar Yadav R, Tomar BS (2021) Genomic tools to accelerate improvement in okra (*Abelmoschus esculentus*). Intech Open. https://doi.org/10.5772/intechopen.97005

Li J, Li J, Ma Y, Kou L, Wei J, Wang W (2022) The complete mitochondrial genome of okra (*Abelmoschus esculentus*): using nanopore long reads to investigate gene transfer from chloroplast genomes and rearrangements of mitochondrial DNA molecules. BMC genomics 23(1):481

Mallela RR, Vutukuri PR, Kanuri M (2009) *In vitro* plant regeneration and genetic transformation of okra (*Abelmoscus esculentus* [L.] Moench). Fruit Veg Cereal Sci Biotech 3:1–6

Mangat BS, Roy MK (1986) Tissue culture and plant regeneration of okra (*Abelmoshus esculentus*). Plant Sci 47(1):57–61

Manickavasagam M et al (2015) Assessment of factors influencing the tissue culture-independent *Agrobacterium*-mediated *in planta* genetic transformation of okra [*Abelmoschus esculentus* (L.) Moench]. Plant Cell Tiss Org 123:309–320

Martinello GE, Leal NR, Amaral Jr AT, Pereira MG, Daher RF (2001) Comparison of morphological characteristics and RAPD for estimating genetic diversity in *Abelmoschus* spp. Acta Horticulturae 546:101–104

Mishra GP, Singh B, Seth T, Singh AK, Halder J, Krishnan N, Tiwari SK, Singh PM (2017) Biotechnological advancements and *Begomo* virus Management in Okra (*Abelmoschus esculentus* L.): status and perspectives. Front Plant Sci 8:360. https://doi.org/10.3389/fpls.2017.00360
Muthaiah G, Elangovan D, Mottaiyan P, Ravishankar KV (2023) Comprehensive Transcriptome Analysis in Okra (*Abelmoschus esculentus* L. Moench): Analysis of LncRNA and Transcription Factors Involved in Abiotic Stress. Russ J Plant Physiol 70(3):47
Narendran M, Deole SG, Harkude S, Shirale D, Nanote A, Bihani P, Parimi S, Char BR, Zehr UB (2013) Efficient genetic transformation of okra (*Abelmoschus esculentus* (L.) Moench) and generation of insect-resistant transgenic plants expressing the *cry1Ac* gene. Plant Cell Rep 32(8):1191–1198. https://doi.org/10.1007/s00299-013-1415-4
Patel KG, Mandaliya VB, Mishra GP, Dobaria JR, Radhakrishnan T (2016) Transgenic peanut overexpressing *mtlD* gene confers enhanced salinity-stress tolerance via mannitol accumulation and differential antioxidative responses. Acta Physiol Plant 38:181. https://doi.org/10.1007/s11738-016-2200-0
Priyavathi P, Nagesh S, Kavitha VVK, Johnson C, Gopal P (2018) Comprehensive leaf transcriptome of a non-model plant, Abelmoschus esculentus for the functional genomics studies. J Genet Genome Res 5:036
Puneeth PV, Yadav RK, Lata S, Ghosh A, Chaudhary H, Tomar BS, Tomer A (2022) Vulnerability studies of okra genotypes to bhendi yellow vein mosaic virus (BYVMV). Indian J Hortic 79(2):186–193
Puneeth PV, Lata S, Yadav RK, Wankhede DP, Tomar BS, Choudary H, Talukdar A (2023a) Exploring the genetic diversity using CAAT box-derived polymorphism (CBDP) and start codon targeted (SCoT) markers in cultivated and wild species of okra (*Abelmoschus esculentus* (L.) Moench). Genet Resour Crop Evol 70(3):749–761
Puneeth PV, Yadav R, Lata S, Asif M, Talukdar A, Choudhary H, Tomar A (2023b) Development of EST-SSR markers using transcriptome data in okra (Abelmoschus esculentus L.): a genetically orphan crop. Indian J Genet Plant Breed 83(01):95–102. https://doi.org/10.31742/ISGPB.83.1.12
Ravishankar KV, Muthaiah G, Mottaiyan P, Gundale SK (2018) Identification of novel microsatellite markers in okra (*Abelmoschus esculentus* (L.) Moench) through next-generation sequencing and their utilization in analysis of genetic relatedness studies and cross-species transferability. J Genet 97:39–47
Rizwan HM, Irshad M, He BZ, Liu S, Debnath B, Mitra S, Li M, Lu XC, Sun YT, Qiu DL (2018) Silver nitrate (Agno3) boosted high-frequency multiple shoot regeneration from cotyledonary node explants of okra (*Abelmoschus esculentus* (L.) Moench). Appl Ecol Environ Res 16(3):3421–3435
Rohit Menon RM, Sarao NK, Mamta Pathak MP (2018) In planta Agrobacterium-mediated genetic transformation in okra {*Abelmoschus esculentus* (L.) Moench}. Appl biol res 221–227
Roy MK, Mangat BS (1989) Regeneration of plants from callus tissue of okra (*Abelmoschus esculentus*). Plant Sci 60:77–81. https://doi.org/10.1016/0168-9452(89)90046
Santhiya S, Yadav RK, Lata S, Sharma BB, Talukdar A, Ghosh A, Janani R (2022) Screening of wild okra (*Abelmoschus esculentus*) against yellow vein mosaic and enation leaf curl diseases. Indian J Agric Sci 92(10):1268–1273
Schafleitner R, Kumar S, Lin CY, Hegde SG, Ebert A (2013) The okra (*Abelmoschus esculentus*) transcriptome as a source for gene sequence information and molecular markers for diversity analysis. Gene 517(1):27–36
Sharma BR (1982) Punjab Padmini-a new variety of okra. Prog Farm 82:15–16
Shi D, Wang J, Bai Y, Liu Y (2020) Transcriptome sequencing of okra (*Abelmoschus esculentus* L. Moench) uncovers differently expressed genes responding to drought stress. J Plant Biochem Biotechnol 29(2):155–170
Thakur MR, Arora SK (1986) Okra In: Vegetable Crops in India. Eds. Bose, T.K. and Somm, M.G. pp. 606–622. Noya Prokash Calcutta. India
Wang J, Shi D, Bai Y, Zhang T, Wu Y, Liu Z, Liu Y (2022) Comprehensive proteomic and metabolomic analysis uncover the response of okra to drought stress. Peer J 10:14312

Wang R, Li W, He Q, Zhang H, Wang M, Zheng X, Xing L (2023) The genome of okra (*Abelmoschus esculentus*) provides insights into its genome evolution and high nutrient content. Hortic Res 10:uhad120

Yadav RK, Badiger M, Lata S (2019) Studies of genetic variability and association of yield traits and YVMV disease in *Abelmoschus* species. Indian J Agric Sci 89(4):688–694

Yildiz M, Furan MA, Koçak M, Sadık G, Kuzğun C, Al-Khafaji M, Bahjat NM (2022) Peroxidase gene markers revealed genetic diversity and population structure in okra germplasm. Plant Biotechnol Rep 16(2):195–204

Yu C, Wu Q, Sun C, Tang M, Sun J, Zhan Y (2019) The phosphoproteomic response of okra (Abelmoschus esculentus L.) seedlings to salt stress. Int J Mol Sci 20(6):1262

Yuan CY, Zhang C, Wang P, Hu S, Chang HP, Xiao WJ, Guo XH (2014) Genetic diversity analysis of okra (*Abelmoschus esculentus* L.) by inter-simple sequence repeat (ISSR) markers. Genet Mol Res 13(2):3165–3175

Yuan CY, Wang P, Chen PP, Xiao WJ, Zhang C, Hu S, Guo XH (2015) Genetic diversity revealed by morphological traits and ISSR markers in 48 Okras (*Abelmoschus esculentus* L.). Physiol Mol Biol Plants 21:359–364

Zhan Y, Wu T, Zhao X, Wang Z, Chen Y (2021) Comparative physiological and full-length transcriptome analyses reveal the molecular mechanism of melatonin-mediated salt tolerance in okra (*Abelmoschus esculentus* L.). BMC Plant Biol 21:1–16

Zhang C, Dong W, Gen W, Xu B, Shen C, Yu C (2018) De novo transcriptome assembly and characterization of the synthesis genes of bioactive constituents in *Abelmoschus esculentus* (L.) moench. Genes 9(3):130

Zhu Z, Yu J, Tang X, Xiong A, Sun M (2023) Selection and validation of reference genes in different tissues of okra (*Abelmoschus esculentus* L.) under different abiotic stresses. Genes 14(3):60

Part V

Agronomic Interventions and Nutritional Quality

Agronomic Management for Enhancing Productivity in Okra: An Overview 17

S. K. Singh and T. K. Behera

17.1 Introduction

Okra (*Abelmoschus esculentus* L. Moench) belongs to the Malvaceae family and is best known as a fruit vegetable crop of the tropical and subtropical regions of the world. Okra, commonly known as lady's fingers or bhindi, is grown for its tender pods. By virtue of its high nutritive and culinary values, okra, or lady's finger, is considered the most important vegetable crop in India as well as in other tropical countries like Nigeria, Pakistan, Cameroon, Iraq and Ghana. Okra is usually not grown in Europe and North America, but people in these countries also liked this vegetable because it contains a good amount of vitamin A and folic acid, besides carbohydrates, phosphorus, magnesium and potassium. This crop is suitable for cultivation as a kitchen garden crop as well as on large commercial farms. Okra has a good potential as one of the foreign exchange earner crops and accounts for about 60% of the export of fresh vegetables, excluding potatoes, onions and garlic from India. Burgeoning population and ever-increasing urbanisation have boosted the cultivation of vegetables in peri-urban areas in an intensive way. India ranked top in area and production in the world, accounting for 60.2% of the global okra production. It is grown in all agro-ecological zones of India. In India during the year 2020–21, the acreage of okra cultivation was 0.531 million (m) hectares (ha), with a production of 6.466 m tonnes (t) and a productivity of 12.17 t/ha (Anonymous 2021). The major okra-producing states are Andhra Pradesh, Assam, Bihar, Chhattisgarh, Gujarat, Madhya Pradesh, Orissa, Tamil Nadu, Uttar Pradesh and West Bengal.

Okra production and productivity are significantly affected due to the use of low-yielding susceptible varieties, suboptimal plant density, inappropriate planting

S. K. Singh (✉) · T. K. Behera
ICAR-Indian Institute of Vegetable Research, Varanasi, Uttar Pradesh, India

S. K. Tikoo et al. (eds.), *Okra: Status, Challenges and Opportunities*,
https://doi.org/10.1007/978-981-97-9963-3_17

dates, poor soil fertility management, injudicious use of fertiliser and irrigation, attack of various pathogens, insect pests, weeds, etc. Quality seed and proper agronomic management play a major role in raising the production and quality of okra. Since the okra crop is input-intensive, it requires greater skills to get high productivity. The principles for developing cultural practices for okra as well as the different cultural practices have been discussed hereunder touching cultivars, climate, soil, tillage, planting time, seed rate, plant population and geometry, method of sowing, intercultural operations, fertiliser management, weed management, water management, harvesting and cropping systems.

17.2 Principles for Developing Agrotechniques

The yield is the outcome of genotype X environment interaction. The environment constitutes both aerial and soil. The aerial environment includes solar radiation, temperature, rainfall and humidity, which cannot be easily altered; however, it can be manipulated to some extent through the selection of a suitable season, time of sowing, etc., to attain the highest productivity from the available climatic resources. On the other hand, the soil environment, which includes soil air, soil water, soil pH, nutrient concentration and living organisms, is amenable for manipulation through tillage, fertiliser application, irrigation, etc. The development of optimum cultural practices requires an understanding of the crop phenology and physiological requirements during each phenological phase. Okra is an annual, erect, herbaceous plant 1–2 m in height. A flower bud appears in the axil of each leaf above the 6th to 8th leaf, depending upon cultivar and season. The flowering is continuous and produced individually at the leaf nodes near the top of the stem. As the stem elongates, the lowermost flower bud opens in to flower. There may be a period of 2, 3 or more days between the time of development of each flower, but never dose more than one flower open on a single node stem. Under normal conditions, crop duration extends from 90 to 110 days.

The flower bud initiation generally takes place 22–26 days after sowing. The flower bud takes another 22–26 days from initiation to full bloom. The first flower opens 41–48 days after sowing and continues for another 40–60 days. Sowing to germination, emergence to true leaf stage, true leaf appearance to vegetative growth phase, flowering, fruiting and senescence are the major phases that could be delineated. Okra has an indeterminate growth habit and continues to produce new leaves from the main stem and on axillary branches until late in the growing season. Due to this indeterminate growth habit, the stem elongation, plant growth, flowering and fruiting occur simultaneously and continue till the senescence of the plant.

The germination or pre-emergence phase starts with sowing, and the duration of this phase depends upon the soil temperature and the moisture content of the soil. Genotype, day length, temperature and nutrition, irrigation and weed intensity affect the period from germination to vegetative growth. Varieties differ with respect to the time taken by them for flowering, i.e. some of them initiate flowers early at the lower nodes while others initiate flowers late at the higher nodes. Temperature and humidity are important weather factors that affect this phase (flowering). Nutrition is another important factor affecting the duration of this phase. The period of flowering and fruiting is mainly affected by nutrition and moisture status. High nitrogen nutrition increases the growth duration. Good nutrition ensures good early canopy cover and delays the onset of senescence, and this increases the duration of growth and fruiting. After the linear profuse flowering and fruiting phase, the rate of flowering and fruiting starts decreasing. This period coincides with the onset of senescence of foliage. Therefore, delaying the onset of senescence of foliage ensures a longer period of linear growth of fruiting, resulting in a high yield. This is achieved through the application of proper nutrition and irrigation.

17.3 Cultivar

Cultivars play an important role in enhancing the okra yield. Since okra is an important crop in India, different varieties are used by farmers in order to meet the demand of okra by consumers. Okra is a self-pollinated crop. The open-pollinated varieties give a reasonable yield in the range of 10–14 t/ha; however, many hybrids that have been developed give a yield in the range of 17–22 t/ha. Farmers grow both types of varieties, depending on the availability of seeds. These varieties have different responses to a given soil and season. Different cultivars of okra require different sowing times, plant spacing and other inputs like fertiliser and irrigation. Some cultivars are sensitive to excessive soil moisture. Others are slightly tolerant of salt. Most cultivars are adapted to high temperatures throughout the growing period with little seasonal fluctuation. The okra crop is sensitive to certain diseases, such as yellow vein mosaic, which not only decreases the yield considerably but also deteriorates the quality of the produce. There is a need to develop more varieties resistant to YVMV disease to boost the yield, particularly during the wet season (kharif) in the northern plains. Apart from this, mites and thrips are the major pests causing considerable damage to okra crops, particularly during the spring–summer season.

17.4 Climate

Okra has specific climate requirements. Okra is basically a hot-weather crop and requires a long, warm and humid growing period. It thrives best during warm and moist seasons. The okra crop is highly sensitive to low temperatures and frost. The seeds do not germinate when the temperature is below 17 °C. Although the optimum temperature for seed germination is 29 °C, for uniform and quick seed germination, optimum soil moisture and a temperature between 25 and 35 °C are needed. The soil temperature really makes a difference: at temperatures of about 16.5 °C, germination takes 27 days, whereas at about 24.0 °C, germination may take only 13 days, and in ideal conditions perhaps a week (Anonymous 2012). Seed germination percentage and speed of seedling emergence are optimal at 30–35 °C (Akande et al. 2003).

Normal plant growth and development occurs above 20 °C temperature (Lamont 1999). The optimum temperatures are in the range of 20–30 °C, with a minimum temperature of 18 °C and the maximum of 35 °C (Benchasri 2012). A monthly average temperature range of 21–30 °C is considered appropriate for growth, flowering and pod developments (Tindall 1983; Nonnecke 1989). It can be successfully grown at a temperature ranging between 25 °C and 30 °C. It grows fairly well in the hottest summer, but when day temperature exceeds 42 °C, flower drops (Chauhan 1972), causing yield losses. The plant grows shorter under high temperatures and low humidity conditions. Flower initiation and flowering are delayed with increasing temperatures (Lamont 1999). Adequate sunshine is very important for the growth and yield of crops. Crop is sensitive to frost, low temperature, waterlogging and drought. Adjustment of climatic factors helps in taking at least one (summer) crop in hills, two or even three (summer, kharif and late kharif) crops in the east, west and north Indian plains, and almost year-round cultivation under moderate climate in south India.

17.5 Soil

The soil structure, texture and fertility affect the cultural practices of okra. A soil with good structure and texture is necessary for okra because a proper environment is required for the growth and development of roots. Soils with poor structure have poor aeration, which affects the respiration and growth of the roots. These problems are especially aggravated under relatively warmer and wet conditions, particularly during the kharif season. Well-drained sandy loam soil, rich in organic matter, is the most suitable for the okra crop. However, sandy to clay soils, so long as they are properly fertilised (manured), supplied with enough organic matter and have good drainage facilities, are fit for okra cultivation. Soil texture affects the water retention and movement, hence affects the number and frequency of irrigation. Soil texture also determines the number and type of tillage operations. Light soils are easier to prepare than heavy soils. Ridging is also easy in light soils. Planting on ridges or raised beds becomes necessary during the rainy season. Okra prefers soils with a pH range between 6.0 and 6.8. It is very sensitive to alkalinity, and soils with a pH >7.5 should be avoided. However, the cultivar Pusa Sawani has some tolerance to salts and thus also to a higher pH range.

17.6 Tillage

The tillage requirements of okra are a fine tilth for easy placement of seeds at the required depth, proper penetration of roots and checking weeds in the early stages of growth. The tillage can be done either with bullock-drawn or tractor-drawn implements. The type and number of tillage operations would depend upon the soil type, previous crop, etc. Normally, one to two ploughings and one harrowing followed by planking are advocated. In heavier soils, additional tillage may be required, while in very light soils, one ploughing may be done away with. During the land preparation, all soils need to be pulverised and well decomposed farmyard manure/ compost may be incorporated. The post-planting tillage operations are confined to weeding and, in certain cases, earthing up for making the ridges to ensure proper irrigation and drainage. The use of rotravator tillage to pulverise the soil may also be resorted.

17.7 Sowing Time and Season

Sowing time depends upon variety (short, medium or long duration), season (early or spring crop, summer, kharif, late kharif) and type of crop, i.e. vegetable or seed crop. Sowing times have a great impact on the germination, growth, flowering, production and quality of okra. Besides, the time of seed sowing has a direct bearing on the incidence of YVMV diseases. Saikia and Bhagabiti (1994) reported more incidence of YVMV in rainy season crops (73%) than summer (25%).

The sowing time determines the length of favourable growing period available to the crop. This in turn influences the fruit size and number of fruits/plants, harvesting frequency and yield. The criteria for working out the optimum sowing time involve the availability of a suitable temperature for germination and growth and that, after emergence, the available growing period should be more than 80–90 days so that reasonable and economic yields could be obtained. In south India, where the winter is mild, it is possible to grow okra throughout the year (Singh and Bhagchandani 1967). In the northern plains, usually two crops, i.e. spring–summer and rainy season crops, are raised. The spring–summer crop is sown from the beginning of February to the end of March. It can, however, be sown a little earlier in Assam, West Bengal, Orissa and Bihar. The rainy season crop is sown during May–July (Table 17.1). In the hills, crop is sown in May–June (Anonymous 1978, 1983). Okra sown during the second week of June in the northern and eastern plains minimises the incidence of both Jassids and shoot and fruit borer (Hazara et al. 2011).

Table 17.1 Planting schedules for okra in different areas

Region	Season	Planting time	Season	Planting time	Season	Planting time
Eastern India	**Spring**–Summer	January-early February	Kharif	May–July	Rainy autumn	Mid-August
Western India	**Spring**–Summer	February–March	Kharif	May–July	Rainy autumn	Mid-August
South India	**Spring**–Summer	Feruary	Kharif	May–July	Winter	November
Hills	Summer	May–June				
Plateau regions			Rain-fed (*Kharif*) crop	May–June	Irrigated winter (*rabi*)	Sept-Oct

17.8 Plant Population and Planting Geometry

The optimisation of plant density (plant population) and crop geometry is one of the most important subjects of crop production management because it affects the seed cost, plant development, yield and quality of the crop/produce (Bussan et al. 2007). The main objective of the farmer involved in the production of the okra crop is to achieve the highest yield and quality simultaneously, and this could be achieved by adopting proper crop management practices. One of the important and essential crop management practices is the establishment of required plant density. The final plant density influences total yield, fruit size and quality. Total yield influences the profit; fruit size is important to meet market requirements, as some markets call for large fruits/pods (processing/seed production) while others need small fruits (canning/export). The quality determines the acceptability of the produce in the market/consumers and its use and price.

Plant density influences okra yield, which is determined by the number and size of fruits. The yield per unit area is increased up to a certain level of increase in plant density due to utilisation of growth factors. And after a point, it reaches a plateau and subsequently decreases with an increase in density. As the plant densities increase, the number of fruits per plant decreases, but the number of fruits per unit area generally increases. Thus, though yield per plant decreases gradually as plant density per unit area is increased, the total yield is increased up to a point. The size of fruit depends on the availability of adequate nutrition, soil moisture and light. The size of fruits may be limited by competition among plants for these growth factors. When the density is higher, individual plants get narrower space, leading to competition for growth factors between plants, resulting in a reduction in fruit size. Therefore, plant population, or the number of plants per unit area, should be such that all plants get equal opportunity to utilise the resources for their optimum growth and development. Maximum yield and best quality of an okra crop in a given environment can be obtained when competition among the plants is less or minimum.

The size and shape of space available to plants (plant geometry) play a role in manipulating the microclimate and hence help in enhancing the yield and quality of crop. Thus, it is evident that plant density and plant geometry decide the spacing to be followed in the okra crop. The desired crop geometry is achieved through manipulation of inter- and intra-row plant spacing. Proper/optimum spacing improves the availability of nutrients, aeration and light intensity and also avoids excessive competition among the plants, which means better crop growth and response to the inputs and thereby enhanced yield and quality (Ravindran et al. 1986). Hence, optimum spacing induces the plant to achieve its potential yield. However, the spacing varies according to location, season, variety and market requirement. Different cultivars of okra may require different plant spacing. Suitable plant spacing can lead to optimum yield, but incorrect plant spacing could result in relatively low yield and poor-quality fruits (Absar and Siddique 1982; Moniruzzaman et al. 2007). A good cultivar that is sown at an improper spacing will give a very poor yield. Proper plant spacing and suitable cultivars are critical to increased production of okra.

The effect of row spacing on the yield and yield components of okra showed that the productive nodes increased with increasing row spacing. Close spacing increased yield and reduced the number of lateral branches per plant (Moterran et al. 1963). Plants at the closer spacing gave the highest yield, but harvesting was difficult (Taj and Gardner 1968). The closest row spacing (30 cm) suppressed weeds better and had low fruit yield/plant when compared to other plant densities (Ibeawuchi et al. 2005).

Plants at the wider spacing were generally shorter and leafy than at the closer spacing. Randhawa and Panna (1969) reported that the number of branches, flowers and fruits per plant in okra increased with the wider row spacing at 60 cm, but the highest yield was obtained from the closer row spacing of 30 cm. Similarly, Kamalanathan et al. (1970) observed that spacing of 60 × 30 cm had significantly and consistently produced higher yield per hectare of okra, while spacing of 60 × 20 cm recorded higher yield over 60 × 40 cm and 60x30 cm in terms of number of fruits per hectare, but the yield per plant was higher in the wider spacing. The fruit weight was higher in the wider spacing of 60 × 30 cm as compared to 60 × 20 cm spacing.

In a flat bed system, Rizvi and Jugirdar (1973) found row spacing of 45 cm is the best practice for a higher yield of okra. Although total yield per hectare increases with adoption of closer spacing, fruit size was decreased with increasing plant density of okra (Albregts et al. 1974). Even within okra varieties, the number of fruits and marketable yield per unit area were affected by plant density; the number of fruits and marketable yield per plant decreased with increasing plant density (Albregts et al. 1976).

Based on the research results and information available in okra, a plant distance of 60 × 30 cm accommodating 55,500 plants/ha is generally recommended for branching types, while 45 × 30 cm accommodating 66,000 plants/ha for non-branching types. During the spring–summer season, the vegetative growth is relatively less; hence, with less plant growth, the spacing is kept at 45 × 20 cm for the non-branching type and 45 × 30 cm for the branching type. For harvesting smaller fruits for fresh fruit export, a strip accommodating 2–3 rows at 20 cm distance and keeping 60 cm between the strips could be grown by maintaining 20–30 cm plant-to-plant spacing in the rows. This will make harvesting easy and check branching. On average, 50–70 thousand plants can be accommodated on one hectare of land. The spacing may vary due to mechanisation in planting as well as the use of modern methods of irrigation and the architecture of plants.

The depth of sowing is also very important for early and uniform germination, which in turn affects the growth and yield of okra. The depth of sowing of 2.5–3.0 cm is found to be optimal as it gives 80–85% germination in 6–7 days, which culminates in giving higher biomass and yield of okra.

The plant spacing influenced not only the quality and yield of the green pod but also the seed. Spacing determines the available area for a plant to source for growth resources, such as water, light and nutrients. Absar and Siddique (1982) noted that plant density is an important factor that affects okra seed production. Suitable plant spacing can lead to optimum seed yield, while too high or too low plant spacing could result in relatively low yield and quality. To maximise seed yield of okra, various planting geometry/spacing were evaluated by different workers. Some advocated closer spacing and some wider spacing for getting a higher seed yield of okra; however, it depends on various factors like variety, agro-climatic condition, edaphological condition, agronomical practices, particularly fertiliser and plant protection measures. Plant spacing for okra seed production suggested by different authors ranges from 20 to 40 cm and 30 to 60 cm between rows (Thakur and Arora 1986; Rastogie et al. 1987; Khan and Jaiswal 1998; Hossain et al. 1999). The germination percentage and seed vigour index were unaffected due to different plant spacing.

Cultural operations like thinning and gap filling are done in a field after the crop has emerged to keep optimum density. Thinning is done to reduce density due to overseed rate or more seeds/hole. Gap filling is done by resowing with seeds to fill the gaps that exist due to low germination, either because of poor-quality seed, very shallow or very deep placement of seeds or poor moisture availability. Sometime crust breaking shallow tilling is resorted to for breaking soil crusting formed due to rain immediately after sowing in silt-reach soils to improve germination.

17.9 Methods of Sowing

The method of sowing depends upon the availability of implements, power source, irrigation method, type of soil and sowing depth. Various practices of sowing, like line sowing, dibbling, the ridge and furrow method and bed planting, are followed in various parts of the country. Seeds are sown by hand dibbling, behind the plough or by seed drill. Broadcasting is not recommended due to the high seed rate as well as inconvenience in cultural operations and harvesting. For manual planting, seeds are sown/dibbled in a line on the properly prepared land and covered with soil. In another method, a ridge is made using bullock/tractor-drawn plough, and seeds are dibbled using hand tools at recommended spacing on the ridges. The seed should be sown on the sides of ridges facing the sun during February–March in the northern and eastern plains. For mechanised planting, beds are made with a tractor-drawn bedmaker, and the seeds are sown/dibbled using hand tools on the beds. For drip irrigation in crops, okra is sown on raised beds to economise on drip laying to some extent. Normally, the depth of sowing depends upon the prevailing temperature and moisture. Under ambient conditions, shallow planting is advocated for quick and uniform emergence. It is better to sow the seeds in pre-irrigated, moist soil than irrigate the field after sowing.

In recent years, bed planting techniques have been developed as an alternative to the conventional system for producing higher yields with less water. Change over from growing crops in a flat/ridge-furrow system of planting to a raised bed alters the land configuration and crop geometry and offers more effective control over irrigation and drainage. This bed planting method of tillage and crop establishment also provides opportunity for crop diversification as intercropping of cowpea and other crops with okra can be done in the beds.

The technique of paired row planting is one way of accommodating the whole population of the base crop and creating inter-spaces wide enough to accommodate one or two rows of intercrop. Therefore, in paired row planting, two adjacent rows of the base crops are paired to reduce the inter-row space to create some space between pairs of base crop rows but wide enough to minimise undue competition among plants of the base crop. The paired row technique of planting is generally used for intercropping two or more crops. However, in the case of okra, this technique can be used to obtain the desired plant population through changing planting geometry as well as for intercropping of other crops to maximise the return. The aim is not to sacrifice the yield of okra but to get some additional yield/income from intercrop. Hence the land equivalent ratio (LER) is always >1, as the combined yield/income from intercropping is more than sole okra yield.

17.10 Seed Rate and Seed Treatment

Quality of seed is the most important attribute that affects the germination, plant population and the initial vigour of the plants. Seed rate depends on the pant density and the crop geometry to be adopted, which in turn is influenced by the season and growth habit of the cultivar used. It requires about 18–22 kg seed/ha during the summer season, whereas 8–10 kg seed/ha is required for the rainy season crop (Nath 1976). During the summer season, the vegetative growth is relatively less, and hence the seed rate is higher than in the rainy season crop. The seed average weight is 4.7–8.1 g/100 seeds. One ounce (28.35 g) of seed contains about 500 seeds with an average germination of 50–60% (Sundaraj et al. 1965). Magagula and Ossam (2011) reported variations in emergence rate with seed size, being higher in bolder seeds and lower in smaller seeds. Delayed and erratic emergence is a serious problem in okra (Anderson et al 1960) that creates problems with fertiliser utilisation, post-emergence weed control and uniform harvesting. Temperature, salinity and the hard seed coat of okra seeds interfere with water uptake. The seed should be soaked overnight in water for better germination. Soaking the seed with 400 ppm GA or 20 ppm IAA enhances germination. Soaking of seeds in 45 °C water for 1.5 h before sowing results in rapid field emergence (Onwueme 1975). Hydropriming of okra seeds for 12 h and solid matrix (SM) priming with calcium aluminium silicate (1:0.4:1; seed: SM: water) for 24 h significantly increased the seed germination, seedling vigour, mean germination time and fruit yield up to 55% in comparison to control in Cv, Hissar Unnat in northern India (Sharma et al. 2014). Similarly, Osmo-priming of seeds with 5% polyethylene glycol for 24 h enhanced germination percentage (94.39%) and better seedling growth in Cv. Punjab-8 (Kaur et al. 2015). Soil application of vermicompost at 5 t/ha before 10 days of sowing increases the germination percentage (100%) emergence speed index (15.17 plants/day) and mean emergence time (6.4 days) (Sarma and Gogoi 2015). Soaking the seeds in 0.2% carbendazim 50% overnight protects the seedlings from wilt disease.

17.11 Manuring and Fertiliser Application

The availability of essential plant nutrients in a balanced form not only has a direct influence on crop growth and development but also on the quantity and quality of produce. Okra plant has a well-developed root system and is a good feeder. Okra has a tap root system that goes to a depth of 45–100 cm The plants form a deeply penetrating taproot with dense shallow feeder roots reaching out in all directions in the upper 45 cm of the soil. Okra responds well to the application of both Farm Yard Manure (FYM) and mineral fertilisers. On the other hand, uptake of fertiliser nutrients (NPK) by okra per unit area is high due to the faster rate of early growth and high yield. It has been reported that for 100 kg of pod yield, okra removes 0.30, 0.13, and 0.45 kg/ha of nitrogen (N), phosphorus (P) and potassium (K) from soil (Hazra et al. 2011). Hence, application of nutrients to the soil in adequate amounts taking into account the soil nutrient status is necessary for meeting the crop's nutritional requirement as well as to maintain optimum soil fertility status. The efficiency of utilisation of the applied fertiliser is low in okra due to frequent irrigations during the early part of the crop in the summer season, frequent rain during the kharif season, short growing season, etc. The plant needs to produce 5 tonnes of dry matter to produce 10 tonnes of green fruit, equivalent to 1 tonne of dry matter in fruits. Hence, the okra crop requires high fertiliser input.

Nitrogen contributes to the yield by enhancing plant height, number and size of fruits, fruiting rate, fruiting period and advanced flowering and delaying the senescence. Phosphorus is involved in a wide range of plant processes, including the development of root systems and enhanced crop maturity. Application of K increases plant height and crop vigour and imparts resistance against drought and diseases. The response of okra to nitrogen is more pronounced when applied in conjunction with P and K. Combined application of the optimum dose of NPK maximises the growth and yield of okra and increases the efficiency of each nutrient. Application of P, K and P + K with N increased the yield by 5.6%, 6.2%, and 26% compared to application of N alone.

The fertiliser needs of okra have been found to vary with the soil and season, variety, climatic conditions, plant population, purpose of crop, use, etc. Okra responds well to nutrient application and increased fruit yield can be obtained with application of 150 kg N, 75 kg P_2O_5 and 75 kg K_2O per hectare (Dadema and Dongale 2004) in the open-pollinated improved varieties. The optimum rates of NPK fertilisation based on pod (fruit) yield response under eastern plains conditions have been found to be 125–150: 60–75: 60–75 kg/ha. Full doses of phosphorus and potassium are applied as basal doses at the time of sowing. Nitrogen being highly mobile is prone to losses and is therefore applied in splits, 1/3rd at the time of planting, and the remaining is top dressed in two splits, first at 30 and the second at 50 days after sowing. A general fertiliser dose of 200 kg N, 100 kg P_2O_5 and 100 kg K_2O per hectare is recommended for the hybrids, as the growth period and yield potential of hybrids are higher than those of open-pollinated varieties. Half N along with a full dose of phosphorous and potassium should be given as a basal dose at the time of sowing, and the rest N should be topdressed in two equal split doses at a

30-day interval after sowing. In spring–summer crop application in splits after every 3–4 pickings increases crop duration and yield. For the seed crop, foliar sprays of 0.3% zinc sulphate and 0.2% ammonium molybdate in the standing crop enhance fruit yield. Soil application of calcium is beneficial for the seed crop. Vermicompost at 4–6 t/ha can be applied as broadcasting during final land preparation.

At planting, using diammonium phosphate or mixed NPK fertiliser combined with urea for basal application and top dressing through urea of the remaining dose of N is economical, efficient and safe. Readily soluble sources of P, such as single superphosphate, triple super phosphate and diammonium phosphate, are most suitable for okra. Sources of K like muriate of potash and sulphate of potash are equally efficient to okra. In order to obtain maximum yields in okra, proper supply of nutrients during the growing period should be ensured. Since the efficacy of fertilisers applied to soil is low due to various losses and fixation problems, it is difficult to achieve the desired crop response to fertilisers applied to soil. Split application of N is essential for maximising efficiency. Spraying 2% urea solution 40–50 days after sowing corrects visual deficiency symptoms of N, if any. Foliar application of nitrogen fertiliser has been recommended by various workers for enhanced yield and higher nutrient efficiency. The rate of nitrogen for foliar spray varies from one fourth to three fourths of the applied dose. Split application of K is advantageous only in light-textured loamy sand soil. No benefits from a split dose of P are reported because it is required mostly for early root and shoot growth. Band placement of P fertiliser invariably is better than broadcast because of the fixation of P in most soils. Methods of placement of K fertiliser in bands at sides or above or below seed or broadcast are equally efficient. Therefore, combined basal application of a full dose of P & K along with one third to half of N at planting is economical, efficient and convenient. Supplying nutrients to crop plants, particularly NPK, through foliar application of water-soluble fertilisers has potential advantage in eliminating the problem of fixation and immobilisation of nutrients in soil. Foliar feeding can increase productivity and also help in reducing production costs. By spraying commercially available water-soluble NPK (19:19:19) fertiliser at a 0.5% concentration three times at a 10-day interval starting at 30 days after sowing (DAS), maximum yield of okra (298.83 g/plant) could be obtained in cultivar Makhmali (Sundaram and Kanthaswamy 2005).

Nutrient elements involved in shoot-specific processes such as photosynthesis and chlorophyll biosynthesis like Mg, Mn, Fe, Cu and N are good for foliar application. Moderation in the ratio of fertilisation is possible based on soil test values after calibration, depending upon the soil, climate, season and variety. Soil and plant test-based fertiliser recommendation results in fertiliser economy through balanced application.

Intensive cropping systems and decline in organic manuring have laid the groundwork for a wide spread deficiency of secondary and micronutrients such as Zn, Fe, Mn, Cu, B and Mo in soil. Application of such nutrients is essential for sustaining higher yields and quality of produce. The beneficial effects of micronutrients in okra are reported widely in literature. In the semi-arid region of Rajasthan, foliar application of zinc up to 30 kg $ZnSO_4$/ha and boron up to 20 kg borax/ha was found to significantly improve the yield attributes and yield of okra (Kumar and Sen 2005). Application of manganese (Mn—0.3% to 0.6%), magnesium (Mg—0.5% to 1.0%) and zinc (Zn—0.25% to 0.5%) as foliar spray at 40 and 60 days after sowing (DAS) increased the fruit yield in okra (Surendra et al. 2006). In the semi-arid zone of Uttar Pradesh, application of Zn, boron (B) and molybdenum (Mo) significantly increased the yield of okra (Singh et al. 2009). In B-deficient acid soils of Uttarakhand and West Bengal, soil application of boron (0.3%) 50% as basal along with Di Ammonium Phosphate (DAP) and the rest in two equal splits at 30 and 45 days after emergence is recommended for getting optimum yields of okra (Srivastava et al. 2009). Severe deficiency of Zn, Fe, Mn, Cu, B and Mo requires soil application at 25, 50, 25, 25, 2, & 2 kg/ha of zinc sulphate, ferrous sulphate, manganese sulphate, copper sulphate, sodium borate and ammonium molybdate, respectively. Foliar spray of solution of 0.2% zinc sulphate, 0.3% ferrous sulphate, 0.2% manganese sulphate, 0.2% copper sulphate, 0.1% sodium borate and 0.1% ammonium molybdate at 40–50 DAP is beneficial if crop suffers from visual deficiency symptoms of the corresponding micronutrients. Foliar spray should be avoided during bright sun shine hours between 11 and 15 h to prevent scorching of tender young leaves.

The crop benefits from the application of the farmyard manure not only from the amount of nitrogen, phosphorus, potassium and other nutrients that it contains but also from its improving effect on the tilth and the moisture holding (retaining) capacity of the soils. Application of farmyard manure also meets the secondary and micronutrient needs of the crop. Therefore, well-decomposed FYM at 20–25 t/ha should be uniformly applied in the field during land preparation. Several workers have reported the increase in okra yield due to the combined use of organic manures such as vermicompost's and non-edible oil cakes in a 50:50 ratio with inorganic fertilisers (Usha Kumara et al. 1999; Bharadiya et al. 2007).

The biofertilisers, namely, *Azotobacter*, *Azospirillum* and Phosphobacteria, can also help in reducing the dose of N and P. Phosphorus-solublising microorganisms not only solublise inorganic phosphate present in the soil and/or added through either crop residues or organic manures but also mineralise fixed phosphorous. Thus, phosphate solubilising bacteria inoculants play an important role in increasing P availability to plants. The use of *Azotobacter* and/or *Azospirillum* as organic fertiliser in okra increases the yield of okra by 6–14% and brings saving in the nitrogenous fertiliser by 10–15% (Mishra and Patjosji 1995; Ganeshe et al. 1998). Specific plant beneficial microbes may be systematically deployed as part of an integrated crop fertilisation management strategy. The results from a large number of experiments on manures and fertilisers reveal that neither chemical fertilisers alone nor organic sources used exclusively can sustain the productivity of soils under highly intensive cropping systems (Appireddy et al. 2008). Integration of various sources of fertilisers is more appropriate because this not only reduces the use of inorganic fertiliser but is also an environment-friendly approach. A number of studies on INM conducted in okra have demonstrated the beneficial effects of integrated use of inorganic fertilisers, organic manures and biofertilisers on the yield of okra. Better growth and yield of okra with the combined use of NPK, vermicompost and micronutrient (Zn) was reported in the northwestern plain region. In

West Bengal, Phonglosa et al. (2015) observed that application of phosphate solubilising bacteria (PSB) and FYM to supplement 75% of N, P and total K produced the highest okra yield and the best economic return. Similarly, in Orissa, maximum yield and fruits per plant of okra could be obtained by the application of 75% NPK + FYM at 10 t/ha + Vermicompost at 5 t/ha + biofertilisers (*Azotobacter*, *Azospirillum* and PSB) (Mal et al. 2013).

17.12 Weed Management

Weeds in okra fields compete with the crop for nutrients, moisture, space, and light and often serve as alternate hosts for several insect pests and diseases. *Solanum nigrum* is a host for fruit and shoot borer (*Leucinodes orbonalis*). *Ageratum* spp. and *Lantana* spp. provide shelter to white fly, which transmits the yellow mosaic virus. Both during the spring–summer and kharif seasons, weeds start growing in okra fields at different times, right from the sowing of okra till its final harvesting. The reduction in yield in okra due to weeds has been estimated in the range of 10–80% under different agro-climatic conditions.

Wider spacing, frequent irrigations/rains and liberal use of manures and fertilisers provide favourable conditions for an early growth of the weeds, well before the crop emergence. It creates a growth margin in favour of weeds and, if unchecked, causes serious harm to the crop. Early weed competition usually reduces crop yields more than late-season weed growth. The most critical period of crop-weed competition is the first 4–7 weeks after sowing, when the crop must be kept free from weeds.

Understanding the type of weed flora in the okra crop and devising suitable control measures is of prime importance. Like in other vegetable crops, manual and mechanical methods are most commonly practiced for weed control. Herbicide options are very limited in okra, unlike cereal crops. It may be kept in mind that the weed control strategy should not depend heavily on a single method but rather an integration of several measures, starting from preventive, cultural, mechanical, chemical and integrated, may be applied.

Depending upon the agro-ecological zone, growing season, cropping system, cultivation practices, etc., different kinds of weeds infest the okra crop. For effective weed management, it is essential to have knowledge of weed flora infesting the crop. The dominant weeds found in okra fields are *Setaria glauca, Commelina benghalensis, Ageratum conyzoides, Physalis minima, Amaranthus viridis* and *Cyperus rotundus,* recorded in abundance in the Madhya Pradesh region (Bhalla and Parmar 1986). In Andhra Pradesh, occurrence of *C. rotundus* along with *Cynodon dactylon, Panicum* sp., *Dactyloctenium aegyptium, Cleome viscosa, Tridax procumbens, Euphorbia hirta* and *Acanthospermum hispidum* (Basha and Reddy 2001; Reddy et al. 2001). In the Akola region of Maharashtra, the major weeds recorded in the okra crop were *Parthenium hysterophorus, Commelina benghalensis, Digera arvensis, Amaranthus viridis, Dinebra retroflexa, Cynodon dactylon and Cyperus rotundus* (Jadhao et al. 2001). At Anand in Gujarat state, *Eleusine indica* was the most dominant weed, followed by *Digitaria sanguinialis, Dactyloctenium aegyptium,*

Phyllanthus niruri, Boerhavia repanda, Digera arvensis, Cyperus iria and *Cyperus rotundus* (Sharma and Patel 2009).

Pulling-out of weeds by hand or removal by hand trowel, hand hoe, spades, etc. is the oldest and still widely practiced and most efficient method for reducing weed growth in okra crops. It is very effective against annual and biennial weeds. This method is also useful for the destruction of weeds growing within the rows where they cannot be controlled by cultivation. At least two hand weedings are required till the crop canopy covers the soil surface. Animal-drawn three-tyne cultivator (*Triphali*) is a useful implement for weeding and tilling operations in okra. A tractor-mounted cultivator consisting of nine spring tynes with narrow reversible shovels fitted to a tractor tool bar cultivates three okra rows at a time.

Mulching is also an effective practice for reducing the growth of annual weeds in okra. It has a smothering effect on weeds and also provides an effective barrier to weed emergence. In addition, mulching conserves soil moisture, resulting in the quick emergence of okra. Mulching can be done with dry straw, grasses, and other plants, as well as plastic sheets and polythene films. Organic mulch should be thick enough to prevent transmission of light that will suppress photosynthesis by weeds.

Effective weed control can also be achieved through herbicides, as it is quicker and much less labourious; hence, large areas can be covered in a short period of time with limited labour. Application of herbicides reduces the number of weeds to one in summer and two during kharif season. Chemical weed control in the seed production programme minimises the spread of contagious viruses by omitting some of the cultural operations during the crop growth. Several herbicides have been tested for the weed management in okra, which may be used depending on their availability, cost, weed flora and the cropping pattern followed. The important herbicides for okra are pendimethalin (0.75–1.0 kg a.i./ha), fluchloralin (1.0–1.5 kg a.i./ha), alachlor or nitrofen (1–1.5 kg a.i./ha), metribuzin (0.7–1.0 kg a.i./ha) and Prometryn (1.6 kg a.i./ha). Pre-sowing incorporation of 2.5-litre fluchloralin controlled weeds effectively for 20 days after sowing okra, as reported by Kumar et al. (1986). Pendimethalin 30 EC at 3 L/ha as post sowing, but pre-emergence soil surface spray with a flat fan nozzle followed by one-hand weeding at 30 days after sowing was the best to control weeds to get higher yield and economic returns (Singh et al. 2010). Herbicide alachlor 1.0–1.5 kg/ha immediately after sowing was found to be as effective for controlling weeds in okra as manual weeding, whereas in *kharif* season, the addition of one hand weeding with alachlor applied as pre-emergence was found essential for getting the higher yield (Singh 1978). Prometryn 1.6 kg/ha as a pre-emergence application was the most effective herbicide for weed control in okra. Post-emergence application of quizalofop-ethyl (75 g/ha), followed by one hand weeding 30 days after sowing, significantly reduced the plant population. Pre-emergence application of oxyflurofen at 0.15 kg a.i./ha + one hand weeding 30 days after sowing effectively controlled the weed population and influenced the crop growth attributes such as plant height, fruit yield and leaf area index.

Growing a competitive crop of okra with the appropriate agronomic practices for the rapid growth of okra plants, practicing proper crop rotation, summer cultivation and combining cultural and chemical methods of weed control are some of the practices that can be integrated to reduce the weed infestation in the okra crop. Clean cultivation, keeping bunds, channels, etc., free from weeds and destroying weeds before planting produce the best results.

17.13 Intercultural Operations

The purpose of intercultural operations is (a) to remove weeds, (b) to incorporate the supplemental N so that growth of the crop is boosted and (c) to ensure full canopy cover is attained early. The first 30 days after planting are critical for crop weed competition. The intercultural operations should be started about 25 days after sowing. After weeding, interculturing or hoeing may be done, and after hoeing, the field may be left for a day or two to provide aeration to the roots of plants. The top dressing of N should be applied between the rows. The crop is then earthed up, and the field is irrigated after a day or two. For ratoon crop from quick branching varieties, pruning the spring–summer crop 20–25 cm above the ground after summer fruiting with the onset of monsoon rains, followed by manuring the crop and plant protection cover, ensures good establishment of the crop. For the ratoon crop, 125 kg N, 75 kg P_2O_5 and 60 kg K_2O/ha are applied. Basal dressing should be done immediately after pruning, followed by irrigation and plant protection measures.

17.14 Water Management

An optimal supply of water is the key factor in the production of the okra crop. Both oversupply and deficiency of water are harmful to the crop. The water requirement of the crop has been found to vary from 450 to 550 mm depending on soil type, crop duration, variety and climatic conditions. The yield and quality of the okra crop depend upon a proper balance between soil air and soil moisture, which can be achieved by adopting a scientific water management programme.

The seed should be sown when the soil has sufficient moisture. Pre-planting irrigation is usually given for land preparation, and one-light irrigations are also given to ensure quick plant emergence and establishment. Irrigation should be given at the initiation of the first true leaf during the spring–summer season and at its expansion during the kharif season.

Knowledge of critical stages of water stress is essential for judicious use of irrigation water. In okra, flowering and fruit initiation/setting stages are the most critical. An inadequate supply of water during these stages affects the yield (70%) more than the water deficit at other stages. Therefore, plants should be irrigated during critical stages of growth. Irrigation according to the critical stages of water stress would therefore help in improving water use efficiency and economy. In okra, flowering and fruit setting are continuous processes that start from 40 days after sowing and continue up to 80–100 days. Therefore, irrigation needs to be applied at a 5–6-day interval during the summer and as and when required during the kharif season, depending on rain.

The irrigation method used in the okra crop must ensure adequate and uniform application of water with minimum losses from deep percolation and runoff. When okra crop is sown on ridges, furrow irrigation is a method for irrigating the crop. For efficient irrigation, the furrows should have a uniform and gentle slope down their length. The length of furrows varies with the water intake rate of soil and percentage slope. In furrow irrigation, water should never overflow the ridges, as this seals the surface soil pores, thereby raising its bulk density and leading to anaerobic conditions. The ridges may be wetted only up to about two-thirds of height, leaving the rest to be moistened gradually by capillary movement of water. A proper balance between soil water and soil air in the root zone is essential for proper growth and development. However, the okra crop is also sown in flat beds, where flood irrigation methods are commonly practiced. For efficient irrigation, the field should have a uniform and gentle slope down its length. To achieve this, land is carefully levelled and graded with plankers while land preparation. Field channels of about 70–100 cm in width serve bilaterally in each bed.

Irrigation efficiency under the flood irrigation method is lower than furrow irrigation. However, even in furrows, irrigation efficiency seldom exceeds 50%. Moreover, a lot of water percolates down in the channels, particularly when their number is large. To increase irrigation water use efficiency, modern irrigation methods such as sprinklers and drip irrigation may be used. About 20–50% saving in irrigation water is observed under drip irrigation systems over conventional methods, with a 20–30% increase in yield. The increase is contributed by the increase in number and size of fruits. Ideally, the quantity of water applied at each irrigation should be just sufficient to restore the soil moisture in the root zone to field capacity. The significant influence of plastic mulch with drip irrigation on yield was obtained over the corresponding level of irrigation under drip alone. Drip irrigation coupled with plastic mulching increases the okra crop yield significantly over furrow irrigation to the tune of 45–72%. Under fertigation, 50% N and K of the recommended dose through WSF or even 75% of RDF through conventional fertiliser applied as fertigation gives yield equivalent to 100% RDF through soil application. Fertigation produced a 42.4% higher yield (Padmanabha et al. 2018). Drip irrigation coupled with polythene mulch, fertigation and fertigation along with black silver polythene mulch have been found to increase the okra yield significantly over ridge and furrow irrigation and also save water to the tune of 50–65% and nitrogen fertiliser by 20–25%. As reported by different research workers throughout the world, drip fertigation in okra saves 20–60% water, increases yield by 13–76%, saves fertiliser from 15% to 30% and results in high use efficiency from 30.5% to 50.8% compared with traditional.

There have been different approaches to scheduling the irrigation in the okra crop. Under flood irrigation or ridge and furrow irrigation, the depth of irrigation water should be around 5–6 cm, depending on the method of irrigation used and other related factors. In sandy and loamy sand soils, irrigation at 15–25 mm CPE or 0.3 bar tension gave the highest yield, while on clay loam soils, irrigation at 40 mm CPE gave good yields. The irrigation at IW/CPE ratio 1:1 produced a significantly higher yield of okra on loam soils. Maximum number of fruits, fruit length, fruit weight and fruit yield have been recorded with fertigation daily at 100% of CPE (Venkadeswaran et al. 2014).

Use of paddy straw or organic mulch should be encouraged, as it is reported to increase irrigation water use efficiency by 15–25%. Further, the crop should be raised with the adoption of optimal cultural practices with optimum levels of other inputs like manure and fertilisers, herbicides, etc., as they improve the water use efficiency of the crop. Under conditions of water scarcity, 38% saving in irrigation water was achieved by applying water to alternate furrows in turns, at the expense of a 10.7% decrease in yield. This practice seems promising where irrigation water is a major limiting factor. There should be proper facilities of drainage in the field, particularly in heavy-textured soils, in order to avoid excess water either due to rainfall or over irrigation. Excess water causes physiological wilting to the okra plants. Water management, besides influencing the yield of okra, has also special significance in pest management. Adequate soil moisture helps in reducing the attack of cutworms and mites. The moist soil conditions decrease the mobility of cutworms and thus prevent losses from their attack.

The current trend in advanced irrigation technique is to switch from manual to automatic-operated systems in drip irrigation with an appropriate approach to irrigation and fertigation scheduling. An automated system is integrated with various types of sensors and electronic controllers and uses microclimate data to schedule irrigation water. The less amount of water loss due to infiltration, seepage, evaporation and overwetting of the plant root zone takes place in sensor-based automated drip fertigation to enhance water savings.

17.15 Harvesting

The greatest increase in fruit weight, length and diameter occurs during the 4th to 6th day after pollination. Fibre formation starts from the 5th to 6th day of fruit formation. The time of harvesting for the production of maximum yield of edible fruit is 7–8 days after pollination, 6–7 days after fruit set or 6th day of flowering onwards. Harvesting of fruits between 4 and 8 days of flowering was ideal in terms of fresh weight length, free from thick pericarp, maximum ascorbic acid and phenol content, thus ideal for consumption. This may depend on cultivar used and management practices followed. Fruits harvested the 8th day after flowering or 7 days after fruit set become fibrous with prominent seeds and are not suitable for consumption. The fruits harvested at marketable maturity vary in terms of nutritive value, sensory value and crude fibre content. On the other hand, optimum seed yield and seed quality in terms of germination percentage and vigour index were obtained when fruits were harvested 38 days after flowering.

The tender young fruits 7–10 cm long should be harvested. The fruits are 9–15 g by weight at the marketable stage and take 40–50 days from sowing to reach marketable maturity, depending on variety, season and other cultural factors. However, small-sized green tender, 6–8 cm long, 4–5 ridged, slender and roundish with pointed tips fruits get a premium price in the market and are ideal for export purposes. There are 16–25 fruits per plant; however, in hybrids, sometime it may go up to 35–40 fruits/plant. Picking interval can also influence the yield of okra since it is a quick-growing crop (Talukder et al. 2003). Earlier picking depresses yields because of low fruit weight, but delayed picking depresses marketable yields because overaged fruits become fibrous (Charles 1977). Okra yield was gradually decreased with increasing picking interval, and the highest pod yield was obtained at a 2-day picking interval. Short picking intervals encouraged the plant to produce a higher number of fruits per plant. The total number of fruits per plant increased with increases in picking frequency (Talukdar et al. 2003). Therefore, the picking should be done every second day to get tender fruits. Besides, frequent picking promotes fruit development, and hence yield is increased.

Harvesting is invariably done manually. Harvesting of okra is a laborious process, and the hair-like bristles on the fruit cause irritation to the fingers. The tender fruits are harvested by bending the pedicle with a jerk or with the help of a knife. The use of cotton cloth gloves to protect fingers and harvesting in the morning when hairs are soft is the best practice for nearby markets. For distant markets, the fruit should be harvested in the late evening and transported in a cool condition at night. The fruit should be harvested in plastic crates. This will prevent the fruits from getting bruised while harvesting and grading. Mechanical harvesting with tractor-drawn robotic harvesters/pickers may be used in large fields. Proper care of the produce has to be taken while harvesting; otherwise, many pests and diseases infect the fruits. Harvesting is scheduled depending upon the market situation also.

17.16 Cropping Systems

Okra, in India, is cultivated in almost all the states under very diverse conditions. This crop has special significance since it can be successfully cultivated throughout India in the summer and rainy seasons and during the winter season in southern and western parts of the country where the temperature is relatively congenial to its cultivation. Besides, okra is grown as a cash crop and gives high net returns in a short period as it gives high yields/unit area/unit time. Okra is a short-duration crop and highly responsive to fertiliser and irrigation. There is flexibility in planting and harvesting time (depending on choice of varieties), and therefore it is suitable for inclusion in multiple, relay and intercropping systems. It can be easily included in various cropping systems prevalent in different regions of the country to increase the productivity, income and resource use efficiency of these cropping systems. The growers of western and southern regions of the country start their cultivation in October as an off-season-early crop to get higher prices. It has compatibility of cultivation with other crops, and picking/harvesting of fruits/pods can also be started at pre-mature stages. Spring–summer season okra gives economic yields even at 75–90 days after sowing, while kharif season varieties require 90–100 days. With the availability of different duration varieties, the time of planting and harvesting of okra can be easily adjusted as per the need of other crops in sequences or as intercrop. For example, to fit in a cropping system, okra varieties like Kashi Kranti and

VRO-6 harvesting can be completed in 75 days after sowing as it gives economic yields even at this short duration, while the seed crop of these varieties requires 120 days before harvesting. Thus, okra can play an important role in intensifying the cropping system through sequential and relay-intercropping systems that utilise the resources efficiently and produce high yields to foster food and nutritional security and health for all. Several studies have shown that inclusion of okra in cropping systems is usually more profitable than cereal-based cropping systems.

17.16.1 Intercropping Systems

Intercropping provides insurance against low returns and weather vagaries and gives stability to the system. The inter- and intra-row spacing of the okra crop can be adjusted without sacrificing much yield for successful intercropping with cowpea, amaranth, maize, onion, etc. Intercropping of okra with the above-mentioned crops at marginally extra cost and with a minor loss in yield gives increased return and stability to the system than the sole crop of these crops. Sugarcane is an important crop in the northern plains and is planted either in the autumn or spring season. The short duration and fast growth of okra make it an ideal intercrop in sugarcane/maize during the spring season. The weed population is also considerably less when intercropping with okra. As regards yield, the total productivity of the intercropping system was considerably 15–20% higher than the individual sole crops. Short-duration varieties of okra should be preferred for the intercropping system.

17.17 Conclusion

At present, the per capita availability of okra in the country is approximately 5.388 kg/head/year. The projected target of okra by 2050 would be 8.082 mt at the present level of per capita availability. To achieve this target, we need to make a concerted effort at all fronts of okra development. This includes the development of cultivars suitable for every niche and pocket of the region, making available good-quality seed to every grower, and following improved agrotechniques and plant protection measures in okra cultivation. Besides, for sustaining the growth of vegetables in general and okra in particular, we need substantial growth in refrigerated pack houses, centres for perishable cargoes, integrated post-harvest handling facilities, state-of-the-art cold chain infrastructure and quality assurance measures and cold chain transportation facilities and processing. There is a need to develop strong market intelligence based on information technology. It is the responsibility of every stakeholder to do their bit for the development of okra in the region. In recent times, apart from large investments pumped in by the private sector, the public sector has also taken the initiative with APEDA for setting up several of these facilities in the country.

References

Absar N, Siddique MA (1982) Influence of plant density on the yield of three varieties of okra. Bangladesh J Agric 7:15–21

Akande MO, Oluwatoyinbo FJ, Adediran JA, Buari KW, Yusuf IO (2003) Soil amendments affect the release of P from rock phosphate and the development and yield of okra. J Veg Crop Prod 9:3–9

Albregts BB, Howard CM (1974) Response of okra to plant Density and Fertilization. Hort Sci 9(4):400

Albregts BB, Howard CM (1976) Response of four cultivars on mulched beds to plant density. Hort Sci 11(3):242

Anderson WH, Carolus RI, Watson DP (1960) The germination of Okra seed as influenced by treatment with acetone and alcohol. Proc Am Soc Hort Sci 62:427–432

Anonymous (1978) Package of practices for kharif crops of Himachal Pradesh, Himachal Pradesh Krishi Vishwavidyalaya, Palampur

Anonymous (1983) Package of practices for kharif crops of Punjab. Punjab Agric Univ, Ludhiana

Anonymous (2012) Okra. (*Abelmoschus esculentus*) http://growingtaste.Com/vegetables/okra.shtml

Anonymous, (2021). https://agricoop.nic.in › sites › default › files › PDF Okra/Ladyfinger 2020–21 (Final)

Appireddy GK, Saha S, Mina BL, Kundu S, Selvakumar G, Gupta HS (2008) Effect of organic manures and integrated nutrient management on yield potential of bell pepper (Capsicum annuum) varieties and on soil properties. Arch Agron Soil Sci 54(2):127–137

Basha GK, Reddy KB (2001) Integrated weed management in summer irrigated okra (Abelmoschus esculentus L. Moench). Madras Agric J 88(10–12):678–682

Benchasri S (2012) Okra [*Abelmoschus escusentus* (L.) Moench] a vegetable of the world. RatarPovrt 49:105–112

Bhalla LP, Parmar RP (1986) Effectiveness of pre-emergence herbicides on weed control and seed yield of okra (Abelmochus esculentus (L) Moench). Seed Farms 12(2):36–43

Bharadiya PS, Kalalbandi BM, Shinde VN (2007) Effect of inorganic and organic fertilizers on growth and yield of okra. Asian J Hort 2:199–201

Bussan AJ, Mitchel PD, Copas ME, Drilias MJ (2007) Evaluation of the effect of density on potato yield and tuber size distribution. Crop Sci 47:2462–2472

Charles WB (1977) Okra Growing, vol 28. Extension Bulletin. University of the West Indies, p 9

Chauhan DVS (1972) Vegetable production in India, 3rd edn. Ram Prasad and Sons, Agra

Dademal AA, Dongale JH (2004) Effect of manures and fertilizers on growth and yield of okra and nutrient availability in latertic soil of Konkan. J Soil Crop 14(2):278–283

Ganeshe RK, Pandey RP, Rawat AK (1998) Comparative performance of biofertilizers and chemical nitrogen on yield of okra. Hort J 11:53–58

Hazara P, Chattopadhya A, Karmakar K, Duta S (2011) Modern technology in vegetable production. New India Publishing Agency, New Delhi

Hossain MO, Salam MA, Islam MS, Masud MAT (1999) Yield and quality of Okra (BARI Dherosh-1) seed as influenced by time of sowing and plant spacing. Bangladesh J Seed Sci Technol 3(1–2):83–87

Ibeawuchi II, Obiefuna JC, Ofoli MC (2005) Effects of row spacing on yield and yield components of okra (Abelmoschus esculentus) and mixture groundnut (Arachis hypogaea). J Agron 4(4):304–307

Jadhao BJ, Patil BM, Karunakar AP, Joshi PS, Mahorkar VK (2001) Weed management in seed crop of okra (Abelmoschus esculentus L.). J Soil Crop 11(1):106–108

Kamalanathan S, Sunderajan S, Thamburaj S (1970) Studies on optimum spacing and manuring of okra (Abelmoschus esculentus (L.) Moench). Madras Agric J 57:10–17

Kaur H, Chawla N, Bassia G, Pathak M (2015) Effect of different seed priming treatments on germination of okra (Abelmoschus esculentus L.). Int J Curr Sci 15:51–58

Khan AR, Jaiswal RC (1998) Effect of nitrogen, plant spacing and green fruit pickings in the seed production of okra (Abelmoschus esculentus (L.) Moench). Vegetable Sci 15(1):8–14
Kumar JC, Singh C, Singh C (1986) Evaluation of recommended herbicide at the farmer's fields in okra (Abelmoschus esculentus L. Moench). J Res-PAU 23(1):44–48
Kumar M, Sen NL (2005) Effect of zinc, boron and gibberellic acid on yield of okra Abelmoschus esculentus L. Indian J of Hort 62(3):308–309
Lamont W (1999) Okra – a versatile vegetable crop. Hort Tech 2:350–352
Magagula P, Ossam E (2011) Effect of seed size on seedling vigor of okra (Abelmoschus esculentus L.) in Swaziland. Adv Environ Biol 5:180–187
Mal B, Mahapatra P, Mohanty S, Mishra HN (2013) Growth and yield parameters of okra (Abelmoschus esculentus) influenced by Diazotrophs and chemical fertilizers. J Crop Weed 9(2):109–112
Mishra M, Patjoshi AK (1995) Effect of biofertilization on Okra (Abelmoschus esculentus L.) Environ Ecol 13:732–733
Moniruzzaman M, Uddin MZ, Chaudhury AK (2007) Response of okra seed crop to sowing time and plant spacing in south eastern hilly region of Bangladesh. Bangladesh J Agril Res 32(3):393–402
Moterran J, Badley GA, Bowden HL (1963) Production clemoson spindan okra. Arkans Em Res 12(2):10
Nath P (1976) Vegetables for the tropical region. ICAR, New Delhi
Nonnecke IL (1989) Vegetable production. Van Nostrand Reinhold AVI Publishing, pp 608–609
Onwueme IC (1975) Enhancement of okra (Hibiscus esculentus) seedling growth and emergence by heat stress conditions. J Agric Sci 85:233–239
Padmanabha K, Lingaiah HB, Jayappa J, Anjanappa M, Anilkumar S, Hanumanthappa DC (2018) Effect of fertigation in okra (Abelmoschus esculentus L.). Trends Biosci 11(13):2353–2355
Phonglosa A, Bhattacharyya K, Ray K, Mandal J, Pari A, Banerjee H, Chattopadhyay A (2015) Integrated nutrient management for okra in an inceptisol of eastern India and yield modeling through artificial neural network. Sci Hortic 187:1–9
Randhawa GS, Panna MS (1969) The effects of row spacing and level of nitrogen on the growth and yield of okra (Abelmoschus esculentus (L.) Moench). J Res Punjab Agric Univ Ludhiana 9(6):320–324
Rastogie KB, Sharma PP, Singh NP, Korla BN (1987) Effect of different levels of nitrogen and plant spacing on seed yield of okra (Abelmoshus esculentus). J Farming Syst 2(3–4):13–15
Ravindran DLV, Rao R, Reddy N (1986) Effect of spacing and nitrogen levels on growth, flowering and yield of African marigold. South Indian Hort 345:320–323
Reddy MD, Reddy CN, Devi MP (2001) Integrated weed management in okra (Hibiscus esculentus). Indian J Weed Sci 33(3–4):217–219
Rizvi IK, Jugirdar SAP (1973) Effect of irrigation interval, spacing and planting methods on the yield of lady's finger. Agric Pak 24(2):151–157
Saikia AK, Bhagabiti KN (1994) In: Rishi N, Ahuja KL, Singh BP (eds) Virology in the tropics. Malhotra publishing House, New Delhi, pp 303–308
Sarma B, Gogoi N (2015) Germination and seedling growth of Okra (Abelmoschus esculentus L.) as influenced by organic amendments. Cogent Food Agric 1:1030906
Sharma S, Patel BD (2009) Weed management in okra grown in kharif season under middle Gujarat conditions. Indian J Weed Sci 43(3&4):226–227
Sharma AD, Rathore SVS, Srinivasana K, Tyagi RK (2014) Comparison of various seed priming methods for seed germination, seedling vigour and fruit yield in okra (Abelmoschus esculentus L. Moench). Sci Hortic 165:75–81
Singh BP (1978) Comparative efficiency and selectivity of herbicides in okra. Indian J Weed Sci 10(1&2):63–71
Singh HB, Bhagchandani PM (1967) Bhindi cultivation in India. ICAR, New Delhi
Singh M, Prabhukumar S, Sairam CV (2010)Integrated Weed Management in Okra (Abelmoschus esculentus (L.) Moench). Ann Plant Prot Sci 18(2):481–483

Singh DP, Dabbas MR, Prakash HG (2009) Effect of micro nutrients on growth, yield attributes and pod yield of okra (Abelmoschus esculentus (L.) Moench) in semi-arid zone of Uttar Pradesh. Asian J Hort 4(2):488–490
Srivastava PC, Singh MV, Tyagi AK, Pachauri SP (2009) Proceedings of the International Plant Nutrition Colloquium XVI. UC Davis. http://escholarship.org/uc/item/91k19559
Sundaraj DD, Balasubramanian G, Soundrapandia G (1965)Standard of germination in Abelmoschus Esculentus Linn.(Bhendai or Okra). Madras Agric J 52:10–12
Sundaram V, Kanthaswamy V (2005) Response of okra to foliar feeding of water soluble fertilizers. Veg Sci 32(1):92–93
Surendra P, Nawalagatti CM, Chetti MB, Hiremath SM (2006) Effect of plant growth regulators and micronutrients on yield and yield components in okra. Karnataka J Agric Sci 19(2):264–267
Taj K, Gardner B (1968) A spacing and fertilizer trials with okra. Poe Trop Reg Amer Soc Sci 12:39–46
Talukder MAH, Munnaf MA, Alam MK, Salam MA, Amin MMU (2003) Influence of Sowing Time, Plant Spacing and Picking Interval. Pak J Biol Sci 6(18):1626–1630
Thakur MR, Arora SK (1986) In: Bose TK, Som MG (eds) Vegetable crops in India. Naya Parkash, 200, Bidhan Sarani, pp 613–622
Tindall HD (1983) Vegetables in the tropics, vol 33. McMillan AVI, pp 325–327
Ushakumari K, Prabhakumari P, Padmaja P (1999) Efficiency of vermicompost on growth and yield of summer crop okra (Abelmoschus esculentus L.). J Trop Agric 37:87–88
Venkadeswaran E, Sundaram V, Sankar R (2014) Influence of trickle fertigation on yield and quality of hybrid okra [Abelmoschus esculentus (L.) Moench]. Asian J. Hort 9(2):285–290

18 Optimization of Seed Yield and Quality in Okra (*Abelmoschus esculentus* L. Moench)

B. S. Tomar, V. R. Yalmalle, and Paresh Chaukhande

18.1 Introduction

Okra (*Abelmoschus esculentus L.* Moench) is one of the most important vegetables of enormous economic value to farmers in the tropical and subtropical regions of the world. For centuries, okra cultivation has been practiced as a garden and large-scale commercial crop. Cultivated okra is an African-originated crop, and due to its polyploidy nature, wide variability is seen around the various tropical and subtropical parts (Zeven and Zhukovsky 1975). Okra is cultivated commercially in India, Turkey, Iran, Western Africa, Yugoslavia, Bangladesh, Afghanistan, Pakistan, Burma, Japan, Malaysia, Brazil, Ghana, Ethiopia, Cyprus, and the Southern United States (Fig. 18.1). According to FAOSTAT (2022), the global okra production is 10.5 million tonnes with a major share from Asia (68.9%), and India is the leading okra-producing nation with more than 6350 thousand tonnes of production (more than 60% share in the global production) and a productivity level of 12.0 t/ha (Fig. 18.1).

According to NHB 2018–19, the leading five okra-producing states in India (Fig. 18.2) are Gujarat (921 thousand tonnes), followed by West Bengal (914 thousand tonnes), Bihar (787 thousand tonnes), Madhya Pradesh (638 thousand tonnes), and Odisha (566 thousand tonnes). In North India, *bhendi* is mainly cultivated during the spring-summer (March–June) and Kharif (July–September) seasons, while in temperate regions, it is cultivated during the summer. Besides, okra is also cultivated as a winter crop in Gujarat, Andhra Pradesh, Karnataka, and Tamil Nadu.

B. S. Tomar (✉) · V. R. Yalmalle · P. Chaukhande
ICAR-Indian Agricultural Research Institute, New Delhi, India

S. K. Tikoo et al. (eds.), *Okra: Status, Challenges and Opportunities*,
https://doi.org/10.1007/978-981-97-9963-3_18

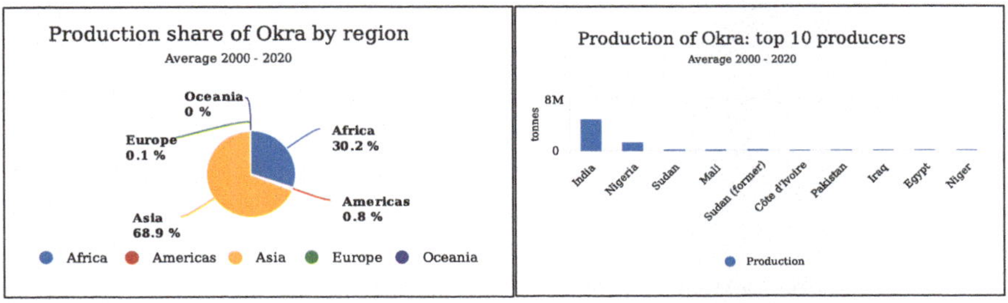

(A) (B)

Fig. 18.1 World production share of okra (**a**) and top 10 okra-producing countries (**b**) in two decades (2000–2020) as per FAOSTAT 2020

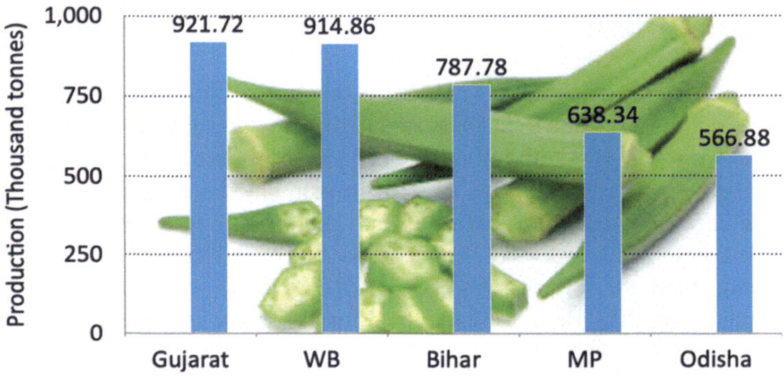

Fig. 18.2 Top five okra-producing Indian states

18.2 Nutritional Profile of Okra

Okra has nutritional and medicinal value (Elkhalifa et al. 2021). The tender green pod contains oxalic acid, thiamine, riboflavin, nicotinic acid, and vitamins A, B, and C. The calcium content in okra is very high (around 66 mg per 100 g) as compared to other vegetables. Okra is an excellent source of iodine necessary for resistance to throat diseases such as goiter. Besides, okra is also suggested for people suffering from heart failure. Okra ripe seeds are roasted, ground, and used as a substitute for coffee in some countries. Mature pods and stems containing crude fiber are used in the paper industry. Nowadays, attention has been paid to the use of okra seeds as a source of proteins (about 20% of dry matter) and vegetable oil (about 14% of dry matter). The greenish-yellow edible oil has a pleasant taste and smell and is rich in monounsaturated fatty acids such as oleic acid and palmitic acid (Martin and Rhodes 1983) and high lysine levels (Al-Wandawi 1983). In India, okra or *bhendi* is cultivated for its tender green fruits during spring-summer and in the rainy season and utilized by people in diverse ways; immature pods are consumed in the form of boiled vegetables; also dried and used as soup thickeners or in stews, which are rich sources of vitamins, calcium, potassium, and other minerals.

18.3 Floral Biology of Okra

Okra flowers are solitary, axillary with epicalyx (up to 10). There are 5 yellow petals with crimson-spotted claws (Fig. 18.3). The staminal column consists of numerous stamens united at the base of petals. Stigma is 5–9 lobed. Botanically, okra fruit is a capsule. Dehiscence usually begins in the morning hours (8–10 am), and the flower remains open for a shorter duration and withers in the afternoon. Okra being protogynous (stigma mature first than anthers) in nature affects the pollination process at the bud stage. Okra is a self-pollinated crop; however, cross-pollination up to 12–19% occurs, and hence okra is categorized as an often-cross-pollinated crop.

18.4 The Technique of Hybridization

Emasculation is easy in okra, and buds likely to open the next day are given a slight ring cut at the base of the flower bud with the help of a blade. Petals along with calyx sheath are removed, and staminal column and stigma are exposed. Undehisced anthers are removed with the pair of forceps, and emasculated buds are bagged. On the next day, bags are removed, and the emasculated buds are pollinated with the

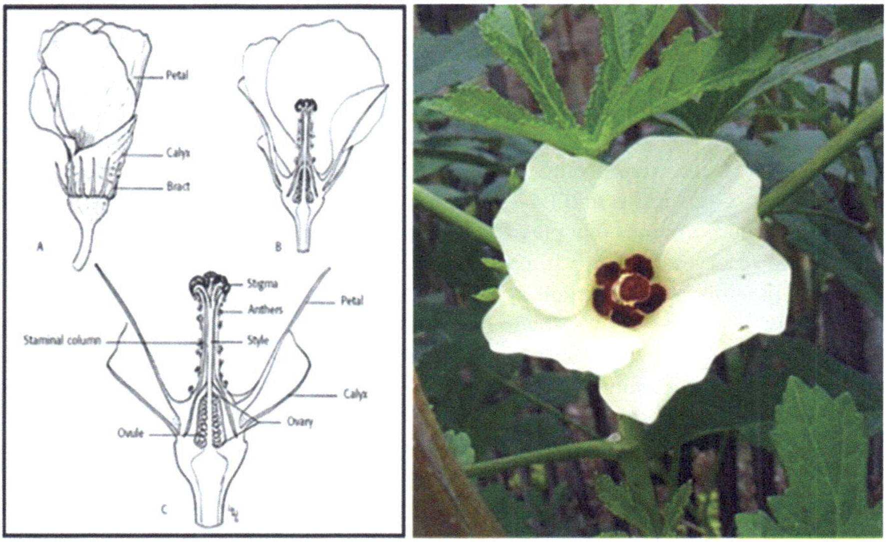

Fig. 18.3 Okra floral morphology

help of male flowers that were bagged a day earlier. The pollinated flowers are bagged and labeled. After a few days, the bags can be removed. The hybridization process can continue for up to 8 weeks depending upon the okra's genetic and other environmental factors. Pollination time and the ratio between female and male flower crossing are to be optimized to get an increased seed set and yield in the okra female parent. For hybrid seed production, a planting ratio of 8:1 (female:male parents) is considered appropriate. Under optimum management practices, a seed yield of 1–1.2 tonnes/ha is expected.

18.5 Node Position and Fruit Retention Affect Seed Quality

The node position and fruit retention have considerable influence on seed yield and quality. Hedau et al. (2010) studied the seed quality from the fruits harvested from different nodes: lower (1st–3rd) and middle (4th–8th). Maximum seed germination and seed quality were obtained from the fruits harvested from lower and middle fruits. Lowest seed quality. The node position and fruit retention have considerable influence on seed yield and quality. Recommended the retention of fruits borne on 1–3 nodes for seed purposes and harvesting of remaining fruits for vegetable purposes as a superior combination for higher seed quality and higher economic returns, whereas Jakhar and recommended retention of only middle nodes (6–10 nodes) for higher seed quality.

18.6 Use of Plant Growth Regulators (PGRs) in Quality Okra Seed Production

Plant growth regulators (PGRs) are organic compounds used in small quantities that can alter the physiological processes of plants. PGRs use has gained a separate field of study besides varietal, manurial, and cultural methods in vegetable improvement. Among the several growth substances, gibberellic acid (GA_3) and indole-3-butyric acid (IBA) are found very promising, and these are used in vegetable crops for various purposes. Gibberellic acid (GA_3) or gibberellins comprise a group of naturally occurring plant hormones that play a key role in early seed germination by mobilizing storage reserves and activating enzyme production. Furthermore, gibberellins foliar application stimulates and synchronizes flowering and fruit sets as well as enhances photosynthesis and growth (Yuan and Xu 2001). Okra seeds often exhibit physical dormancy due to the hardness present in the seed, which complicates crop management due to nonuniform germination after sowing, resulting in a lack of crop uniformity (Mohammadi et al. 2011). On the flip side, Pal and Hossain (2001) found that the GA_3 effect was lacking in okra seed germination and seed weight, but a significant effect of this hormone was found on plant height and the number of pods per plant. The positive effect of GA_3 on plant height has been widely reported in okra (Ilias et al. 2007). Additionally, foliar application of GA_3 (20–40 ppm) and naphthalene acetic acid (NAA; 50–100 ppm) on okra was found to increase the number of pods per plant and the number and weight of seeds (Abduljabbar et al. 2007). Besides, Dhankhar and Singh (2009) noticed that with GA_3 application, the okra pod length and diameter have increased. These results could be attributed mainly to the stimulatory effect of GA_3 on plant growth, which resulted in higher rates of biosynthesis and therefore higher amounts of assimilates available for distribution to the pods. Further investigation by various workers on PGR studies in okra is presented in Table 18.1.

Table 18.1 Use of growth regulators for yield enhancement in okra

Growth regulator	Mode and dosage	Morphological trait improvement	References
Cycocel (CCC)	Foliar (400 ppm)	50% flowering (34.33 days), nodes per plant (30.83), picking per plant (15.66), and pod yield per hectare (132.32 quintals) in Pusa Sawani	Maulana and Bahadur (2020)
Gibberellic acid (GA_3)	Foliar (100 ppm)	Pod and seed yield (3.12 t/ha)	
	Foliar (150 ppm)	Improved different growth and yield attributes of okra, such as fruit girth, fruit length, number of pods per plant, pod length, pod fresh and dry weight, seed yield, and number of seeds per fruit	
	Foliar (50–100 ppm)	Plant growth, pod, and seed characteristics (number of seeds per pod)	
	Foliar (100 ppm)	Number of leaves & branches (18.13 & 3.48), plant height (104.17 cm), number of fruit per plant (23.39), fruit fresh weight (17.95 g) and fruit yield (130.88q/ha), 1000 seed weight (78.49 g) chlorophyll-a (1.84), chlorophyll-b (0.81), zinc content (30.81 ppm), boron (22.55 ppm), copper (18.24 ppm), and germination (85.48%)	
	Foliar (200 ppm + 200 ppm naphthalene acetic acid [NAA])	Number of pods per plant, pod length, pod fresh and dry weight, seed yield, and seed quality (germination percentage and 1000-seed weight)	
Indole acetic acid (IAA)	Foliar (200 ppm)	Pod yield (214 q/ha)	

18.7 Hard Seeds in Okra

The hard seed coat is one of the major problems in okra, which results in slow and erratic germination. To combat the problem of hardheadedness, the growers are forced to use a high seed rate. Hard seededness is influenced by several factors, such as time of harvest, seed moisture content, and drying rate during seed development (Hosamani et al. 2012). Purquerio et al. reported the difference in the onset of hard seededness in angular pods and cylindrical pods and that angular pods (45 days after anthesis [DAA]) must be harvested earlier than cylindrical pods (45–55 DAA), and delay beyond this period will lead to poor seed quality. The physical dormancy imposed on impervious hard seed coats can be broken by several treatments. Hot water treatment is an eco-friendly and farmer-friend on-field approach for breaking physical dormancy in okra. Demir (2001), treated okra seeds at 50 °C for 48 h to reduce hardheadedness in seeds harvested having different maturity levels. Seeds harvested at 55 days after anthesis (DAA) had 84% hard seeds, which was reduced to 4% in treated seeds. Velempini et al. (2003), recommend hot water treatment of wild okra (*Corchorus olitorius*) seeds at 80 °C for 10 min or use of concentrated H_2SO_4 for 5 min for breaking physical seed dormancy. Mohammadi et al. (2012) treated hard seeds with different dry and wet heat treatments. The results show that among hot water treatments, 75 °C for 5 min was effective, whereas among dry heat treatments, 100 °C for 24 h resulted in higher seed germination. Hedau et al. (2010) found an inverse relationship with respect to dead and hard seed percentage; seeds from farther fruits and mature fruits of the turning stage produced more dead and hard seeds.

18.8 Seed Quality Enhancement in Okra

Seed quality enhancement through various priming treatments has been undertaken in okra. Seed priming in okra is essential because okra seeds have problems of hard seeds and low field emergence. Seed priming is one of the important seed quality enhancement techniques. Seed priming is a controlled hydration treatment that results in enhanced field emergence and increases tolerance to various biotic and abiotic stresses. Different seed priming techniques include hydro priming, osmo priming, halo priming, and solid matrix priming. The different studies on seed quality enhancement in okra have been summarized in Table 18.2.

18.9 Other Challenges

Seed is the prime factor that determines the quantitative and qualitative characteristics of the crop that is going to be harvested later on. Nowadays only 20% of quality seed has been produced in India, and it should be increased to 45% by managing other key inputs (Seednet India 2020). Therefore, more attention should be directed

Table 18.2 Treatments for breaking physical dormancy in okra

S.no	Treatment	Duration/ chemical	Traits improved	References
1.	Hydropriming	Water for 24 h	Seed germination, seedling vigor, mean germination time, and marketable fruit yield	Sharma et al. (2014)
2.	Hydropriming	Priming with hot water at 50 °C for 5 min	Seed germination, seedling vigor	Tania et al. (2020)
3.	Solid matrix priming	Calcium aluminum silicate (1:0.4:1; Seed:SM: Water) for 24°	Seed germination, seedling vigor, mean germination time, and marketable fruit yield	Sharma et al. (2014)
4.	Solid matrix priming	Sand matricpriming (3 h in 60% WHC of sand)	Seed germination, seedling vigor	Kuppusamy and Ranganathan (2014)
5.	Hormonal	100 mg\L salicylic acid for 4 h at 25 °C	Seed germination percentage, germination speed index (GSI), and seedling vigor index (SVI),	Hussein (2015)
6	Hormonal	Seed priming with 200 ppm GA3 solution for 24 h	Seed germination, seed vigor index, mean germination rate	Lamichhane et al. (2021)
7	Osmopriming	Soaking of seeds in polyethylene glycol (−1 MPa) solution for 7 days	Increase in plant height and 30% increased yield	Maiti et al. (2011)

toward increasing seed yield with good quality. Being a profitable enterprise, successful okra seed production is conditional on certain agricultural practices. The seed of okra possesses distinct properties or values that can be harnessed in diversified ways. For successful crop production in okra, innovative technologies may help to reap the benefits of achieving higher yield and quality seed production. The implementation of decision support system (DSS) tools in okra has been reported for irrigation (El-Gafy and El-Ganzori 2012), nutrient management (Phonglosa et al. 2015), and yield prediction (Ojesanmi et al. 2016), which has opened new vistas of crop management. Furthermore, the DSS tools should be inclined toward field level that can help farmers implement in okra production.

Okra seed production is affected by several diseases, pests, and other environmental stresses. The crucial disease that hinders successful okra production is the yellow vein mosaic virus (YVMV), which can cause yield loss of up to 100% (Mishra et al. 2017). Okra seed yield was highly influenced by YVMV incidence, and a higher yield was obtained during a disease-free period (Chattopadhyay et al. 2011). Studies have also been conducted on different sowing dates (Fig. 18.4) on okra for various yield-attributing traits in okra (Abduljabbar et al. 2007).

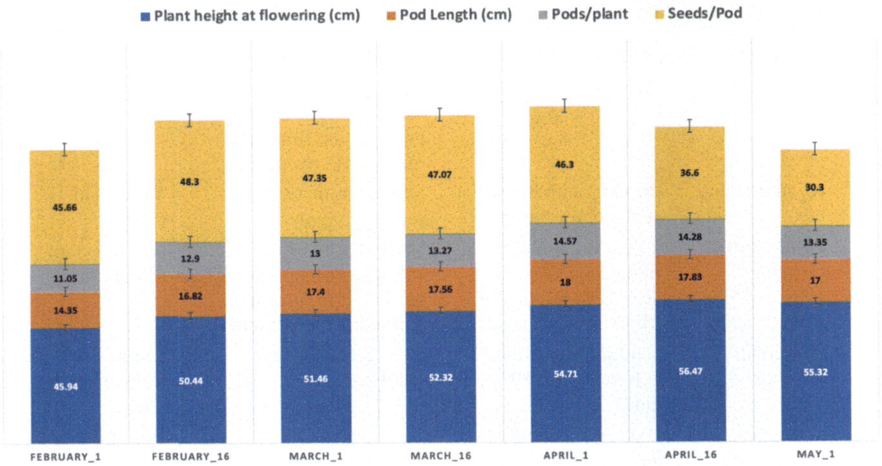

Fig. 18.4 Pooled 4 years (2006–10) data shows the different sowing dates effect on okra for yield-attributing traits (Chattopadhyay et al. 2011)

18.10 Conclusion

Seed production of okra requires a lot of skill and consistent efforts in raising a successful crop. Proper planning by sourcing authenticated/pure seeds along with following the field standards for okra seed producers can reap a quality seed harvest.

References

Abduljabbar IM, Abduljabbar I, Shukri HS (2007) Effect of sowing date, topping, and some growth regulators on growth, pod, and seeds yield of okra (Abelmoschus esculentus L.M.). In: African crop science conference proceedings, vol 8, pp 473–478

Al-Wandawi H (1983) Chemical composition of seeds of two okra cultivars. J Agric Food Chem 31(6):1355–1358

Bairwa HL, Shukla AK, Mahawer LN, Kaushik RA, Shukla KB, Ameta KD (2009) Response of integrated nutrient management on yield, quality and physico-chemical characteristics of okra cv Arka Anamika. Indian J Hortic 66(3):310–314

Chattopadhyay A, Dutta S, Chatterjee S (2011) Seed yield and quality of okra as influenced by sowing dates. Afr J Biotechnol 10(28):5461–5467. https://doi.org/10.5897/AJB10.979

Demir İ (2001) The effects of heat treatment on hardseededness of serially harvested okra seed lots at optimum and low temperatures. Sci Hortic 89(1):1–7. https://doi.org/10.1016/S0304-4238(00)00216-8

Dhankhar BS, Singh R (2009) Okra hand book, global production, processing and crop improvement. HNB Publishing, p 475

El-Gafy I, El-Ganzori A (2012) Decision support system for economic value of irrigation water. Appl Water Sci 2(2):63–76

Elkhalifa A, Alshammari E, Adnan M, Alcantara JC, Awadelkareem AM, Eltoum NE, Mehmood K, Panda BP, Ashraf SA (2021) Okra (*Abelmoschus esculentus*) as a potential dietary medi-

cine with nutraceutical importance for sustainable health applications. Molecules (Basel, Switzerland) 26(3):696. https://doi.org/10.3390/molecules26030696
Food and Agriculture Organization of the United Nations (2022) FAOSTAT statistical database. FAO, Rome. https://www.fao.org/faostat/en/#data/QCL/visualize
Hedau NK, Singh G, Mahajan V, Singh SRK, Gahlain A (2010) Seed quality and vigour in relation to nodal position and harvesting stage of okra under mid hills of North-Western Himalayas. Indian J Hortic 67(4):251–253
Hosamani J, Pandita VK, Tomar BS (2012) Seed development and acquisition of desiccation tolerance during maturation of okra seed. Indian J Hortic 69(3):353–359
Hussein HJ (2015) Effect of seed priming treatment with salicylic acid on viability of okra (Abelmoschus esculentus L.) seeds. Euphrates J Agric Sci 7(2):1–9
Ilias I, Ouzounidou G, Giannakoula A, Papadopoulou P (2007) Effects of gibberellic acid and prohexadione-calcium on growth, chlorophyll fluorescence and quality of okra plant. Biol Plant 51(3):575–578. https://doi.org/10.1007/s10535-007-0126-5
Kuppusamy N, Ranganathan U (2014) Storage potential of primed seeds of okra (*Abelmoschus esculentus*) and beet root (*Beta vulgaris*). Aust J Crop Sci 8(9):1290–1297. https://doi.org/10.3316/informit.675827684140779
Lamichhane A, Mamata KC, Shrestha M, Baral B (2021) Effect of seed priming on germination of okra (*Abelmoschus esculentus* var. Arka Anamika). Malaysian J Sustain Agric 5(2):111–114. https://doi.org/10.26480/mjsa.02.2021.111.114
Maiti RK, Vidyasagar P, Rajkumar D, Ramaswamy A, Gonzalez Rodriguez H (2011) IJbSM 2(1) seed priming improves seedling vigor and yield of few vegetable crops. Int J Bio-Resource Stress Manag 2(1):125–130
Martin FW, Rhodes AM (1983) Seed characteristics of okra and related *Abelmoschus* species. Plant Foods Hum Nutr 33(1):41–49
Maulana S, Bahadur VPV (2020) Effect of different levels of cycocel (CCC) on two different cultivars of okra under Prayagraj agro-climatic conditions (*Abelmoschus esculentus* L.). IJCS 8(4):133–136
Mishra GP, Singh B, Seth T, Singh AK, Halder J, Krishnan N et al (2017) Biotechnological advancements and begomovirus management in okra (*Abelmoschus esculentus* L.): status and perspectives. Front Plant Sci 8:360
Mohammadi G, Khah EM, Bannayan AM (2011) Differential responses for harvesting times and storage on hardness of different varieties of okra. Notulae Scientia Biologicae 3(4):117–122
Mohammadi G, Khah E, Shirkhani A (2012) Effects of seed hardness breaking techniques on Okra (Abelmoschus esculentus L.). Int J Agric Crop Sci 4(6):264–273
National Horticultural Board (NHB) Databases 2018–19. https://www.nhb.gov.in/StatisticsViewer.aspx?enc=CaSOU6CTRN/Ty9tOdH8Lkgol2QGWss6940QLGSpy2T0m6tZbw4/y0R84pftyqIGN
Ojesanmi OA, Adekoya AD, Awoseyi AA (2016) Adaptive neuro-fuzzy inference system for okra yield prediction. J Nat Sci Eng Technol 15(2):95–102
Pal AK, Hossain M (2001) Effect of seed soaking on growth, pod yield, and seed yield in okra (Abelmoschus esculentus (L.) Moench.). Hortic J 14:61–65
Phonglosa A, Bhattacharyya K, Ray K, Mandal J, Pari A, Banerjee H, Chattopadhyay A (2015) Integrated nutrient management for okra in an inceptisol of eastern India and yield modeling through artificial neural network. Sci Hortic 187:1–9
Seednet India Portal (2020). https://seednet.gov.in/material/IndianSeedSector.htm
Sharma AD, Rathore SVS, Srinivasan K, Tyagi RK (2014) Comparison of various seed priming methods for seed germination, seedling vigour and fruit yield in okra (*Abelmoschus esculentus* L. Moench). Sci Hortic 165:75–81. https://doi.org/10.1016/j.scienta.2013.10.044
Tania SS, Rhaman MS, Hossain MM (2020) Hydro-priming and halo-priming improve seed germination, yield and yield contributing characters of okra (*Abelmoschus esculentus* L.). Trop Plant Res 7(1):86–93. https://doi.org/10.22271/tpr.2020.v7.i1.012

Velempini P, Riddoch I, Batisani N (2003) Seed treatments for enhancing germination of wild okra (*Corchorus olitorius*). Exp Agric 39(4):441–447. https://doi.org/10.1017/S0014479703001388
Yuan L, Xu DQ (2001) Stimulation effect of gibberellic acid short term treatment on leaf photosynthesis related to the increase in Rubisco content in broad bean and soybean. Photosynth Res 68:39–47. https://doi.org/10.1023/A:1011894912421
Zeven AC, Zhukovsky PM (1975) Dictionary of cultivated plants and their centers of diversity. Centre of Agricultural Publishing and Documentation, Wageningen, p 219

Challenges in Seed Quality Assurance in Okra

19

Vinod K. Pandita and Jagadish Hosamani

Quality seed is the basic input for commercial agriculture production systems. Assessment of seed quality is an important mission of seed regulatory agencies, and based on the reports of seed testing laboratories, the commercial seed of okra (*Abelmoschus esculentus* L. Moench) varieties and hybrids is distributed to the farmers. In India, seed testing rules/protocols are available in the Seed Testing Manual (Chalam et al. 1967) and Handbook of Seed and Planting Material Testing Manual for Horticultural Crops. In the United States and Canada, seed-testing rules have been developed and standardized by the Association of Official Seed Analysts (AOSA) and in other countries around the world by the International Seed Testing Association (ISTA). To meet the demands of new technology and new crops, the rules and procedures are re-examined and revised annually. For example, the development of seed-coating technology has led to changes and additions to the "Rules for Testing Seeds" as published by the AOSA (1983) and the "International Rules for Seed Testing" published by the ISTA (1985).

Testing for genetic purity, physical purity, and germinability is done while evaluating the quality of seeds. This information is also provided on the seed label along with the name of the variety or cultivar. The vigor test results are not part of the regulatory labeling requirement under the Indian Seed Act. However, vigor tests are extensively used by some commercial seed-producing agencies as a part of in-house quality control. Seed quality assurance (QA) is the combination of efforts and activities undertaken by public and private seed-producing agencies to ensure that the

V. K. Pandita (✉)
ICAR-Indian Agricultural Research Institute, Regional Station, Karnal, India

J. Hosamani
UAS, Dharwad, Dharwad, India

Department of Seed Science & Technology, College of Agriculture, Vijayapur, India

S. K. Tikoo et al. (eds.), *Okra: Status, Challenges and Opportunities*,
https://doi.org/10.1007/978-981-97-9963-3_19

seed produced conforms to minimum seed quality standards. Generally, two types of seed quality control, i.e., internal and external seed quality control measures, are undertaken. Internal control is carried out by the internal seed quality control professionals/experts of the seed-producing agencies. External control is undertaken by the concerned state seed certification agency. The inspection by the seed certification agency is carried out during the production and processing stages.

It takes a lot of time and investment to develop new okra varieties or hybrids with desirable traits like high productivity, resistance to biotic and abiotic stresses, or unique quality traits. Only when farmers have access to high-quality seed of these improved varieties and hybrids at reasonable prices, the results of such investments are realized. The system of seed quality assurance in okra also relies on testing the main four components of seed quality, namely, genetic purity, physical purity, germination percentage, and freedom from seed-borne pathogens of designated diseases. Hence, the seed quality assurance in okra varieties/hybrids is based on seed quality control measures as envisaged in the Indian Minimum Seed Certification Standards. The seed quality control in okra is performed in the following steps based on field (Tables 19.1 and 19.2) and seed (Table 19.3) standards:

A. *Field Standards*

Table 19.1 Isolation distance requirements for each class of seeds

	Minimum distance				
	IMSCS		AOSCA		
Contaminants	FS	CS	FS	RS	CS
Field of other varieties	500 m	250 m	1320 ft (402 m)	1320 ft (402 m)	825 ft (251 m)
Field of same variety not conforming to varietal purity, wild okra *A. ficulneus*, *A. manihot*, *A. moschatus*	500 m	250 m	–	–	–

[a]*FS* Foundation seed, *CS* Certified seed, *RS* Registered seed, *IMSCS* Indian Minimum Seed Certification Standards, *AOSCA* Association of Official Seed Certifying Agencies

Table 19.2 Specific requirements (permissible off-types)

	Minimum permitted (%)			
	IMSCS		AOSCA	
Factor	FS	CS	FS	CS
Off-type				
Other varieties: definite	0.10%	0.20%		
	(1:1000)	(1:500)	None	0.08%
Other varieties: doubtful				(1:1250)
			None	0.2%
				(1:500)
Objectionable weed plant	None	None	None	None

B. *Seed Standards*

Table 19.3 Seed standards for each class of seeds

	Standard for each class			
	IMSCS		AOSCA	
Factor	FS	CS	FS	CS
Pure seed (min.)	99%	99%	99%	98%
Inert matter (max.)	1.0%	1.0%	1.0%	2.0%
Other crop seed (max.)	None	None	None	None
Total weed seed	None	None	None	None
Objectionable weed seed (max.)	None	None	None	None
Other distinguishable variable (max.)	10/kg	20/kg	0.2%	10.5%
Germination including hard seed (min.)	65%	65%	65%	65%
Moisture (max.)	10%	10%		
For vapor-proof containers (max.)	8%	8%		

For people working in seed certification programs, quality assurance means being sure that the seed is inspected for genetic variability in the field, quality testing in the laboratory, and after certification is complete through grow-out tests, it is fit for marketing. For seed analysts, it means conducting tests to assure trueness-to-type, freedom from contamination, and the ability to produce normal, healthy plants in the field. Several state seed certification agencies have set up specialized programs for quality assurance and are providing this service to the seed producers. Thus, quality assurance means different to different people, but the ultimate aim is to supply high-quality seed to the farmers as per the standards laid out (IMSCS 2013). Some of the challenges encountered in quality assurance and control measures are discussed below:

19.1 Maintenance of Genetic Purity of Varieties and Hybrids

Most quality assurance programs begin during seed production in the field. The use of high-quality, genetically pure seed, proper isolation distances and roguing operations, along with proper seeding rates, row spacing, fertility levels, and irrigation scheduling, is essential in producing high-quality seed. Absence of obligate autogamy in okra continues to impede progress in both the maintenance of genetic stocks and the breeding of new varieties. According to reports, the extent of natural outcrossing in okra varies from 0% to 60% depending on the genotype, cropping season, pollinator population, and region (Chandra and Bhatnagar 1975; Martin 1983; Tanda 1985). The contamination of genotypes during evaluation and seed multiplication will result in progressive adulteration (loss of varietal distinctions). Maintaining varietal purity is therefore crucial for any successful okra seed production program. The production of genetically pure breeder seed from the variety or paternal line is the first step in the seed production chain. If the breeder seed is not highly pure, the contaminants become multiplied by several times in subsequent generations of the production of foundation and certified seeds. These off-types may even cause the variety or hybrid to lose all of its improved traits. Therefore, the variety development organizations must implement an appropriate systematic maintenance breeding program for okra (Fig. 19.1).

Before conditioning and marketing the seed, the seed quality assurance program should be carried out. A seedling grow-out is one of the often-used genetic purity tests. This evaluation is used routinely for vegetable crops. Many seed companies now employ electrophoretic fingerprinting in addition to grow-out tests. These assessments will become more essential for a quality assurance program as cultivars become more specific in the environmental conditions they require for growth. The official measure for ensuring the seed lot's genetic purity is the grow-out test. It acts as both a "post-control" test and a "pre-control" test to prevent genetic contamination. It is a requirement for seed certification of Foundation seeds and all hybrids, according to Indian statutory regulations. According to the regulations of the Seeds Act of 1966, the test must be conducted to verify the information on the seller's label regarding the genetic purity status of the seed lot. Grow-out tests could also be employed as a yardstick for assessing the effectiveness of the certification agency or

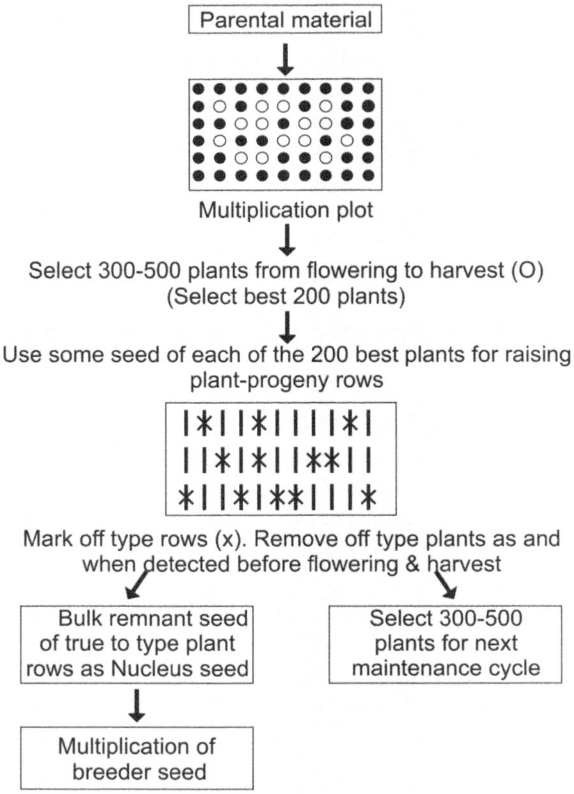

Fig. 19.1 Variety maintenance for nucleus seed production in okra (rest/reserve seed method)

Table 19.4 The size of the submitted sample

Size of the sample (g)	Crops
1000	Maize, cotton, groundnut, soybean, and species of other genera with seeds of similar size

Table 19.5 Specifications for the grow-out test

Crop	Row length (m)	Plant-to-plant distance (cm)	Space between rows (cm)	Space between plots (cm)	No. of replications
Okra	5	10	45	45	2

Table 19.6 Minimum standards for genetic purity in okra

Class of seed	Percentage
Foundation	99.00
Certified	
Varieties	98.00
Hybrids	90.00

the inspector. In accordance with the recommended sampling procedures, the samples for the grow-out test must be drawn at the same time as the samples for the other seed quality tests (Tables 19.4, 19.5, and 19.6).

19.2 Occurrence and Development of Hard Seeds

The hard seed coat dormancy present in okra is a major physiological constraint to proper stand establishment in the field. Several studies reported an interrelation between seed moisture content and hard seededness, and the percentage of hard seeds showed a progressive decline with increasing moisture content. Ellis et al. (1985) reported little or no hard seededness at 13% but a considerable amount at 4–6% moisture content. It is seen that seed producers often leave the seeds in pods until they dry completely to below 10% moisture level, resulting in a higher proportion of hard seeds in the seed lot. According to Hosamani et al. (2012), maximum hard seeds were detected when the color of the seed coat changed from white to gray at 40 days after anthesis (DAA) with the advancement of seed development in okra (Table 19.7). The moisture content of seeds dramatically decreased as seed development advanced and hard seeds started to occur. Maximum hard seeds were found in freshly developed and rapidly dried seeds when seed moisture content was minimum at 40 DAA (Fig. 19.2). Another significant contributing factor to the slow and erratic germination of okra is the presence of a hard seed coat (Demir 1997; Marsh 1993; Standifer et al. 1989). It is important that the seed analyst should identify and report the presence of hard seeds in the seed lot for taking corrective measures.

Table 19.7 Seed coat color of freshly harvested seeds during seed development and its relationship with occurrence of hard seededness in okra

Days after anthesis	Seed coat colour	Percentage	Hard seed (%)
10	White group 155	100	0
17	White group 155A	100	0
24	White group 155D	95	0
	Yellow-white group 158B	5	
28	Yellow-white group 158B	78	8.75
	Grey-yellow group 161B	22	
32	Grey group 200C	100	11.75
36	Grey group 200C	100	14.75
40	Grey group 200C	100	20.0

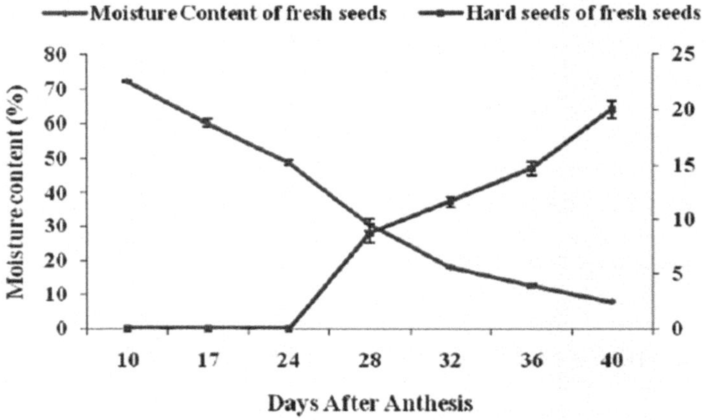

Fig. 19.2 Relationship between seed moisture content and occurrence of hard seeds in okra

19.3 Improvement in Planting Value of Seed Through Seed Enhancement and Processing

Proper and uniform stand establishment is essential for better plant growth and yield in vegetable crops. The major problem with okra growers is difficulty in obtaining proper seedling emergence under field conditions. Okra seeds require high temperatures for germination, and poor germination is common under low temperatures during early spring-summer sown crops. Okra seed germinates in soil at a temperature between 21 and 35 °C, with 29 °C being the optimal temperature (Lorenz and Maynard 1980). Okra seed germination can be erratic due to environmental conditions such as inadequate and excessive soil moisture, low soil temperature, and pre-emergence damping off caused by *Pythium* spp. (Conway et al. 2001).

Several seed enhancement treatments have been used to increase percentage and uniformity of emergence in many vegetable crops (Nagarajan et al. 2005; Pandita and Nagarajan 2000; Parera and Cantiliffe 1994; Pill 2020). The solid matrix priming (SMP) approach developed by Taylor (1990) and Eastin (1999) is one of several priming techniques that can be used to improve the rate and uniformity of germination in a wide range of crops (Khan 1992; Lin and Sung 2001; Pandita et al. 2007). In order to control the effects of water, oxygen, and temperature on germination, this procedure combines solid matrix materials, water, and seed. Hydration of the seed is limited to a level that permits "pre-germination" activity but prevents the emergence of radicles (Harman and Taylor 1988). The matric potential of the seed regulates water uptake (Khan et al. 1992b; Kubik et al. 1988). The properties of the solid matrix system reduce aeration problems and facilitate the incorporation of fungicides (Baird 1994; Harman and Taylor 1988; Khan et al. 1992a) or biological agents (Conway et al. 2001; Harman et al. 1989).

Solid matrix priming in combination with *Trichoderma viridae* improved field emergence by 19% in conditions of suboptimal temperatures and by 11% in optimal temperatures compared with the non-primed control (Table 19.8). Final marketable pod yield was improved by SMP alone or in combination with *Trichoderma viridae* under suboptimal temperatures but not at optimal temperatures. In either environment, priming had no impact on the quantity of pods per plant or pod yield per plant. The findings imply that SMP in combination with *Trichoderma viridae* can be utilized effectively to improve seedling emergence and productivity of okra under low temperatures (Hosamani et al. 2020; Pandita et al. 2010). Low and erratic emergence of okra seed is a major problem under low temperatures. Improper plant stand establishment leads to low okra productivity, and to achieve a proper plant stand during early spring plantings under northern-plain conditions, the okra growers prefer to use a high seed rate to achieve good plant stands. The use of solid matrix priming technology can be safely used to improve uniformity of field emergence in okra under suboptimal temperatures.

Table 19.8 Effect of solid matrix priming in combination with bio-agents and fungicides on laboratory germination and field performance in okra seeds

Treatment	Germination (%)	Speed of germination	Field emergence (%)	Field emergence index	Seedling ht. (cm)	Seedling fresh wt (g)	Seedling dry wt (g)
SMP	76.3^{b}	62.4^{c}	65.0^{b}	85.1^{a}	6.5^{c}	3.48^{c}	0.50^{c}
SMP + TV	85.3^{a}	79.8^{a}	73.0^{a}	85.6^{a}	6.9^{b}	3.67^{a}	0.62^{a}
TV + SMP	70.3^{d}	52.2^{d}	50.6^{d}	72.0^{bc}	7.6^{a}	3.42^{d}	0.41^{e}
SMP + TH	83.0^{a}	71.0^{b}	64.0^{bc}	77.0^{ab}	5.8^{d}	3.59^{b}	0.59^{b}
TH + SMP	74.3^{bc}	54.8^{d}	48.6^{d}	65.4^{cd}	5.9^{d}	3.42^{d}	0.46^{d}
SMP + T	74.3^{c}	63.1^{c}	58.0^{c}	77.9^{ab}	5.9^{d}	2.73^{e}	0.24^{f}
SMP + C	72.3^{cd}	60.1^{c}	48.6^{d}	67.3^{cd}	5.4^{e}	1.74^{f}	0.24^{f}
Control	66.6^{e}	34.1^{e}	40.0^{e}	60.0^{d}	5.4^{e}	1.59^{g}	0.20^{g}
C.D ($p = 0.05$)	3.01	4.49	6.42	8.08	0.118	0.013	0.009

$^{a\text{-}g}$Mean in the columns having same superscript letters are not significantly different

Seed conditioning is an essential step in making seeds of genetically superior cultivars available to crop producers. Some objectives of seed conditioning are to remove contaminants, upgrade quality, improve plantability, apply seed treatment, and package the seed. Okra seeds are extracted after the hand-harvested pods become dry and brittle. The most efficient method of hand seed extraction is to twist the pods open. Alternatively, the pods are either flailed or the seeds are extracted with a stationery thresher.

19.4 Development of Appropriate Vigor and Viability Tests

Vegetable seed quality is also important due to their high price resulting from high quality control, maintenance, and high production cost. Due to the high cost of hybrid seed in okra, it is important that the seed lot have the highest percentage of germination and vigor. In order to perfectly align crop production schedules to market demand, vegetable growers are showing increasing interest in laboratory-based seed tests that can aid in predicting emergence. Farmers commonly use the standard germination test to assess the quality of their seed and determine the quantity of seed to be sown. Standard germination tests may be used to predict seedling emergence in the field under optimal field conditions. Vigor tests offer additional information because field conditions are often suboptimal.

Many seed companies use one or more vigor tests to predict seed planting value in their quality control system. Many vigor tests for different crop species have been proposed (AOSA 1983; ISTA 1993, 1995, 1999, 2006, 2013), but accelerated aging (Hampton and TeKrony 1995) and controlled deterioration (Powell and Matthews 2005) have been successfully used to predict the field emergence potential of seed lots of various crops such as onions, carrots (Matthews 1980), and melons (Mavi and Demir 2007). A number of vigor tests have been used to assess the vigor status of seed lots, but no single test has yet been identified for rating the seed vigor across the crops. Pandita et al. (2014) demonstrated that controlled deterioration and paper-piercing tests can be successfully used to predict seedling emergence in okra seed lots (Fig. 19.3).

The germination test is the preferred technique for evaluating the physiological quality of seeds. However, this test has limits for okra seeds, especially given the relatively long 21-day period needed to gather data and make conclusions (Anonymous 2009). Specifically for species with dormancy or slower germination, like okra seeds, the tetrazolium test has been widely utilized in quality control programs to assess seed viability in a quick and efficient manner (Anonymous 2009; Marcos Filho 2015). However, a number of factors, including solution concentration, temperature, seed exposure time, and seed preparation before staining, interfere with performing a tetrazolium test (De Paiva et al. 2017). According to Souza et al. (2018), okra seeds can be tested with a 0.075% tetrazolium solution for 90 min at 35 °C to determine their viability.

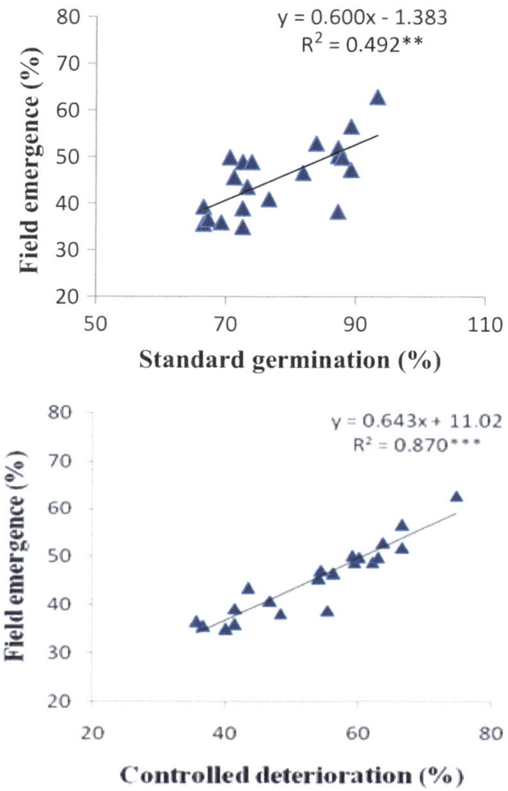

Fig. 19.3 Determination coefficient of standard germination and controlled deterioration for field emergence in okra seed lots

19.5 Detection of Seed-Borne Pathogens

Okra is one of the most significant economic crops in India, but it is also one of the most vulnerable to rot, decay, and deterioration by mycoflora, viruses, insects, and other organisms, which lowers the quality of crop seeds. With appropriate seed health testing, precautions must be taken to lower the risk of damage caused by these infections. Literature shows that numerous pathogenic and saprophytic fungi are found on okra seeds. *Alternaria alternata, Aspergillus flavus, A. niger, Chaetomium globosum, Curvularia lunata, C. pallescens, C. robusta, Drechslera hawaiiensis, D. rostrata, Fusarium moniliforme, F. oxysporum, Penicillium* sp., *Rhizopus* sp., *R. nigricans, sclerotium* sp., *and Stachybotrys atra* have been reported on okra seed. The pathogens may accompany, adhere to, or be inside the seed during the marketing of the seed and finally may lead to low crop productivity. Therefore, detection of these pathogens is important for disease risk management. For the accurate detection of many seed-borne pathogens, it is necessary to scale up the use of modern tools such as conducive environmental incubation and identification (for seed-borne fungi), liquid plating assays (for seed-borne bacteria), enzyme-linked immunosorbent assays (for seed-borne viruses), and rolling circle replication (RCR) technology.

19.6 Conclusion

Quality assurance (QA) programs require systematic sampling and testing of the seed during production, conditioning, and storage. Quality assurance and control are keys to the production and supply of certified seed to the farmers. Most of the private seed agencies are supplying truthfully labeled (TL) seed of vegetables, including okra, to the farmers, and the QA system plays an important role in building trust among the farmers for the procurement of TL seed. It is therefore important for seed testing laboratories in the public and private sectors to harmonize the testing protocols with the procedures presented by ISTA. The challenges for seed testing laboratories are to maintain the highest level of consistency, repeatability, and accuracy during the seed testing process.

References

Anonymous (2009) Rules for seed analysis. Ministry of Agriculture, Livestock and Supply, Secretary of Agricultural Defense, General Coordination of Laboratory Support, Brasília, p 395

AOSA (1983) Seed vigour testing handbook, 1st edn. Association of Official Seed Analysts, East Lasing, p 32

Baird RE (1994) Evaluation of seed treatments on *Shrunken-2* sweet corn. Plant Dis 78:817–821. https://doi.org/10.1094/PD-78-0817

Chalam GV, Singh A, Douglas JE (1967) Seed testing manual. Indian Council of Agricultural Research and United States Agency for International Development, I.M.H, Press private, New Delhi

Chandra S, Bhatnagar SP (1975) Reproductive biology of *Abelmoschus esculentus* (L.) Moench, reproductive behaviour, floral morphology, anthesis and pollination mechanism. Acta Bot Indica 3(2):104–113

Conway KE, Mereddy R, Kahn BA, Wu Y, Hallgren SW, Wu L (2001) Beneficial effects of solid matrix chemo-priming in okra. Plant Dis 85(5):535–537. https://doi.org/10.1094/pdis.2001.85.5.535

De Paiva EP, Torres SB, Nobre De Almeida JP, Da Silva Sa FV, Torquato Oliveira RR (2017) Tetrazolium test to evaluate the viability of gherkin seeds. Rev Cienc Agron 48(1):118–124. https://eurekamag.com/research/070/790/070790890.php#:~:text=10.5935/1806%2D6690.20170013

Demir I (1997) Occurrence of hardseededness in relation to seed development in okra (*Abelmoschus esculentus* (L.) Moench). Plant Var Seed 10(1):7–13

Eastin JA (1999) Solid matrix priming of seed with microorganisms and selected chemical treatment. U. S. patent no. 5,974,734

Ellis RH, Hong TD, Roberts EH (1985) Handbook of seed technology for gene banks. In: Principles and methodology, vol I. International Board for Plant Genetic Resources, Rome, p 210

Hampton JG, TeKrony DM (1995) Handbook of vigour test methods. International Seed Testing Association, Zurich

Harman GE, Taylor AG (1988) Improved seedling performance by integration of biological control agents at favorable pH levels with solid matrix priming. Phytopathology 78:520–525
Harman GE, Taylor AG, Stasz TE (1989) Combining effective strains of *Trichoderma harzianum* and solid matrix priming to improve biological seed treatments. Plant Dis 73:631–637. https://doi.org/10.1094/PD-73-0631
Hosamani J, Pandita VK, Tomar BS (2012) Seed development and acquisition of desiccation tolerance during maturation of okra seed. Indian J Hortic 69(3):353–359
Hosamani J, Pandita VK, Tomar BS (2020) Solid matrix priming and bio-agents improve seed emergence in okra [*Abelmoschus esculentus* (L.) Moench]. J Pharmacogn Phytochem 9(4):3087–3090
IMSCS (2013) Indian minimum seed certification standards, 2013. The Central Seed Certification Board, Department of Agriculture and Co-operation, Ministry of Agriculture, Government of India, New Delhi
ISTA (1985) International rules for seed testing, 1985. International Seed Testing Association, Zurich
ISTA (1993) International rules for seed testing, 1993. Seed Sci Technol 21(Suppl):141–168
ISTA (1995) In: Hampton JG, TeKrony DM (eds) Handbook for vigour test methods, 3rd edn. International Seed Testing Association, Zurich
ISTA (1999) International rules for seed testing, 1999. Seed Sci Technol 27(Suppl):27–32
ISTA (2006) International rules for seed testing, 2006. International Seed Testing Association, Zurich
ISTA (2013) Seed vigour. In: International rules for seed testing. International Seed Testing Association, Bassersdorf
Khan AA (1992) Preplant physiological seed conditioning. Horti Rev 13:131–181. https://doi.org/10.1002/9780470650509.ch4
Khan AA, Abawi GS, Maguire JD (1992a) Integrating matriconditioning and fungicidal treatment of table beet seed to improve stand establishment and yield. Crop Sci 32(1):231–237. https://doi.org/10.2135/cropsci1992.0011183X003200010047x
Khan AA, Maguire JD, Abawi GS, Ilyas S (1992b) Matriconditioning of vegetable seeds to improve stand establishment in early field plantings. J Am Soc Hortic Sci 117(1):41–47. https://doi.org/10.21273/JASHS.117.1.41
Kubik KK, Eastin JA, Eastin JD, Eskridge KM (1988) Solid matrix priming of tomato and pepper. In: Proceedings of international conference on stand establishment in horticultural crops, Lancaster, pp 86–96
Lin JM, Sung JM (2001) Pre-sowing treatments for improving emergence of bitter gourd seedlings under optimal and sub-optimal temperatures. Seed Sci Technol 29(1):39–50
Lorenz OA, Maynard DN (1980) Knott's handbook for vegetable growers, vol 144, 2, edn, Aufl. John Wiley & Sons, New York, Chichester, Brisbane, Toronto, Kartoniertes, Ringbuch, Pflanzenernaehr, Bodenk, pp 122–123. https://doi.org/10.1002/jpln.19811440118
Marcos Filho J (2015) Physiology of seeds of cultivated plants, 2nd edn. Abrates, Londrina, p 660
Marsh L (1993) Moisture affects cowpea and okra seed emergence and growth at low temperatures. Hortic Sci 28(8):774–777
Martin FW (1983) Natural outcrossing of okra in Puerto Rica. J Agric Univ Puerto Rica 67(1):50–52. https://doi.org/10.46429/jaupr.v67i1.7690
Matthews S (1980) Controlled deterioration: a new vigour test for crop seeds. In: Habblethwaite PD (ed) Seed production. Butterworths, London, pp 647–660
Mavi K, Demir I (2007) Controlled deterioration and accelerated ageing tests predict relative seedling emergence potential of melon seed lots. Hortic Sci 42(6):1431–1435. https://doi.org/10.21273/HORTSCI.42.6.1431
Nagarajan S, Pandita VK, Joshi DK, Sinha JP, Modi BS (2005) Characterization of water status in primed seeds of tomato (*Lycopersicon esculentum* Mill.) by sorption properties and NMR relaxation times. Seed Sci Res 15(2):99–111. https://doi.org/10.1079/ssr2005200
Pandita VK, Nagarajan S (2000) Osmopriming of fresh seed and its effect on accelerated ageing in Indian tomato (*Lycopersicon esculentum* Mill.) varieties. Indian J Agric Sci 70(7):479–480
Pandita VK, Anand A, Nagarajan S (2007) Enhancement of seed germination in hot pepper following presowing treatments. Seed Sci Technol 35(2):282–290

Pandita VK, Anand A, Nagarajan S, Seth R, Sinha SN (2010) Solid matrix priming improves emergence and crop performance in okra. Seed Sci Technol 38(3):665–674. https://doi.org/10.15258/sst.2010.38.3.14

Pandita VK, Patil P, Tomar BS, Seth R (2014) Controlled deterioration and paper- piercing tests predict seedling emergence potential in okra seed lots. Sci Hortic 179:21–24. https://doi.org/10.1016/j.scienta.2014.09.011

Parera CA, Cantiliffe DJ (1994) Presowing seed priming. Horti Rev 16:109–142

Pill WG (2020) Low water potential and presowing germination treatments to improve seed quality. In: Seed quality, pp 319–359. https://www.sciencegate.app/app/redirect#aHR0cHM6Ly9keC5kb2kub3JnLzEwLjEyMDEvOTc4MTAwMzA3NTIyNi0xMA

Powell AA, Matthews S (2005) Towards the validation of the controlled deterioration vigour test for small seeded vegetables, vol 129. Seed Testing International, ISTA News Bulletin, pp 21–24

Souza AAT, Nogueira NW, Torres SB, Benedito CP, Neta MLS, Paiva EP (2018) Tetrazolium test to estimate okra (*Abelmoschus esculentus* (L.) Moench) seed viability. Rev Caatinga 31(4):817–822. https://doi.org/10.1590/1983-21252018v31n403rc

Standifer LC, Wilson PW, Drummond A (1989) The effect of seed moisture content on hardseededness and germination in four cultivars of okra (*Abelmoschus esculentus* (L.) Moench). Plant Var Seed 2(3):149–154

Tanda AS (1985) Floral biology, pollen dispersal and foraging behaviour of honeybees in okra (*Abelmoschus esculentum*). J Apic Res 24(4):225–227. https://www.sciencegate.app/app/redirect#aHR0cHM6Ly9keC5kb2kub3JnLzEwLjEwODAvMDAyMTg4MzkuMTk4NS4xMTEwMDY3Nw==

Taylor AG (1990) Solid matrix priming of seed. U S patent no 4,912,874

Agronomic Interventions to Enhance Yield in Okra 20

Vijay Kumar Kori, Mahipal Sharma, Belgamwar Prashant, Arun Pandey, and Venu Gopal Reddy

20.1 Introduction

Okra (*Abelmoschus esculentus* L.), a member of the family Malvaceae, originated in the Abyssinian centre of origin of cultivated plants, an area that includes present-day Ethiopia, the mountainous portion of Eritrea, and the eastern, higher part of the Anglo-Egyptian Sudan. If we look back into the history of okra cultivation, in the 1950s, it was grown like a field crop under high-density planting with seed drills from homegrown seed. Okra is adapted to a wide range of soils. A well-drained, fertile soil with adequate organic matter content is suitable for okra cultivation. During the decade of 2000–2010, the okra cultivation moved from open-pollinated to large-scale adoption of hybrids having high resistance to Yellow Vein Mosaic Virus (YVMV) and good fruit quality. After 2010, another virus called Okra Enation Leaf Curl Virus (OELCV) has been posing significant economic yield loss under severe infestation. In 2014, Advanta Seeds (UPL Ltd.) introduced the first combined resistance hybrid for both the viruses YVMV and OELCV along with good fruit quality, which helped significantly increase the area of its cultivation with a high return on investment to the farmers.

The new generation of hybrids having high genetic potential for productivity require customized levels of cultivation practices to optimize their yield through better expression of their attributes and fruit quality. Among the entire input cost of cultivation to the farmers, seed value is one of the major components, and hence it is necessary to grow an optimum number of plant populations in a given acreage to get the highest yield potential. This is largely dependent on spacing between row-to-row and plant-to-plant. Different hybrids of okra require different sowing windows and spacing. A good hybrid sown in improper season and spacing will perform

V. K. Kori (✉) · M. Sharma · B. Prashant · A. Pandey · V. G. Reddy
Advanta Seeds UPL Ltd., Hyderabad, Telangana, India
e-mail: vijay.kori@advantaseeds.com

S. K. Tikoo et al. (eds.), *Okra: Status, Challenges and Opportunities*,
https://doi.org/10.1007/978-981-97-9963-3_20

poorly and lead to lower yields. Selection of proper plant spacing with an optimum package of practices is very crucial for exploiting the yield potential of new generation hybrids. The objective of this study was to determine the most suitable hybrid variety and its corresponding plant spacing for better yield potential in okra in Haryana state.

20.2 Materials and Methods

Three hybrids named Navya (Advanta Seeds), along with two other popular hybrids of this area referred to as Checks 1 and 2 here, were used for the experiment. The effect of sowing with different spacings, namely, 30 × 5 cm, 6,66,6666 plants/ha, 30 × 15 cm, 2,22,222 plants/ha, 60 × 30 cm, and 55,555 plants/ha, on these three hybrids was studied to know the optimum spacing and plant density in targeted geography.

The experiment was carried out in March 2020 at Advanta's Technology Development farm located at village Badshahapur of Karnal district in Haryana state, which comes under the major sowing season of okra in Haryana, that is, summer. In this area, temperatures range from 3 °C minimum to 46 °C maximum during the winter and summer seasons, respectively. Major soil type in the area is sandy loam. The experiment was laid out in a randomized block design with three replications by sowing two seeds per hole per treatment and thinning out to one seedling per hole after 15 days to keep the best seedlings for experimental studies. The observations were made for major yield-contributing traits like plant height, number of branches, internodal distance, virus pressure, and fruit yield.

20.3 Results and Discussion

The experiment resulted in a significant difference among the treatments (hybrids) with different spacing for yield level. Among 3 hybrids, namely, Navya (Advanta Seeds), Hybrid1 and Hybrid 2, referred to here as Checks 1 and 2, with 3 different spacings, namely, 60 × 30 cm, 30 × 5 cm, and 30 × 15 cm; the closer row spacing of 30 × 5 cm with 6,66,666 plants/ha has resulted in better yields when compared to the other two spacings 30 × 15 cm with 2,22,222 plants/ha and 60 × 30 cm with 55,555 plants/ha, respectively. The yields obtained are given in Table 20.1. The response obtained due to diverse treatments and spacing showed enhanced yield effect in closer spacing.

Effect of Hybrids: A considerable change in plant attributes, like plant growth and development, number of branches, health of the secondary branches, and finally yield was noticed in three experimental hybrids. Hybrid Navya from Advanta Seeds resulted in significant high yield potential, followed by Checks 1 and 2, respectively. The performance of hybrid Navya proved incremental for yield with a potential of 46.74 mt/ha, which was far superior over Check 1, which produced 42.82 mt/ha, and Check 2, with 41.14 mt/ha.

Table 20.1 Effect of treatment, spacing, and interaction on yield (t/ha)

Cultivars	Spacing			
	30 × 05 cm	30 × 15 cm	60 × 30 cm	Cultivar mean
Navya	46.74	40.79	31.38	39.64
Check 1	42.82	39.08	29.52	37.14
Check 2	41.14	37.54	29.15	35.94
Spacing mean	43.57	39.14	30.02	
C.D. for cultivars: 5%–0.887	C.D. for spacing: 5%–0.887		C.D. for interaction: 5%–1.537	

Effect of Spacing: With three different spacings, namely, 30 × 5 cm, 30 × 15 cm and 60 × 30 cm used in the experiment, spacing 30 × 5 cm resulted in the highest yield potential, clearly demonstrating that close spacing with a high density plant population of a hybrid with suitable genetics for the geography could enable growers to get the fullest yield potential and a high return on investment. Even the general mean yield in the closest spacing of 30 × 5cm was significantly higher than the other two spacings.

In conclusion, we recommend that hybrid Navya be grown at the closest spacing of 30 × 5cm for its optimum yield. It is also clear that in okra, closer spacing should be adopted for maximizing yield potential.

21 Nutritional Quality of Okra: A Global Overview

Neena Chawla, Simarjot Kaur, and Sandeep Shakya

21.1 Introduction

The okra is a treasure of multiple biochemical nutritional compounds that constitute to make it a vegetable with tremendous health benefits. It is widely distributed across Africa, Asia, Southern Europe, the Mediterranean region, and the United States of America. The per 100 g edible portion of okra comprises of 0.70 g ash, 90.17 g water, 31 Kcal energy, 3.2 g total dietary fibre, 7.03 g carbohydrate, 1.2 g total sugar, 0.40 g sucrose, 0.13 g glucose, 0.21 g fructose, 0.34 g starch, 0.10 g total lipid, 0.026 g total saturated fatty acids, 0.017 g total mono-unsaturated fatty acids, 0.027 g total polyunsaturated fatty acids, 24 mg phytosterols, 2 g protein, and minerals including calcium 81.0 mg, copper 0.1 mg, iron 0.8 mg, potassium 303 mg, magnesium 57.0 mg, manganese 1.0 mg, sodium 8.0 mg, phosphorus 63.0 mg, selenium 0.7 µg, and zinc 0.6 mg, respectively (Roy et al. 2014; USDA 2010).

Historically, okra pods have been utilized for various purposes such as food, appetite stimulants, astringents, and even as an aphrodisiac. This versatile crop, with its fresh leaves, buds, flowers, seeds, pods and stems serves multiple purposes. The unripe green pods of okra are commonly consumed as vegetables, and the extract from these pods can be used to thicken soups and sauces in various dishes (Dhaliwal 2010; Kumar et al. 2013). The okra seeds are rich in protein and oil. These can be roasted and ground to create a caffeine-free coffee alternative (Calisir and Yildiz 2005; Martin 1982; Roy et al. 2014). The crop is cultivated for economic purposes in many countries, including India, Japan, Turkey, Iran, Western Africa, Bangladesh, Pakistan, Myanmar, Afghanistan, Malaysia, Brazil, Ethiopia, Cyprus,

N. Chawla (✉)
Department of Vegetable Science, Punjab Agricultural University, Ludhiana, Punjab, India
e-mail: chawlaneena@pau.edu

S. Kaur · S. Shakya
Department of Biochemistry, Punjab Agricultural University, Ludhiana, Punjab, India

S. K. Tikoo et al. (eds.), *Okra: Status, Challenges and Opportunities*,
https://doi.org/10.1007/978-981-97-9963-3_21

Thailand, Yugoslavia, and various South American nations (Qhureshi 2007). The extensive studies conducted throughout the time determined numerous active compounds that have consistently demonstrated health benefits supported by both in vitro and in vivo research. It has recently been employed for its nutraceutical and therapeutic characteristics in addition to its nutritional benefits due to the presence of numerous significant bioactive chemicals and their related bioactivities that are becoming more desirable as functional food ingredients nowadays.

21.2 Nutritional Value of Okra

According to Wayne et al. (1984), the fruit quality of okra is determined by the length, diameter, chlorophyll, mucilage and fibre content. The okra is an excellent source of protein, vitamins, and minerals, with more than half of its composition consisting of soluble fibre like gums and pectin, which help in the reduction of serum cholesterol and lower the risk of heart disease. The rest is made up of insoluble fibre, which helps maintain a healthy digestive system (Shalau 2002) while being low in calories. Having great nutritional value, it also has high water content, averaging 85% (Kumar et al. 2010; Siemonsma and Kouame 2004). In addition to this, the glycans are found in abundance in okra that are primarily responsible for the viscosity of the aqueous suspension (Kumar et al. 2010). The protein content determined in okra is mainly traced in seeds, which is known to be quite higher than that reported in other fresh vegetables. Sami et al. (2013) reported that analysis of amino acids detected the presence of ten essential amino acids, among which aspartic acid is the predominant one found in okra-associated proteins. Although the essential amino acid content was lower than the non-essential amino acid content, the ratio of essential to non-essential amino acids was similar to that found in soybean.

The okra pods are low in calories, rich in nutrients, mucilage, and an excellent source of ingestible fibre. These consist of significant bioactive compounds like thiamine, folic acid, carotene, amino acids, riboflavin, vitamin C, oxalic acid and niacin. In pods, which contain about 17% seeds, the primary mineral components are potassium, sodium, magnesium, and calcium. Moyin-Jesu (2007) reported the presence of iron, zinc, manganese, and nickel. Okra is an excellent source of zinc, which is crucial for the proper functioning of the human immune system (Cook et al. 2000). Moreover, okra is abundant in dietary fibre, especially viscous fibre (Kendall and Jenkins 2004).

Along with the supply of optimum quantities of nutrients, these are low in cholesterol and saturated fat (Lim 2012; USDA 2010). Okra also exhibits antioxidant properties. Shui and Peng (2004) reported that quercetin derivatives and (−)-epigallocatechin are the primary antioxidant compounds found in okra. The quercetin derivatives include quercetin 3-O-xylosyl (1″→2″) glucoside, quercetin 3-O-glucosyl (1″→6″) glucoside, quercetin 3-O-glucoside, and quercetin 3-O- (6″-O-malonyl)-glucoside that contribute to 70% of the total antioxidant activity. Liao et al. (2012) identified a new flavonol glycoside, 5, 7, 3′, 4′-tetrahydroxy-4″-O-methyl flavonol-3-O-β-D-glucopyranoside, along with another previously reported compound, 5, 7, 3′, 4′-tetrahydroxy flavonol-3-O-[β-D-glucopyranosyl-(1→6)]-β-D-glucopyranoside in okra.

Okra is abundant in polysaccharides and serves many different biological purposes. According to several studies, the primary monosaccharides in the polysaccharides of okra include galactose (Gal), rhamnose (Rha), galacturonic acid (GalA), and arabinose (Ara). Furthermore, NMR studies indicated that rhamnogalacturonan-I (RG-I) is a distinct structural component of okra polysaccharides. Pectin, a natural polysaccharide with exceptional gelling and emulsifying properties that can be utilized as an emulsifier or thickener in the food industry, is often found in the mucilage and cell walls of okra. Different okra varieties have varying levels of pectin concentration. The young tissues have greater pectin concentration than older tissues (Liu et al. 2017). According to Zhang et al. (2018), the pectin monosaccharide composition of okra primarily includes rhamnose, galacturonic acid, galactose, arabinose, glucuronic acid, glucose, and a significant amount of branching rhamnogalacturonan-I (RG-I) fragments.

Polyphenols usually comprise a hydroxyl-containing aromatic ring. In order to reduce oxidative damage, they are frequently coupled with polysaccharides located in cell walls (Balasundram et al. 2006). Catechins, anthocyanins, quercetin, and other polyphenols fall within the family of polyphenols known as flavonoids (Cook and Samman 1996). The abilities of polyphenols as antioxidants are intimately linked to their other characteristics. They often provide antioxidant, anti-inflammatory, anti-fatigue and other health benefits by reducing reactive oxygen species (ROS) and enhancing the activity of antioxidant enzymes. The polyphenols in okra peel mainly include quercetin and hydroxycinnamic acid derivatives, while those in the seeds are predominantly quercetin derivatives and catechins (Arapitsas 2008). There is currently less study on polyphenols with significant content in okra, such as catechin and hydroxycinnamic acid, and more research on total polyphenols, total flavonoids, and quercetin. The mesocarp of okra seeds includes hydroxycinnamic and quercetin derivatives (0.2 and 0.3 mg/g of the skin, respectively), oligomeric catechins (2.5 mg/g of seeds), and flavonol derivatives (3.4 mg/g of seeds) (Arapitsas 2008; Manach et al. 2005; Ngoc et al. 2008). According to Moyin-Jesu (2007), the chemical analysis of okra seed flour revealed the following composition: moisture at 6.96%, total carbohydrates at 30.81%, protein at 22.14%, oil at 14.01%, and crude fibre at 27.30%.

Proteins are crucial for human nutrition. The biological value of a protein is assessed based on the amount, composition and digestibility of its amino acids by the human body (Ewa et al. 2011). Okra has been called "the perfect villager's vegetable" because of its resilience, nutritional fibre and unique seed protein balance of lysine and tryptophan amino acids, which sets it apart from the proteins in cereals and pulses (Holser and Bost 2004; Sanjeet et al. 2010). It contains active trypsin, abundant essential amino acids, and a chemical score of 67 (Rao 1985). The amino acid profile of okra seed protein is comparable to that of soybean and it has a higher protein efficiency ratio (PER) (Adetuyi et al. 2012). The amino acid composition of the protein makes it a valuable supplement to diets that rely on legumes or grains

(Ndangui et al. 2010). Among plant protein sources, okra seeds stand out for their high-quality protein content, particularly rich in essential amino acids (Gemede et al. 2016). As a result, it is a crucial component of the human diet (Farinde et al. 2007). The entire seeds and kernels of okra are rich in protein and fat. The seed kernel contains significant amount of protein and fat, whereas the seed coat or hull is primarily composed of crude fibre.

Okra mucilage refers to the thick, slimy substance present in both fresh and dried pods. Mucilaginous compounds are usually found in the pod walls and consist of acidic polysaccharides that are chemically bonded to minerals and proteins (Woolfe et al. 1977). Most of the carbohydrates are present in the form of mucilage (Kumar et al. 2009; Liu et al. 2005). There are various types of polysaccharides, but neutral sugars such as rhamnose, galactose and galacturonic acid are commonly found (Hirose et al. 2004; Sengkhamparn et al. 2009). It consists of long-chain molecules with a molecular weight of around 1,70,000 formed from sugar molecules and amino acids in young fruits. The composition primarily includes galactose (25%), rhamnose (22%), galacturonic acid (27%), and amino acids (11%). It is highly soluble in water with an intrinsic viscosity in aqueous solutions of about 30%. The okra mucilage has potential applications in food, non-food products, and pharmaceuticals. In the food industry, it can be used as an ingredient in flour-based adhesives, a whipping agent for reconstituted egg whites and as a supplement for clarifying sugarcane juice, particularly in India. The non-food applications of okra mucilage include its use as a brightening agent in metal electro-deposition, a deflocculant in paper and cloth production and a barrier to reduce pipe flow friction (BeMiller et al. 1993; Ndjouenkeu et al. 1996). It is suitable for applications in both industry and medicine (Akinyele and Temikotan 2007). It is commonly used in the production of glace paper and is also utilized in confectionery. Additionally, okra has applications in medicine as a blood volume expander or plasma substitute (Adetuyi et al. 2008; Kumar et al. 2010; Lengsfeld et al. 2004; Markose and Peter 1990; Savello et al. 1980).

The okra leaf also holds a large nutritional potential. The per 100 grams of edible okra leaves do contain water (81.50 g), energy (235.00 kJ or 56.00 kcal), protein (4.40 g), fat (0.60 g), carbohydrate (11.30 g), fibre (2.10 g), calcium (532.00 mg), phosphorus (70.00 mg), iron (0.70 mg), ascorbic acid (59.00 mg), β-carotene (385.00 mg), thiamine (0.25 mg), riboflavin (2.80 mg), and niacin (0.20 mg) (Gopalan et al. 2007; Varmudy 2011).

The okra seeds are utilized to make oil. Almost all food kinds benefit considerably from the nutritional and sensory value provided by lipid components of the oil. Generally, vegetable oils typically contain four different kinds of lipids: triacylglycerols, diacylglycerols, polar lipids and free fatty acids. The fatty acid content of lipids influences their physical properties, stability and nutritional value. Triacylglycerols are the most common form of storage lipids found in nature. These compounds are made up of both saturated and unsaturated fatty acids, which vary in acyl chain length along with the number and positions of double bonds. They include saturated, monoenoic, and polyunsaturated fatty acids, each differing in their specific fatty acid composition. The structural difference between monoenoic and polyunsaturated fatty acids is the existence of repeated methylene units. These molecules comprise a significant group that influences health by forming an incredibly flexible chain that quickly changes structural states (Vermerris and Nicholson 2006). The okra seed oil has a potential concentration of up to 20% of glycerol and higher fatty acids that are mostly derived from plant fruits, seeds, and germs (Jarret et al. 2011). The okra seed oil holds significant potential as a supplement or alternative to other seed oils (Clopton et al. 1948; Rao 1985). According to Crossley and Hilditch (1951), okra seed oil and cottonseed oil share some characteristics, making okra seed oil more desirable. It may be produced using traditional methods and can be used in place of cotton seed oil. Numerous studies have also revealed that it is an excellent source of unsaturated fatty acids, particularly linoleic acid (32.22–43.07%), linolenic acid (6.79–12.34%), and oleic acid (4.31–6.98%), with oleic acid and linoleic acid content being greater than that of other species. Linoleic acid is a polyunsaturated fatty acid that makes up the majority of the oil (up to 47.4%) (Andras et al. 2005). Linoleic acid, necessary for human nutrition, is abundant in okra seed oil (Savello et al. 1980). Okra seeds comprise palmitic acid (29.18–43.26%) and stearic acid (6.36–7.73%) as the main saturated fatty acids found in okra seeds, with okra containing a higher concentration of palmitic acid compared to other plants (Jarret et al. 2011; Ogungbenle and Omosola 2015). The *A. esculentus* contains palmitic acid in the range of 10.3 to 36.35% (Jarret et al. 2011). Greece-sourced okra seeds have oil contents that range from 20 to 40%, depending on the technique of extraction (Benchasr 2012; MEF 2013).

21.3 Okra: Health Benefits

The importance of nutrition to human health has received more emphasis. Consuming a diet rich in plant-based foods is associated with a reduced risk of various chronic diseases such as cancer and atherosclerosis (Gosslau and Chen 2004). Flavonoid antioxidants like β-carotene, xanthein, and lutein are abundant in okra pods (Dilruba et al. 2009). The okra has the potential to cure those suffering from leucorrhoea and general weakness. Demir (2001) found that the high iodine content in okra fruit is beneficial for preventing goitre, treating ulcers and alleviating haemorrhoids. It is an important source of calcium and other essential vitamins and minerals that are often lacking in the diets of people in impoverished countries (Kahlon et al. 2007). The positive effects are partly due to the chemicals with antioxidant properties. The primary antioxidants found in vegetables include vitamin C and E, carotenoids and phenolic compounds, particularly flavonoids. These antioxidants neutralize free radicals that prevent the initiation of chain reactions or halt their progression (serving as the second line of defence). Since they quench singlet oxygen, vitamin E and carotenoids also support the initial line of defence against oxidative stress (Krinsky 2001). Flavonoids and vitamin C not only protect α-tocopherol in human LDL but have also been shown to regenerate vitamin E from the chromanoxy radical (Davey et al. 2000). Rossetto et al. (2002) studied that the synergistic action of these dietary antioxidants in foods may reduce reactive oxygen species levels more efficiently. Furthermore, a combination of lipid- and water-soluble antioxidants can neutralize free radicals in both aqueous and lipid moieties (Trombino et al. 2004). Using the liposome oxidation technique, the combined activity of quercetin or catechins with tocopherol was significantly greater than the sum of their individual activities. In systems involving human erythrocyte membrane ghosts and phosphatidylcholine liposomes, combinations of α-tocopherol or vitamin C with phenolic compounds also exhibited synergistic effects (Liao and Yin 2000).

21.4 Development of Okra-Based Nutraceuticals

Nutraceuticals are generally defined as foods or food components that provide additional health benefits. Okra-based nutraceuticals are preferred health foods due to their natural nutritional value and bioactive components, which contribute positively to health. (Fig. 21.1) (Kumar 2014). The efforts are underway to enhance the well-documented hypoglycemic effects of okra fruit by creating formulations with different ratios of okra seeds to peels (10:90, 20:80, 30:70, 40:60, 50:50%, etc.). This is followed by an evaluation of the anti-diabetic and antioxidant effectiveness of these formulations through in vitro testing (Muhammad et al. 2018; Reddy et al. 2012, Fig. 21.1). The okra polysaccharides, majorly pectin and lectin, have demonstrated the ability to inhibit the growth of human cancer cells (Dan-Dan and Gu 2010). These polysaccharides too play the therapeutic role of being anti-diabetic by reducing body weight, glucose levels, and total serum cholesterol while improving glucose tolerance. Pectin triggers apoptosis, suppresses cellular proliferation, enhances cholesterol breakdown and reduces fat production. It lowers bad cholesterol by modifying bile production in the intestine, which aids in removing clots and cholesterol deposits. The polyphenolic compounds quercetin 3-O-glucosyl (1→6) glucoside (QDG), quercetin 3-O-glucoside (QG), vitamin A, vitamin B (B1, B2, and B6), vitamin C, traces of Zn, Ca, folic acid and epigallocatechin act as major

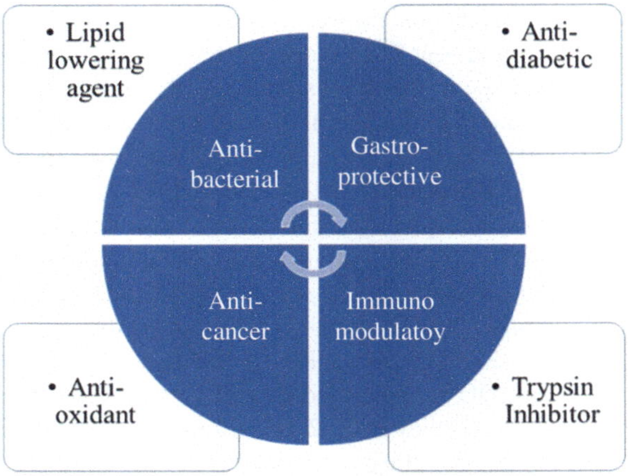

Fig. 21.1 Health benefits of okra-based nutraceuticals

antioxidants with strong reducing power and free radical scavenging properties, including 2,2-diphenyl-1-picrylhydrazyl (DPPH), superoxide anions and hydroxyl radicals. It improves vision by the therapeutic action of beta-carotene (a vitamin A precursor), xanthin and lutein, which have antioxidant properties that help prevent eye issues such as cataracts and glaucoma. These vitamins, along with dietary fibre, benefit pregnancy as folate prevents miscarriages, supports neural tube formation and protection, reduces birth defects like spina bifida and can alleviate constipation during pregnancy. This fibre, along with the mucilage content, treats ulcers with the alkaline mucus that neutralizes acid and coats the digestive tract, aiding in the rapid healing of peptic ulcers. It relieves and prevents constipation by binding toxins and lubricating the large intestine, facilitating smooth and regular bowel movements with its natural laxative effect. The rhamnogalacturonan polysaccharides act as an anti-adhesive by disrupting *H. pylori* adhesion to stomach tissues by targeting outer membrane proteins (Elkhalifa et al. 2021).

21.5 Conclusion

This chapter emphasizes the nutritional value of various edible okra components and their potential as nutraceuticals for enhancing human health. It is a low-fat food rich in minerals and phytochemicals. It is notably high in dietary fibre, potassium, magnesium, manganese, ascorbic acid, folate, thiamine, pyridoxine, vitamin K, and bioactive compounds like flavonoids (especially quercetin) and phytosterols. While knowledge about the disease preventive role and mechanisms of action of edible okra parts is limited, studies suggest that okra components may help reduce the risk of chronic diseases such as obesity, diabetes, atherosclerosis, hypertension, cancer, neurological disorders and neglected tropical diseases (NTDs). The affordability and availability of okra make it a promising source for nutraceuticals, benefiting both low- and high-income groups through nutritional diversification.

References

Adetuyi FO, Osagie AU, Adekunle AT (2008) Effect of postharvest storage techniques on the nutritional properties of Benin indigenous Okra *Abelmoschus esculentus* (L) Moench. Pak J Nutr 7:652–657

Adetuyi F, Ajala L, Ibrahim T (2012) Effect of the addition of defatted okra seed (*Abelmoschus esculentus*) flour on the chemical composition, functional properties and Zn bioavailability of plantain (musa paradisiacal linn) flour. JMBFS 2(1):69–82

Akinyele BO, Temikotan T (2007) Effect of variation in soil texture on the vegetative and pod characteristics of okra (*Abelmoschus esculentus* (L.) Moench). Int J Agric Res 2:165–169

Andras CD, Simandi B, Orsi F, Lambrou C, Khomsug P, Thongjaroenbuangam W, Tatla DM, Panayiotou C, Domokos J, Pakdeenarong N, Suttajit M, Chantiratikul P, Doleschall F (2005) Supercritical carbon dioxide, Antioxidative activities and phenolic extraction of Okra (*Hibiscus esculentus* L.) seeds. J Cont Extracts from Okra Abelmoschus Sci Food Agric 85:1415–1419

Arapitsas P (2008) Identification and quantification of polyphenolic compounds from okra seeds and skins. Food Chem 110:1041–1045

Balasundram N, Sundram K, Samman S (2006) Phenolic compounds in plants and agri industrial by-products: antioxidant activity, occurrence, and potential uses. Food Chem 99(1):191–103

BeMiller JN, Whistler RL, Barkalow DG, Chen C (1993) Aloea, chia, flax seed, okra, psyllium seed, quince seed, and tamarin gums. In: Whistler RL, BeMiller JN (eds) Industrial gums. Academic Press, New York, pp 227–256

Benchasr S (2012) Okra (*Abelmoschus esculentus* (L.) Moench) as a valuable vegetable of the world. Ratar Povrt 49:105–112

Calisir S, Yildiz MU (2005) A study on some physicochemical properties of Turkey okra (*Hibiscus esculenta*) seeds. J Food Eng 68:73–78

Clopton JR, Roberts A, Jeskey HA (1948) Chemical studies on oil bearing seeds. J Am Oil Chem Soci 25(11):401–404

Cook NC, Samman S (1996) Flavonoids—chemistry, metabolism, Cardioprotective effects, and dietary sources. J Nutr Biochem 7(2):66–76

Cook JA, Vander Jagt DJ, Pastuszyn A, Mounkaila G, Glew RS, Millson M, Glew RH (2000) Nutrient and chemical composition of 13 wild plant foods of Niger. J Food Comp Anal 13:83–92

Crossley A, Hilditch TP (1951) The fatty acids and glycerides of Okra seed oil. J Sci Food Agr 2(6):251–255

Dan-Dan R, Gu C (2010) Inhibition effect of Okra polysaccharides on proliferation of human cancer cell lines. Food Sci 31:353–356

Davey MW, Van Montagu M, Inze D, Sanmartin M, Kanellis A, Smirnof F (2000) Plant L-ascorbic acid: chemistry, function, metabolism, bioavailability and effects of processing. J Sci Food Agric 80:825–860

Demir I (2001) Variation in shoot and root characteristics and their association with drought tolerance in lentil landraces. Sci Hortic 89:1–7

Dhaliwal MS (2010) Okra (*Abelmoschus esculentus*) L. (Moench). Kalyani Publishers, New Delhi

Dilruba S, Hasanuzzaman M, Karim R, Nahar K (2009) Yield response of okra to different sowing time and application of growth hormones. J Hortic Sci 1:10–14

Elkhalifa AEO, Alshammari E, Adnan M, Alcantara JC, Awadelkareem AM, Eltoum NE, Ashraf SA (2021) Okra (Abelmoschus esculentus) as a potential dietary medicine with nutraceutical importance for sustainable health applications. Molecule 26(3):696

Ewa C, Agnieszka G, Adametal F (2011) The content of protein and of amino acids in Jerusalem artichoke tubers (*Helianthus tuberosus* L.) of red variety Rote Zonenkugel. Acta Sci Pol Technol Aliment 10(4):433–441

Farinde A, Owolarafe O, Ogungbemi I (2007) An overview of production, processing, marketing and utilisation of okra in egbedore local government area of Osun State, Nigeria. Agric Eng 4:1–17

Gemede HF, Haki GD, Beyene F, Woldegiorgis AZ, Rakshit SK (2016) Proximate, mineral, and antinutrient composition of indigenous okra (Abelmoschus esculentus) pod accessions: implications for mineral bioavailability. Food Sci Nutr 4(2):223–233

Gopalan C, Sastri SBV, Balasubramanian S (2007) Nutritive value of Indian foods, National Institute of Nutrition (NIN). ICMR, India

Gosslau A, Chen KY (2004) Nutraceuticals apoptosis and disease prevention. Nutrition 20:95–102

Hirose K, Endo K, Hasegawa K (2004) A convenient synthesis of lepidimoide from okra mucilage and its growth promoting activity in hypocotyls. Carbohydr Poly 339:9–19

Holser R, Bost G (2004) Hybrid Hibiscus seed oil compositions. J Am Oil Chem Soc 81(8):795–797

Jarret RL, Wang ML, Levy IJ (2011) Seed oil and fatty acid content in okra (*Abelmoschus esculentus*) and related species. J Agric Food Chem 59:4019–4024

Kahlon TS, Chapman MH, Smith GE (2007) *In vitro* binding of bile acids by okra, beets, asparagus, eggplant, turnips, green beans, carrots and cauliflower. Food Chem 103:676–680

Kendall CWC, Jenkins DJA (2004) A dietary portfolio: maximal reduction of low density lipoprotein cholesterol with diet. Curr Atheroscler Rep 6:492–498

Krinsky NI (2001) Carotenoids as antioxidants. Nutrition 17:815–817

Kumar S (2014) Physicochemical, phytochemical and toxicity studies on gum and mucilage from plant *Abelmoschus esculentus*. J Phytopharm 3:200–203

Kumar R, Patil MB, Patil SR, Paschapur MS (2009) Evaluation of Abelmoschus esculentus mucilage as paracetamol suspension. Intl J Pharm Tech Res 1:658–665

Kumar S, Dagnoko S, Haougui A, Ratnadass A, Pasternak D, Kouame C (2010) Okra (*Abelmoschuss pp.*) in West and Central Africa: potential and progress on its improvement. Afr J Agric Res 5:3590–3598

Kumar A, Kumar P, Nadendla R (2013) A review on: Abelmoschus esculentus (Okra). Int Res J Pharm Appl Sci 3:129–132

Lengsfeld C, Titgemeyer F, Faller G, Hensel A (2004) Glycosylated compounds from okra inhibit adhesion of Helicobacter pylori to human gastric mucosa. J Agric Food Chem 52:1495–1503

Liao KI, Yin MC (2000) Individual and combined antioxidant effects of seven phenolic agents in human. J Agric Food Chem 48:2266–2270

Liao H, Liu H, Yuan K (2012) A new flavonol glycoside from the *Abelmoschus esculentus* L. Pharmagnosy Mag 8:12–15

Lim TK (2012) Edible medicinal and non-medicinal plants: Vol 3 fruits. Springer Science+Business Media B V, p 160

Liu IM, Liou S, Lan TW, Hsu FL, Cheng JT (2005) Myricetin as the active principle of Abelmoschus moschatusto lower plasma glucosein streptozotocin-induced diabetic rats. Planta Med 71:617–621

Liu S, Ji H, Li M, Cheng Z, Zhu J, Zhao YK, Guo X, Ye J (2017) Study on the flavonoids and Pectin contents in different Okra (*Abelmoschus escullentus* L.) accessions. J Agric Sci Bot 1(1):12–16

MEF (2013) Biology of Okra. Series of crop specific biology document. Ministry of Environmental and Forest Government of India. pp 1–8

Manach C, Williamson G, Morand C, Scalbert A, Remesy C (2005) Bioavailability and bioefficacy of polyphenols in humans. Am J Clin Nutr 81:230–242

Markose BL, Peter KV (1990) Okra review of research on vegetable and tuber crops. Kerala Agricultural University Press, Kerala

Martin F (1982) Okra, potential multiple-purpose crop for the temperate zones and tropics. Econ Bot 36(3):340–345

Moyin-Jesu EI (2007) Use of plant residues for improving soil fertility pod nutrients root growth and pod weight of okra Abelmoschus esculentum L. Bioresour Technol 98:2057–2064

Muhammad I, Matazu IK, Yaradua IA, Yau S, Nasir A, Bilbis SL, Abbas YA (2018) Development of Okra-based antidiabetic nutraceutical formulation from *Abelmoschus esculentus* (L.) Moench (Ex-maradi Variety). Trop J Nat Prod Res 2:80–86

Ndangui CB, Kimbonguila A, Nzikou JM, Matos L, Pambou NPG, Abena AA, Silou T, Scher J, De Sobry S (2010) Nutritive composition and properties physico chemical of gumbo (*Abelmoschus esculentu s* L.) seed and oil. Res J Environ Earth Sci 2(1):49–54

Ndjouenkeu R, Goycoolea FM, Morris ER, Akingbala JO (1996) Rheology of okra (*Hibiscus esculentus* L.) and dika nut (*Irvinia gabonensis*) polysaccharides. Carbohydr Poly 29:263–269

Ngoc T, Ngo N, Van T, Phung V (2008) Hypolipidemic effect of extracts from *Abelmoschus esculentus* L. (Malvaceae) on Tyloxapol-induced hyperlipidemia in mice. Warasan Phesatchasat 35:42–46

Ogungbenle HN, Omosola SM (2015) The comparative assessment of nutritive values of dry Nigerian Okra (Abelmoschus esculentus) fruit and oil. Int J Food Sci Nutri Eng 5(1):8–14

Qhureshi Z (2007) Breeding investigation in bhendi (*Abelmoschus esculentus* (L.) Moench). Master Thesis, University of Agriculture Sciences, GKVK, Bangalore

Rao PU (1985) Chemical composition and biological evaluation of Okra (Hibiscus esculentus) seeds and their Kernels. Plant Food Hum Nutr 35(4):389–396

Reddy MT, Kadiyala H, Mutyala G, Hameedunnisa B (2012) Heterosis for yield and yield components in Okra (*Abelmoschus esculentus* (L.) Moench). Chil J Agric Res 72:316–325

Rossetto M, Vanzani P, Mattivi F, Lunelli M, Scarpa M, Rigo A (2002) Synergistic antioxidant effect of catechin and malvidin 3- glucoside on free radical-initiated peroxidation of linoleic acid in micelles. Arch Biochem Biophys 408:239–245

Roy A, Shrivastava SL, Mandal SM (2014) Functional properties of Okra *Abelmoschus esculentus* L. (Moench): traditional claims and scientific evidences. Plant Sci Today 1:121–130

Sami R, Lianzhou J, Yang L, Ma Y, Jing J (2013) Evaluation of fatty acid and amino acid compositions in Okra (*Abelmoschus esculentus*) grown in different geographical locations. Biomed Res Int 2013:574283

Sanjeet K, Sokona D, Adamou H, Alain R, Dov P, Christophe K (2010) Okra (*Abelmoschus spp.*) in West and Central Africa: potential and progress on its improvement. Afr J Agric Res 5(25):3590–3598

Savello PH, Martin FW, Hill JM (1980) Nutritional composition of okra seed meal. J Agric Food Chem 28:1163–1166

Sengkhamparn N, Verhoef R, Schols HA, Sajjaanantakul T, Voragen AGJ (2009) Characterization of cell wall polysaccharides from okra (*Abelmoschus esculentus* (L.) Moench). Carbohydr Res 344:1824–1832

Shalau J (2002) Backyard Gardener. www.ag.arizona.edu

Shui G, Peng L (2004) An improved method for the analysis of major antioxidants of *Hibiscus esculentus* Linn. J Chromatograph A 1048(1):17–24

Siemonsma JS, Kouamé C (2004) *Abelmoschus esculentus* (L.) Moench. Internet record from protabase. In: Grubben GJH, Denton OA (eds) PROTA (Plant Resources of Tropical Africa), Wageningen. http://database.prota.org/search.htm

Trombino S, Serini S, Di Nicuolo F, Celleno L, Ando S, Picci N (2004) Antioxidant effect of ferulic acid in isolated membranes and intact cells. J Agric Food Chem 52:2411–2420

U.S. Department of Agriculture, Agricultural Research Service (2010) USDA National Nutrient Database for Standard

Varmudy V (2011) Marking survey need to boost okra exports. Department of economics, Vivekananda College, Puttur

Vermerris W, Nicholson R (2006) Phenolic compound biochemistry. Springer, Dordrecht

Wayne J, Mclaurin WJ, Fontenot RJ, Newsom DW (1984) Effects of nitrogen and quality factors of canned okra. J Hortic Sci 109:524–526

Woolfe ML, Martin FC, Otchere G (1977) Studies on the mucilages extracted from okra fruits (*Hibiscus esculentus* L.) and baobab leaves (*Adansonia digitata* L.). J Sci Food Agric 28:519–529

Zhang T, Xiang J, Zheng G, Ya R, Min X (2018) Preliminary characterization and anti hyperglycemic activity of a Pectic polysaccharide from Okra (*Abelmoschus Esculentus* (L.) Moench). J Funct Food 41:19–24

22 Harnessing Bioactive Phytochemicals in Okra: Nutritional and Therapeutic Perspectives

Pradip Karmakar, Brijesh Kumar Maurya, Himanshu Singh, Vidya Sagar, Sujan Majumder, Rajeev Kumar, P. M. Singh, N. Rai, and T. K. Behera

22.1 Introduction

Okra originated from Africa and can be cultivated in countries with tropical, subtropical, and warm temperate climates (Ndunguru and Rajabu 2004; Naveed et al. 2009). Okra is rich in minerals along with beneficial nutrients and also contains an appreciable amount of carbohydrates, proteins, vitamins, and traces of fatty acids which are useful for boosting and improving the immune system, bones, muscles, and other parts of the human body to protect from diseases (Alessandra et al. 2019). Okra is also rich in several bioactive compounds, including anthocyanin, which is responsible for the red coloration of the okra fruits (Karmakar et al. 2022). Okra has common benefits for the human body and animals due to the chemical components of primary metabolism (Table 22.3), which depends on the type of soil, water, and environment changes (Dias 2012; Ariel et al. 2009).

It is a delectable vegetable crop that is capable of growing in different countries in South Asia, West Asia, including several Arab Nations, West Africa, and North America (Mishra et al. 2017; Singh et al. 2023). Complex bioactive polysaccharides such as hemicellulose, alpha-cellulose, pectin, lignin, waxy substances, numerous crucial vitamins, and essential macro- and micro-nutrients, including phosphorus, potassium, iron, calcium, magnesium, and zinc, are all present in fresh okra. It also contains an appreciable amount of energy in terms of carbohydrates (7%), protein (2%), and ample quantities of dietary fibers (Rakesh and Vishal 2016; Priya et al. 2024). Hence, okra is considered a key vegetable for human health and nutrition. Mucilage present in okra also has industrial use as a turbidity remover of the wastewater (Konstantinos et al. 2009; Monika et al. 2001) and is also being investigated as a biodegradable material for the packaging of food substances (Araújo et al.

P. Karmakar (✉) · B. K. Maurya · H. Singh · V. Sagar · S. Majumder
R. Kumar · P. M. Singh · N. Rai · T. K. Behera
Division of Crop Improvement, Indian Institute of Vegetable Research, Varanasi, India

S. K. Tikoo et al. (eds.), *Okra: Status, Challenges and Opportunities*,
https://doi.org/10.1007/978-981-97-9963-3_22

2018; De-Rosa et al. 2010). The experimentation on biological activity of bioactive compounds isolated from okra fruit has revealed its antioxidant, antidiabetic, neuroprotective, anti-hyperlipidemia, and anti-fatigue properties (Fangbo et al. 2015). Okra polysaccharides have not yet undergone thorough pharmacological investigations.

All parts of the plant, right from the fresh leaves to seeds, have beneficial value and many desirable uses. It can also be used as a salad, stew, fresh or dehydrated, and boiled or fried (Axe 2021). Okra fruit is beneficial against genital-urinary ailments, chronic dysentery, and spermatorrhea. It also can be used during off-season as dried or frozen fruit (Eagri 2023). Okra is rich in protein and edible oil, and its dried fruit contains edible oil 13–22% and protein 20–24% for use as refined edible oil. Dry fruits also have industrial importance and are used for paper manufacture, fibers, and cardboard. During the preparation of jaggery, its vegetative parts (roots and stems) are commonly utilized for clearing cane juice (Eagri 2023).

22.2 Nutritional and Phytochemical Components of Okra

22.2.1 Nutritional Components of Okra

Industrial applications for bast fiber from the plant's stem include the strengthening of polymer composites. Due to its flocculant qualities, okra mucilage it can also be utilized for wastewater treatment. Okra mucilage, which resembles a thick polysaccharide film in composition, is being researched as a biodegradable food container (WIKI 2023). One cup of okra fruits is reported to contain 33 calories of energy, 1.9 g of protein, 3.2 g of dietary fiber, fat at 0.2 g, carbohydrate at 7.5 g, minerals like potassium (299 mg), sodium (7 mg), magnesium (57 mg), calcium 82 mg, and vitamins such as vitamin A at 716 IU, vitamin C at 23 mg, vitamin K at 31.3 mg, folic acid at 60 μg, thiamine at 0.2 mg, and vitamin B_6 at 0.2 mg.

22.2.2 Phytochemical Components of Okra

Greek terms for "plant" and "chemicals" are the origin of the term "phytochemical." Typically, compounds that may have biological relevance but are not recognized as essential nutrients are referred to as phytochemicals. However, in a more restricted sense, the "phytochemical" term refers to the variety of secondary metabolic substances present in plants. Nearly 10,000 distinct phytochemicals are thought to be capable of treating a wide range of diseases. Okra seed contains tannins, terpenoids, and glycosides (Honda et al. 2003; Hung et al. 2014).

In recent times, this crop has also been extensively grown in the northern and southern parts of China (Jiang et al. 2017). Fruits of this tropical vegetable have been used to prevent different diseases and disorders and have a vast range of curative properties (Sabitha et al. 2011). The therapeutic and nourishing properties of it are mainly due to the presence of different biologically active phytonutrients. Okra

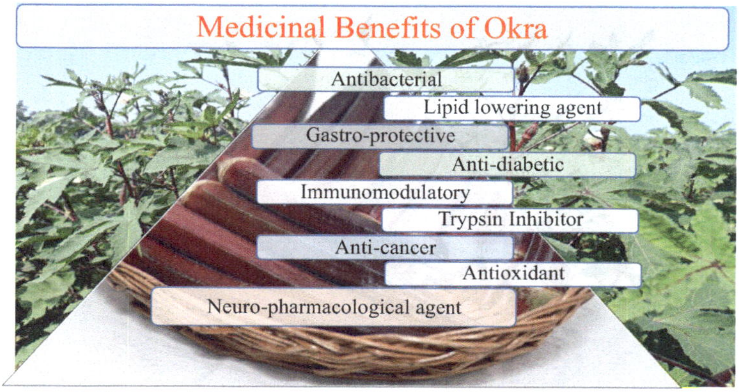

Fig. 22.1 Illustrative representation of the okra-mediated beneficial effects that have been scientifically established to date

fruits play a major role in human nutrition by supplying carbohydrates, protein, fat, vitamins, and mineral nutrients (Benchasri 2012). Okra fruits also consist of huge quantities of vital nutrients such as protein, calcium, dietary fiber, iron, and zinc are also low in anti-nutrients (Gemede et al. 2015). The health-promoting (Fig. 22.1) and bioactive properties of okra are associated with the content of various secondary metabolites of plant origin. The powder from okra fruits, which is extracted using ethyl alcohol along with water, is loaded with carbohydrates, proteins, gums and mucilage, sterols of plant origin, flavonoids, other phenolic compounds, tannins, and essential oils (Saha et al. 2011). Pharmacological experiment has confirmed the various bioactivities, such as antioxidant, antidiabetic, anti-hyperlipidemic, and anti-hyperglycemic effects of okra fruits (Graham et al. 2017; Lu et al. 2016; Sabitha et al. 2011; Zhang et al. 2018). Usually, the phenolic compounds (Jiang et al. 2017; Liao et al. 2012; Lu et al. 2016; Xia et al. 2015; Zeng et al. 2015) are regarded as the key bioactive constituents in okra fruits and are accountable for its various bioactivities. The composition and content of polyphenolic compounds in okra fruits and their bioactive characteristics are affected by varieties, growing environment, and fruit morphology. The nutritional and phytochemical components of okra were documented in Table 22.1.

22.3 Flavonoids in Okra

Okra is rich in flavonoids—a prominent type of polyphenolic compound found in all vascular plants. These flavonoids exist in two forms, namely, aglycones or combined with glycosides and acyl groups. With over 8000 identified varieties, flavonoids make up a vast group of secondary plant metabolites that exhibit their diverse presence in the plant (Gonzales et al. 2015). Flavonoids have a complex structure with two aromatic rings connected by a tetra-hydropyran ring, sharing a similar

Table 22.1 Different bioactive phytochemicals identified in okra

SN	Name	Group	Plant part	Pharmaceutical and health-promoting properties	References
1	Rhamnogalacturonan-I	Carbohydrate (Polysaccharide)	Fruit	Antidiabetic and hypoglycemic properties	Sengkhamparn et al. (2009), Liu et al. (2018)
2	Pectin	Carbohydrate (Polysaccharide)	Fruit	Anticancer, anti-proliferative, promote self-destruction of the cancerous cell, and help to reduce bad cholesterol	Olano-Martin et al. (2003), Vayssade et al. (2009), Das and Ghosh (2019)
3	Lectins	Protein	Leaf and seed	Anticancer activity	Monte et al. (2013)
4	Lysine	Amino acid (Protein)	Seed	Essential amino acids, improve bone health and immune systems, have wound healing properties, and are helpful in reducing the symptoms of dry eyes, cold sores, and schizophrenia	Elkhalifa et al. (2021), Gunarathne et al. (2024)
5	Oleic acid	Fat	Seed	Cardio-protective activity, helpful in reduction of blood cholesterol, and anti-inflammatory activity	Durazzo et al. (2018), Nagy and Tiuca (2017)
6	Linoleic acid	Fat	Seed	Essential for normal growth and development, lowering the risk of heart disease, reducing blood cholesterol, normalizing blood pressure, and improving skin health	Durazzo et al. (2018), Taha (2020), Marangoni et al. (2020)
7	Delphinidin 3-Osambubioside	Anthocyanin	Fruit	Anti-inflammatory properties, regulate lipid metabolism and help to reduce hyperlipidemia	Zhang et al. (2021), Sogo et al. (2015)
8	Cyanidin 3-Osambubioside	Anthocyanin	Fruit	Anti-hypercholesterolemia, antipyretic, antibacterial, antiviral, and anthelminthic properties	Zhang et al. (2021), Lazalde-Cruz et al. (2021)
9	Chlorophyll-a and b	Chlorophyll	Fruit	Antioxidant and anti-inflammatory properties improve liver function and skin health	Karmakar et al. (2022), Martins et al. (2023)

10	Protocatechuic acid	Flavonoids	Fruit	Antioxidant, anti-inflammatory, neuroprotective, anticancer, antiviral, antidiabetic, anti-hyperglycemic, anti-osteoporotic, and anti-aging properties	Wu et al. (2020), Pawar et al. (2023)
11	Quercetin-3-O-gentiobioside	Flavonoids	Fruit	Anti-inflammatory, anti-histamine, anti-fatigue, and vaso-protective properties	Wu et al. (2020), Pawar et al. (2023)
12	Rutin	Flavonoids	Fruit	Antioxidant and anti-inflammatory activity, anti-aging properties, helpful in strengthening and enhancing the flexibility of blood vessels, and plays a protective role against heart disease, diabetes, and various microbial infections	Wu et al. (2020), Semwal et al. (2021)
13	Isoquercitrin	Flavonoids	Fruit	Antioxidant, anti-inflammatory, and anticancer properties	Wu et al. (2020), Macedo et al. (2023)
14	Quercetin	Flavonoids	Fruit	Antioxidant, anti-inflammatory, anti-proliferative, anti-carcinogenic, antidiabetic, and antiviral properties and also improve heart health	Wu et al. (2020), Deepika and Maurya (2022)
15	Iodine	Mineral	Fruit	Helpful in regulation of thyroid hormone and proper bone and brain development during pregnancy and infancy	Adelakun and Oyelade (2011), Bertinato (2021)
16	Selenium	Mineral	Fruit	Boost the immune system, improve thyroid and reproductive health, act as an antioxidant, and reduce the risk of cancer, heart disease, and declining mental health	Elkhalifa et al. (2021), Tóth and Csapó (2018)
17	Zinc	Mineral	Fruit	Improve immune system, helpful in wound healing, control DNA metabolism and repair, and essential for reproduction, vision, taste, and cognition	Elkhalifa et al. (2021), Maret and Sandstead (2006)

(continued)

Table 22.1 (continued)

SN	Name	Group	Plant part	Pharmaceutical and health-promoting properties	References
18	Potassium	Mineral	Fruit	Improve muscle health, control blood pressure, and be helpful in moving the nutrients into cells and waste material out of cells	Elkhalifa et al. (2021), Weaver (2013)
19	Vitamin A	Vitamin	Fruit	Improve vision, cellular growth and differentiation, reproduction, and protect skin by supporting cell production and repair	Elkhalifa et al. (2021), Blomhoff (1994)
20	Vitamin C	Vitamin	Fruit	Antioxidant activity, wound healing properties, improve immune system, reproductive, and skin health	Elkhalifa et al. (2021), Bsoul and Terezhalmy (2004)
21	Vitamin K	Vitamin	Fruit	Vital for blood clotting and injury healing, anticancer activity, improving bone health and menstrual health, and also enhancing the immune system	Elkhalifa et al. (2021)
22	Folic acid	Vitamin	Fruit	Improve reproductive health and be helpful for the growth and development of the fetus.	Elkhalifa et al. (2021)
23	Thiamine	Vitamin	Fruit	Enhance the nervous system and brain health; improve heart, stomach, muscles, and intestine	Elkhalifa et al. (2021)
24	Riboflavin	Vitamin	Fruit	Essential for growth and development and also regulate total good health	Elkhalifa et al. (2021)
25	Niacin	Vitamin	Fruit	Help to maintain our nervous system and improve digestive system, and skin health	Elkhalifa et al. (2021)

phenyl benzopyran skeleton. They are classified into six subgroups based on pyran ring differences, like flavonols (quercetin, rutin, kaempferol, and quercetatagetin), flavanols (epicatechin), flavanones (hesperidin, naringin, and naringenin), flavones (apigenin), iso-flavonoids (genistein and daidzein), and anthocyanins (cyanidins). Quercetin and its derivatives are the major flavonoids and compounds extracted from okra. Okra's flavonoid compounds, primarily quercetin and its derivatives, have been found to have anticancer properties, particularly in inhibiting tumor growth. Quercetin-rich foods have been shown to target glycolysis, leading to anti-cancer effects. Specifically, quercetin can induce cell death in ovarian cancer cells, slow prostate cancer development, and reduce kidney damage and inflammation in obstructive nephropathy.

22.4 Mucilage and Its Potential Benefits in Okra

Okra contains a thick, slimy substance called mucilage found in both fresh and dried pods, which is a mixture of acidic polysaccharides, proteins, and minerals mainly concentrated in the pod walls (Woolfe et al. 1977). The polysaccharides in okra mucilage are diverse, but common components include neutral sugars like rhamnose, galactose, and galacturonic acid, which are frequently found in its composition (Hirose et al. 2004; Sengkhamparn et al. 2009). Okra mucilage can be brought out by different methods to produce a thick, viscous gum. The varying extraction methods appear to influence the chemical composition of the mucilage, leading to differences in its properties and characteristics (Ndjouenkeu et al. 1996). Its mucilage is a recyclable and cost-effective source of biodegradable material. The okra mucilage gum has a unique combination of properties, including being highly soluble in water, flexible, stretchy, and thick, making it a versatile substance with various potential uses (BeMiller et al. 1993). The characteristics of okra can be altered by various factors, including temperature, sugar levels, acidity (pH), salt content, and how long it's stored (Woolfe et al. 1977). Okra mucilage is a versatile substance with various applications. In food, it's used as a whipping agent for egg whites, an additive in flour-based adhesives, and to clarify sugarcane juice in India. Beyond food, it is used to brighten metals in electroplating, improve paper and fabric production as a deflocculant, and reduce friction in pipelines (BeMiller et al. 1993; Ndjouenkeu et al. 1996). Combining polysaccharides with acrylamide can lead to the creation of innovative, biodegradable polymeric materials (Mishra et al. 2008). The mucilage has potential medicinal uses, including extending the action of serum albumin (BeMiller et al. 1993), as tablet binder (Ofoefule et al. 2001), and as a suspending agent in formulations (Kumar et al. 2009). In Asian medicine, okra mucilage is utilized as a protective agent in food to help safeguard against inflammatory gastric disorders and irritations (Lengsfeld et al. 2004).

22.5 Phenolic Compounds in Okra

Phenolics are a vast and diverse group of plant compounds. Phenolics encompass a vast array of over 8000 biologically active compounds, ranging from basic phenol molecules to complex polymeric structures with molecular weights exceeding 3000 Da. Currently, phenolics are categorized into two primary groups: simple phenols and polyphenols, distinguished by the number of phenol subunits they contain. Simple phenols, which include phenolic acids with a carboxyl group that defines their function, are distinct from polyphenols, which have at least two phenol rings. The "French paradox" study, showing reduced cardiovascular disease in moderate red wine drinkers, sparked interest in phenolic compounds as beneficial food components. These compounds exhibit a range of biochemical activities, including antioxidant, anti-mutagenic, and anti-carcinogenic effects, as well as gene expression.

The antioxidant properties of flavonoids and phenolics in okra can vary depending on the cultivar, with significant differences in content observed in different Greek okra cultivars. Phenolics can directly impact digestive enzymes, influencing carbohydrate and fat breakdown in the human diet. Analyzing phenolic compounds in okra fruits is crucial for understanding their biological characteristics. (Nwachukwu et al. 2014) conducted a study to determine the flavonoid content in okra leaves and fruits and revealed the highest phenolic content in young leaves (0.99 mg/g) and the highest flavonoid content in mature leaves (0.79 mg/g). The antioxidant activity of okra is largely due to its flavonoid content. Key phenolic compounds found in okra fruits include quercetin-3-O-gentiobioside, isoquercitrin, rutin, protocatechuic acid, and catechin derivatives, with quercetin-3-O-gentiobioside being the most abundant (Xia et al. 2015; Arapitsas 2008). For instance, Geng et al. (2015) studied the phenolic compounds extracted from okra flowers and demonstrated potent antioxidant properties, including robust 1,1-diphenyl-2-picrylhydrazyl (DPPH) radical scavenging activity and reducing power, making them a valuable functional ingredient with potential applications in the food and pharmaceutical industries.

22.6 Okra Polysaccharide (OP)

From ancient times, mucilage and gums were known as rich sources of polysaccharides. They are widely used in the pharmaceutical and food industries as water retention agents, thickeners and emulsion stabilizers, suspending agents and binders, etc. Newer applications of polysaccharides have been discovered in the creation of textiles, cosmetics, and the paper industry, in addition to their use in finished pharmaceutical products. Due to its ease of use and superior percutaneous (through the skin) absorption, mucilage has good potential for use in pharmaceutical applications such as diluents, binder, and pharmaceutical excipients (an inactive substance that serves as the vehicle or medium for a drug or other active substance). Synthetic excipients, disintegrants in tablets, thickeners in oral liquid, protective colloids in suspensions, film-forming agents in transdermal and periodontal medications (relating to, denoting, or affecting the gums and other tissues surrounding the teeth) films, and gelling agents in gel base for topical application. The main components of this mucilage are galactose (25%), rhamnose (22%), galacturonic acid (27%), and amino acids (11%; Table 22.3) (Sengkhamparn et al. 2010). Previous research has shown that okra Rhamnogalacturonan improves cell proliferation. Savello et al. (1980) have shown the potential application of okra mucilage used for plasma replacement or blood volume expansion. Similarly, the polysaccharides obtained from some red algae families (Rhodophyta), especially Gracilariaceae and Gelidiaceae (Armisen 1995), are widely used in the cosmetic industry, medicine, and as food. Agar is principally a combination of naturally occurring polymer agarose, pyruvated form of agarose, and inter-polymers of sulfated galactans.

The extraction of okra polysaccharide (OP) and its purification methods used to separate and purify the polysaccharides from the plants all affect the degree to which the polysaccharides maintain or lose their structural integrity and biological functions. Huricha et al. (2016) identified three types of okra polysaccharides with different molecular weights (600, 990, and 1300 kDa) and two groups of monosaccharide components (group −1 galactose, rhamnose, galacturonic acid, and glucuronic acid; and group 2, galactose, rhamnose, galacturonic acid, glucuronic acid, and glucose) were identified using the maceration method. They demonstrated that these okra polysaccharides can act as novel anti-inflammatory drugs, which is supported by future research. Benefits of okra polysaccharide (OP) are presented in Table 22.2.

Table 22.2 Benefits of okra polysaccharides

Benefit	Description	References
Nutritional value	Okra polysaccharides are high in dietary fiber, which supports gastrointestinal health and regulates blood glucose levels.	Lobo et al. (2010)
Antioxidant properties	These polysaccharides exhibit antioxidant activity, which helps to reduce oxidative stress and prevent chronic diseases.	Sreerama and Malleshi (2012)
Immunomodulatory effects	Okra polysaccharides can enhance immune function by influencing cytokine production and immune cell activity.	Zhang and Zheng (2014)
Anti-Inflammatory properties	They have demonstrated the ability to reduce inflammation, which is beneficial for managing inflammatory conditions.	Reddy and Yang (2015)
Anticancer properties	Preliminary studies suggest these polysaccharides may inhibit tumor growth and induce apoptosis in cancer cells.	Reddy and Yang (2015)
Industrial applications	Okra polysaccharides are used as thickening agents, gelling agents, and in drug delivery systems due to their biocompatibility.	Kumar and Roy (2018)

22.7 Documented Pharmacological and Therapeutic Properties

Okra fruits are appreciated for their therapeutic use against diabetes, hyperlipidemia, microbes, ulcers, and neurodegenerative diseases. Okra is considered to be loaded with dietary fiber, especially complex polysaccharides, phenolic compound derivaties of oligomeric catechins, and flavonol by-products. It can be utilized as a potent arsenal to combat pathological, chronic, and long-term maladies. Okra is demonstrated to have a significant role in the regularization of glucose content in the blood. This can pave the way for okra-based nutritional drug therapy. Several findings reported that the main health-stimulating characteristics in this crop are associated with the presence of an appreciable number of bioactive polysaccharides, popularly known as okra polysaccharides, or OPs. Usually, the OPs isolated through hot-water extraction contain high amounts of uronic acids, and its key component monosaccharides are reported as galactose, rhamnose, arabinose, and galacturonic acid. Besides, different plant parts of okra have been extensively utilized in various entho-medicinal systems across the globe (Table 22.3).

22.7.1 Anti-inflammatory and Immunomodulatory Properties

Okra extracts possess anti-inflammatory properties by suppressing pro-inflammatory molecules and enzymes, such as cyclooxygenases and cytokines, which could help alleviate inflammation in various conditions. However, a study found that okra lectin administered intravenously to mice at certain doses (0.01, 0.1, and 1 mg/kg) actually induced inflammation (Soares et al. 2012). A recent study discovered that an ethanol extract of okra, given orally to rats at doses of 500, 250, or 100 mg/kg, effectively reduced inflammation in ethanol-induced gastric mucosal injury (Ortaç et al. 2018). Hydrolyzed okra extract and its polysaccharides (25–100 μg/mL) can modulate immune responses in rat bone marrow cells, leading to increased hemoglobin production, enhanced expression of immune cell surface markers (CD80/86), and reduced endocytosis activity (Sheu and Lai 2012). Additionally, the extract stimulated the production of pro-inflammatory cytokines while suppressing anti-inflammatory cytokine production. Similarly, crude okra polysaccharides were found to enhance immune cell proliferation, nitric oxide production, and cytokine secretion in RAW264.7 cells, indicating immunomodulatory effects (Chen et al. 2016).

22.7.2 Analgesic Activity

Okra extracts are reported to have analgesic (pain-alleviating) properties. The mechanisms underlying this potential effect are not well-established but could be related to the modulation of pain pathways or reduction of inflammation, which can contribute to pain relief. It is important to note that most of the research on okra's

Table 22.3 Ethno-medicinal use of okra in various parts of the world

SN	Ethno-medicinal system	Ethno-medicinal application	Plant parts utilized	References
1	Indian	For improving hair health	Leaf	Chowdhury et al. (2019), Odedra and Nathabhai (2009)
2	Indian	To treat diseases related to the respiratory system, such as bronchitis and pneumonia	Flower	Marwat et al. (2011), Dubey and Mishra (2018), Chowdhury et al. (2019)
3	Indian	To treat ailments related to the digestive system (dysentery, diarrhea, and stomach inflammation), urinary system (kidney infection, ardour urinae, dysuria, and as diuretics), fever, gonorrhea to protect the eye as demulcent and emollient and also as poultice	Fruit	Chowdhury et al. (2019), Dubey and Mishra (2018)
4	Indian	Use to relieve cramps or spasms of the stomach, intestines, and bladder; as a cordial and stimulant; for treatment of excessive, involuntary seminal discharge; as a sudorific agent	Seed	Chowdhury et al. (2019), Dubey and Mishra (2018)
5	Indian	For the treatment of syphilis and also use for demulcent and emollient properties	Root	Chowdhury et al. (2019), Dubey and Mishra (2018), Odedra and Nathabhai (2009)
6	Turkish	Treatment of diabetes mellitus as it is capable of enhancing blood glucose level	Seed	Chowdhury et al. (2019), Dubey and Mishra (2018), Calisir et al. (2005)
7	Turkish and Nicaraguan	To improve the function of the stomach and to increase appetite; for the treatment of diabetes and ulcers; and also applied as purgative to cure jaundice	Root	Chowdhury et al. (2019), Barrett (1994), Yesilada et al. (1995)
8	Latin American	Against tumor	Leaf	Chowdhury et al. (2019), Dubey and Mishra (2018)
9	Latin American	For the treatment of tumors	Seed	Chowdhury et al. (2019), Dubey and Mishra (2018)
10	Nepalese	For healing of wounds, cuts, and ulcers	Root	Chowdhury et al. (2019), Dubey and Mishra (2018)

medicinal properties is still in the early stages, and more comprehensive studies, including clinical trials in humans, are needed to confirm and fully understand these effects. Additionally, the concentration and bioavailability of these bioactive compounds in the human body after consuming okra may vary based on factors such as preparation methods and individual differences. If considering okra for its potential health benefits, consult with a healthcare professional to ensure safety and appropriateness for your specific condition and medical history. Always remember that natural remedies should not replace medical treatments prescribed by qualified healthcare providers.

22.7.3 Antioxidant Efficacy

Okra pods, widely consumed as a vegetable globally, are actually immature fruits with antioxidant properties (Gemede et al. 2018; Wang et al. 2018). Research has shown that the seeds of these immature pods contain high levels of polyphenols (29.5%), contributing to their antioxidant efficiency. After consuming food or beverages, flavonoids in the edible matrix must pass from the gastrointestinal tract into the bloodstream to be absorbed. Since most plant flavonoids are present as glycosides, the attached sugar needs to be removed post-consumption for absorption to take place (Williamson et al. 2018). The polyphenols found in immature fruits of okra exhibit antioxidant activity by reducing MDA levels while enhancing the levels of SOD and glutathione peroxidase (GSH-Px) (Shen et al. 2019; Gemede et al. 2015). Quercetin-3-O-gentiobioside is the main contributor to antioxidant capacity and also hinders lipase, α-glucosidase, and α-amylase, which are digestive enzymes (Shen et al. 2019). The ability of okra pods to scavenge free radicals and reduce ferric ions has been highlighted in prior research. In one particular study, extracts from okra, obtained through both cold extraction and boiling in water, were shown to possess significant antioxidant properties (Ansari et al. 2005). Seeds of okra are also a valuable source of phenolic compounds, particularly procyanidin B1 and B2, which contribute to DPPH (1,1-diphenyl-2-picrylhydrazyl) and ABTS (2,20-azino-bis (3-ethylbenzothiazoline-6-sulfonic acid) free radical scavenging activities (Khomsug et al. 2010), not only confirmed the antioxidant and ferric-reducing prowess of okra pods but also identifies two specific glucopyranoside compounds, namely 5,7,30,40 -tetrahydroxy-4”-O-methylflavonol3-O-β-D-glucopyranoside and 5,7,30,40 -tetrahydroxy flavonol−3-O-(β-D-glucopyranosyl- (1→6))-β-D-glucopyranoside. These compounds are responsible for okra’s impressive antioxidant activity (Yuan et al. 2012). Later reports indicated that other parts of the okra plant, such as the flowers, leaves, seeds, and fruits, also demonstrated significant antioxidant activity (Yuan et al. 2012). In a study focused on desirability, okra seed and peel powder were found to enhance hepatic, renal, and pancreatic levels of SOD and glutathione peroxidase in streptozotocin-induced diabetic models.

Additionally, biochemical analyses indicated that the anti-fatigue activity of okra seeds involves lowering blood lactic acid and urea nitrogen levels, enhancing hepatic glycogen storage, and boosting antioxidant capacity by reducing malondialdehyde levels while increasing superoxide dismutase and glutathione peroxidase levels (Xia et al. 2015).

22.7.4 Antidiabetic and Anti-hyperlipidemic Properties

Neither humans nor animals have been specifically studied for okra's anticonvulsant (the ability to prevent or lessen the intensity and frequency of seizures) qualities. Prescribed anticonvulsant drugs are the first medications used to treat epilepsy, and through scientific research and clinical trials, their safety and efficacy are well established. Its seeds have the ability to manage enhanced blood glucose levels. Tomoda et al. (1985) revealed that polysaccharides present in okra exhibit anti-complementary and hypoglycemic effects in normal mice. Okra has shown hypolipidemic activity in both rat models and in mice in in vivo conditions. Its polysaccharide can reduce the blood cholesterol and potentially retard cancer through its capacity to bind bile acids (Kahlon et al. 2007). Cholesterol levels were reduced by 56.45%, 55.65%, 41.13%, 40.50%, and 53.63% in mice groups that were orally administered with dichloromethane and methanol okra plant and fruit extracts and simvastatin as compared to the group injected with tyloxapol (Nagoc et al. 2008). Administering powder of seed and peel at 100 and 200 mg/kg at two different doses to diabetic rats resulted in a considerable ($P < 0.001$) decrease in glucose concentration in blood and an enhancement in body weight compared to the diabetic control rats. A study examined the impact of okra fruit's water-soluble fraction on the absorption of glucose and metformin in the gastrointestinal tract of Long Evans rats. The results showed that this fraction significantly decreased the absorption of glucose in rats that had fasted for 24 hours (Khatun et al. 2011). Thanakosai and Phuwapraisirisan (2013) discovered two significant flavonol glucosides, iso-quercetin and quercetin 3-O-betaglucopyranosyl-($1'' \rightarrow 6''$)-glucoside, in okra seeds, which are α-glucosidase inhibitors. These compounds effectively inhibited enzymes (maltase and sucrase) in the rat intestine, with iso-quercetin being 6–10 times more potent. Additionally, Subrahmanyam et al. (2011) found that okra fruit extract exhibited antidiabetic properties. Okra could therefore be used in the treatment of hyperglycemia and hypertriglyceridemia by diet (Fan et al. 2014).

22.7.5 Enhancement of Nerve Function and Against Neurological Disorders

The nervous system that plays a crucial role in receiving, processing, and transmitting sensations throughout the body organs deteriorates due to a number of circumstances, including aging and environmental toxins. Okra is one of the natural compounds being investigated to prevent nervous system harm. While research on okra's potential to enhance nerve function is still limited, it presents a promising area for exploration as a novel alternative therapeutic approach. Application of a synthetic glucocorticoid receptor agonist, the dexamethasone, causes death in the CA3 layer of the hippocampus, which has been linked with reduced learning and reminiscence. Okra extract and its derivatives, particularly quercetin and rutin, were reported to protect neuronal function and enhance learning and reminiscence discrepancies in rats undergoing this treatment (Tongjaroenbuangam et al. 2011).

Okra treatment provides important in vitro data regarding the effects of okra on various cellular processes associated with AID in H63D variant HFE cells. These results suggest that okra may be beneficial for individuals expressing the H63D variant to reduce the risk of Alzheimer's disease and other neurodegenerative diseases associated with oxidative stress (Mairuae et al. 2015).

22.7.6 Anticancer Activities

Gold nanoparticles derived from okra were employed to assess their effects on three cancer cell lines in vitro, namely, Jurkat (human acute myeloid leukemia), K562 (human chronic myeloid leukemia), and DL (Dalton's lymphoma). The cells were exposed to varying concentrations of okra Au nanoparticles at 0, 1, 5, 10, 25, and 50 μg/ml for a period of 24 hours. The MTT assay was then used to measure the cytotoxicity caused by the Au nanoparticles. The study found that gold nanoparticles derived from okra up to the 50 μg/ml concentration significantly reduced the viability of Jurkat cancer cells in a dose-dependent fashion. Precisely, Au nanoparticle concentrations of 5, 10, 25, and 50 μg/ml reduced the Jurkat cell viability by 45.10%, 48.60%, 81.30%, and 87.20%, respectively. Likewise, K562 and DL cancer cells exhibited substantial viability declines of 38.38% and 50.165%, and 28.51% and 48.165% at 25 and 50 μg/ml doses, respectively (Mollick et al. 2014).

Devanesan and AlSalhi (2021) revealed that the silver (Ag) nanoparticles isolated using okra flower extract showed strong cytotoxic activity against cancer cells. Ag nanoparticles considerably reduced cell viability in a concentration-dependent way in both lung cancer (A-549) and mesenchymal cancer (TERT-4) cell lines. Concentrations of 25 and 50 μg/ml showed substantial cell death, with even more noticeable effects at 100 μg/ml, outperforming the control drug 5-fluorouracil. The IC50 values, indicating the concentration required to inhibit 50% of cell growth, were 1.74 μg/ml and 1.65 μg/ml for A-549 and TERT-4 cells, respectively.

Nevertheless, okra has shown potential anticancer properties through numerous studies. Dried okra seeds reduced tumor necrosis factor-alpha expression in 3 T3-L1, adipocytes (Okada et al. 2010). Okra lectin increased proapoptotic gene expression in breast cancer and skin fibroblast cells, promoting cell death. Okra lectin at 0.0125, 0.025, 0.05, and 0.1 μg/ml increased Bax/Bcl-2 ratio, proapoptotic caspase-3, caspase-9, and p21 gene expression in breast cancer (MCF7) and skin fibroblast (CCD-1059 sk) cells, stimulating cell death (Monte et al. 2013).

22.7.7 Antimicrobial Properties

Previous research has shown that the lipid components, specifically palmitic and stearic acids, found in lyophilized okra pod extracts and their aqueous counterparts, exhibit antimicrobial properties against a range of microorganisms, including *Rhodococcus opacus*, *Rhodococcus erythropolis*, *Mycobacterium aurum*, *Escherichia coli*, *Staphylococcus aureus*, *Pseudomonas aeruginosa*, and

Xanthobacter Py2 (Chen et al. 2016). In addition to its antimicrobial properties, okra extract has also been used to synthesize gold nanoparticles, which have shown significant microbicidal activity against a range of bacteria, including *Bacillus cereus*, *Bacillus subtilis*, *E. coli*, *P. aeruginosa*, and *M. luteus* (De Carvalho et al. 2011). Moreover, carbohydrate-rich fractions of okra pods have been found to exhibit activity against *Helicobacter pylori*, a bacterium associated with stomach ulcers and gastric cancer (Lengsfeld et al. 2004).

Polysaccharide extracts obtained from unripe okra fruit are known to possess antiadhesive properties due to the bacterial adhesion of Helicobacter pylori to the gastric tissue. This study investigates the structure and function of polymers responsible for inhibiting bacterial adhesion to cells. Ammonium sulfate precipitation of the aqueous extract gave two fractions at 60% and 90% saturation with significant protection against *H. pyori*, type J99, after pre-incubation of bacteria at 1 mg/mL (Thole et al. 2015).

22.7.8 Useful During Pregnancy

Okra being a nutrient-dense vegetable is particularly beneficial for pregnant women due to its rich content of vitamins A, B1, B2, B6, and C, as well as zinc, calcium, fiber, and folic acid. These essential nutrients support overall health and well-being during pregnancy while also providing specific benefits like preventing miscarriages and supporting the formation and protection of the neural tube in the fetus. This is useful to avoid birth defects such as spina bifida and can even alleviate constipation throughout pregnancy. Folate played a significant role in women, which prevented the birth defects in developing babies. Ascorbic acid and folate are vital nutrients that hasten the growth and development of the fetus's brain. High quantities of folic acid in okra promote the formation of neural tubes in the fetus during the 4th to 12th week of pregnancy (Zaharuddin et al. 2014).

22.7.9 Gastroprotective Effect

The gastroprotective effect of okra lectin (AEL) on ethanol-induced gastropathy was studied. Fasted mice, pretreated with ethanol (99.9%), were administered AEL (0.01, 0.1, 1.0, 10, or 50 mg/kg), saline (5 ml/kg), or ranitidine (80 mg/kg) in a volume of 0.2 ml per animal. To confirm the stomach injury, the mice were put to death 30 minutes after the ethanol treatment. To recognize the AEL gastroprotective effect, measurements of gastric oxidative stress, tissue hemoglobin concentration, and microscopic characteristics were made. AEL (1 mg/kg) could prevent mucosa damage from ethanol in the presence of two preventive properties of AEL identified through analysis of microscopic characteristics, oxidative stress, and Hb levels. AEL has simultaneously shown an antioxidant impact that is likely connected to its complex defense mechanism (Ribeiro et al. 2016).

22.7.10 Nootropic Potential to Enhance Memory or Other Cognitive Functions

Okra seed extract exhibits antioxidant, antianxiety, and nootropic activities, which should support the medicinal properties of okra. Therefore, pretreatment of rats with okra seeds aqueous extract and methanol seed extract (200 mg/kg, orally) for 7 days significantly ($P < 0.01$) attenuated scopolamine-induced passive cognitive impairment in the avoidance test. These extracts reduce blood glucose, corticosterone, cholesterol, and triglyceride levels in response to stress (Doreddula et al. 2014).

22.7.11 Immunomodulating Activity

Water-soluble polysaccharides (OFPS11) were extracted from okra flowers by aqueous extraction and purified using DEAE-52 cellulose and sephacryl™ S-500 each. Preliminary characterization and studies on its immunomodulatory activity showed that OFPS11 was mainly composed of galactose and rhamnose in a ratio of 2.23:1 and had a molecular weight of 1700 kDa. Pretreatment of RAW264.7 cells with OFPS11 reduced the proliferation of HepG-2 cells. In addition, OFPS11 also improved the phagocytic activity of RAW264.7 cells, increased nitric oxide production, and the release of TNF-α. These findings suggest that OFPS11 may employ its anticancer effects by promoting macrophage activation via the molecular NF-kB pathway (Zheng et al. 2014).

22.7.12 Hepatoprotective Activity

Antioxidant performance was evaluated using DPPH and hydroxyl radical scavenging assays, and the hepatoprotective effect of the extract was evaluated by estimating the levels of liver and antioxidant markers using CCI™-intoxicated HepG™ cell lines in Wistar rats. The IC50 values of okra root extract were 270.99 and 532.86 μg/ml for DPPH and hydroxyl radical scavenging assays, respectively. Incubation of HepG2 cells with CCI™ decreased cell viability and increased aminotransferase leakage. Pretreatment with the essential extract induced 32.25% and 39.04% cell death at concentrations of 200 and 400 μg/ml, respectively. Compound with cells treated with CCI at 100, 200, and 400 μg/ml, ALT leakage was reduced by 18.62%, 38.59%, and 52.15% after treatment, respectively. In in vivo experiments, the treatment also reduced transaminases, ALP, MDA, total bilirubin, elevated TNF-α levels, and increased antioxidant activity in a dose-dependent manner. Histological examination of the liver showed that steatosis (fatty infiltration of liver cells associated with metabolic disease such as alcohol, malnutrition, pregnancy, or drug therapy), necrosis, and inflammation were reduced (Saravanan et al. 2013).

22.7.13 Cardiovascular Disease

Heart problems can be tackled with the consumption of okra in our diet. Okra is rich in pectin which helps lower the serum cholesterol and has the ability to reduce the risk of disease related to the heart by altering the production of bile in the intestines. Fruits of okra also helped in maintaining the level of body cholesterol in humans (Ngoc et al. 2008).

22.7.14 Eye-Sight Improvement and Skin Nourishment

Folate is the natural form of vitamin B9, also found in okra fruit. Folic acid is a water-soluble vitamin and is naturally found in many foods. In vegetable crops, though the leafy vegetables are the best dietary source of folic acid, it is also present in okra. Folic acid is necessary for the synthesis DNA and other genetic material, cell division, producing red and white blood cells in bone marrow, and transforming carbohydrates into energy. Okra pods are useful for eyesight improvement and an excellent source of vitamin A and beta-carotene, both nutrients essential for good vision and healthy skin (Messing et al. 2014).

22.8 Conclusion and Future Perspectives

The nutritional and therapeutic characteristics of the various plant parts, including fruits of okra, documented in this chapter, with special emphasis on human nutrition and health-promoting properties. It is regarded as a low-cost and financially inexpensive vegetable and usually sources several nutrient components such as carbohydrates, proteins, fats, dietary fiber, vitamins, and macro- and micro-elements. The human well-being of okra is attributed due to the presence of numerous bioactive phytochemicals in appreciable amounts. Even though the possible benefits of okra fruit on a number of chronic ailments have been systematically studied and documented fairly, still many aspects, including nutrigenomics, bioavailability, and pharmacokinetics, along with precise modes and mechanisms of function for particular health problems, require additional research and developments. This information might have risen because of the multifaceted and complicated disease etiology and different associated influences for disease development. Owing to its nutritional richness, okra can be practically utilized to alleviate malnutrition and hidden hunger in various underdeveloped and developing nations all around the globe. Additionally, the nutraceuticals formulation based on okra plant parts might substantiate to be useful, because of its easy availability and economical in nature. Nevertheless, okra fruits provide ample opportunities for their application as potent natural antioxidants and natural medicine to control hyperlipidemia and hyperglycemia and to improve heart and gut health. The future lines of research must focus on okra nutrigenomics, the development of nutraceuticals, functional foods, and natural medicines from the okra plant.

References

Adelakun OE, Oyelade OJ (2011) Chemical and antioxidant properties of okra (*Abelmoschus esculentus* moench) seed. In: Nuts and seeds in health and disease prevention. Academic Press, pp 841–846

Alessandra D, Massimo L, Ettore N, Eliana B, Patricia D, Antonello S (2019) Abelmoschus esculentus (L.): bioactive components' beneficial properties focused on antidiabetic role for sustainable health applications. Molecules 24(1):38

Ansari NM, Houlihan L, Hussain B, Pieroni A (2005) Antioxidant activity of five vegetables traditionally consumed by South-Asian migrants in Bradford, Yorkshire, UK. Phytother Res 19:907–911

Arapitsas P (2008) Identification and quantification of polyphenolic compounds from okra seeds and skins. Food Chem 110:1041–1045

Araújo A, Galvão A, Filho C, Mendes F, Oliveira M, Barbosa F, Filho M, Bastos M (2018) Okra mucilage and corn starch bio-based film to be applied in food. Polym Test 71:352–361

Ariel R, Vicente GA, Manganaris GO, Sozzi GO, Crisosto CH (2009) Nutritional quality of fruits and vegetables. Postharvest handling: a systems approach, 2nd edn. Elsevier Inc. ISBN: 978-0-12-374112-7

Armisen R (1995) World-wide use and importance of Gracilaria. J Appl Phycol 7:231–243

Axe J (2021) Okra Nutrition: Improve Heart Health, Eyesight & Cholesterol Levels https://draxe.com/nutrition/okra-nutrition/ Bates, D. M. 1968. Notes on the cultivated Malvaceae 2 Abelmoschus. Baileya.16:99–112

Barrett B (1994) Medicinal plants of Nicaragua's Atlantic Coast. Econ Bot 48:8–20

BeMiller JN, Whistler RL, Barkalow DG, Chen CC (1993) Aloe, chia, flax seed, okra, psyllium seed, quince seed, and tamarind gums. In: Whistler RL, BeMiller JN (eds) Industrial Gums. Academic Press, New York, pp 227–256

Benchasri S (2012) Okra (*Abelmoschus esculentus* (L.) Moench) as a valuable vegetable of the world. Field Veg Crops Res 49(1):105–112

Bertinato J (2021) Iodine nutrition: disorders, monitoring and policies. In: Advances in food and nutrition research, vol 96. Academic Press, pp 365–415

Blomhoff R (1994) Vitamin A in health and disease. CRC Press

Bsoul SA, Terezhalmy GT (2004) Vitamin C in health and disease. J Contemp Dent Pract 5(2):1–13

Calisir S, Ozcan M, Haciseferogullari H, Yildiz MU (2005) A study on some physico-chemical properties of Turkey okra (Hibiscus esculentus) seeds. J Food Eng 68:73–78

Chen H, Jiao H, Cheng Y, Xu K, Jia X, Shi Q, Wang F (2016) In vitro and in vivo immunomodulatory activity of okra (*Abelmoschus esculentus* L.) polysaccharides. J Med Food 19:253–265

Chowdhury NS, Jamaly S, Farjana F, Begum N, Zenat EA (2019) A review on ethnomedicinal, pharmacological, phytochemical and pharmaceutical profile of lady's finger (Abelmoschus esculentus L.) plant. Pharm Pharm 10:94–108

Das S, Ghosh LK (2019) Okra and its various applications in drug delivery, food technology, health care and pharmacological aspects—a review. J Pharm Sci Res 11:2139–2147

De Carvalho CCCR, Cruz PA, Da Fonseca MMR, Xavier-Filho L (2011) Antibacterial properties of the extract of *Abelmoschus esculentus*. Biotechnol Bioprocess Eng 16:971–977

De-Rosa I, Kenny J, Puglia D, Santulli C, Sarasini F (2010) Morphological, thermal and mechanical characterization of okra (*Abelmoschus esculentus*) fibres as potential reinforcement in polymer composites. Compos Sci Technol 70(1):116–122

Devanesan S, AlSalhi MS (2021) Green synthesis of silver nanoparticles using the flower extract of *Abelmoschus esculentus* for cytotoxicity and antimicrobial studies. Int J Nanomedicine 16:3343–3356

Dias S (2012) Nutritional quality and health benefits of vegetables: a review. Food Nutr Sci 03(10):1354–1374

Doreddula SK, Bonam SR, Gaddam DP, Desu BS, Ramarao N, Pandy V (2014) Phytochemical analysis, antioxidant, antistress, and nootropic activities of aqueous and methanolic seed extracts of lady's finger (*Abelmoschus esculentus* L.) in mice. Sci World J 2014(1):519848
Dubey P, Mishra S (2018) Effect of okra seed in treatment of hypoglycemia: a research framework using STZ induced rat. J Med Plants 6(3):85–88
Durazzo A, Lucarini M, Novellino E, Souto EB, Daliu P, Santini A (2018) Abelmoschus esculentus (L.): bioactive components' beneficial properties—focused on antidiabetic role for sustainable health applications. Molecules 24:38
Eagri (2023) Origin, area, production, varieties, package of practices for bhendi (syn: lady's finger, bhindi) (*Abelmoschus esculentus* (l.) moench) (2n = 130). (hindi: bhendi. http://eagri.org/eagri50/HORT281/pdf/lec06.pdf
Elkhalifa AEO, Alshammari E, Adnan M, Alcantara JC, Awadelkareem AM, Eltoum NE, Mehmood K, Panda BP, Ashraf SA (2021) Okra (*Abelmoschus esculentus*) as a potential dietary medicine with nutraceutical importance for sustainable health applications. Molecules 26:696
Fan S, Zhang Y, Sun Q, Yu L, Li M, Zheng B, Wu X, Yang B, Li Y, Huang C (2014) Extract of okra lowers blood glucose and serum lipids in high-fat diet-induced obese C57BL/6 mice. J Nutr Biochem 25:702–709
Fangbo X, Yu Z, Mengqiu L, Qi C, Yonghong L, Xinmin L, Ruile P (2015) Antioxidant and anti-fatigue constituents of okra. Nutrients 7:8846–8858
Gemede HF, Ratta N, Haki GD, Woldegiorgis AZ, Beyene F (2015) Nutritional quality and health benefits of okra (*Abelmoschus esculentus*): a review. J Food Process Technol 6(458):2
Gemede HF, Haki GD, Beyene F, Rakshit SK, Woldegiorgis AZ (2018) Indigenous Ethiopian okra (*Abelmoschus esculentus*) mucilage: a novel ingredient with functional and antioxidant properties. Food Sci Nutr 6:563–571
Geng S, Liu Y, Ma H, Chen C (2015) Extraction and antioxidant activity of phenolic compounds from okra flowers. Trop J Pharm Res 14:807–814
Gonzales GB, Smagghe G, Grootaert C, Zotti MJ, Raes K, Van Camp J (2015) Flavonoid interactions during digestion, absorption, distribution and metabolism: a sequential structure–activity/property relationship-based approach in the study of bioavailability and bioactivity. Drug Metab Rev 47:175–190
Graham JO, Agbenorhevi JK, Kpodo FM (2017) Total phenol content and antioxidant activity of okra seeds from different genotypes. Am J Food Nutr 5(3):90–94
Gunarathne R, Guan X, Feng T, Zhao Y, Lu J (2024) L-lysine dietary supplementation for childhood and adolescent growth: promises and precautions. J Adv Res
Hirose K, Endo K, Hasegawa K (2004) A convenient synthesis of lepidimoide from okra mucilage and its growth-promoting activity in hypocotyls. Carbohydr Poly 339:9–19
Honda AH, Nakagawa S, Ashida H, Kanazawa K (2003) Simultaneous determination of all polyphenols in vegetables, fruits, and teas. J Agric Food Chem 51(3):571–581
Hung NM, Hiep NV, Dung BN, Hai NX (2014) Lead accumulation in different parts of okra plant (Abelmoschus esculentus). ARPN J Agric Biol Sci 6:190–194
Huricha C, Hanwei J, Ying C, Kailian X, Xiaoxiao J, Qiaoyun S, Shiyu G, Manchuriga W, Li D, Fengyang W (2016) In vitro and in vivo immunomodulatory activity of okra (Abelmoschus esculentus L.) polysaccharides. J Med Food 19(3):253–265
Jiang N, Liu C, Li D, Zhang Z, Liu C, Wang D, Niu L, Zhang M (2017) Evaluation of freeze drying combined with microwave vacuum drying for functional okra snacks: antioxidant properties, sensory quality, and energy consumption. LWT Food Sci Technol 82:216–226
Kahlon TS, Chapman MH, Smith GE (2007) In vitro binding of bile acids by okra, beets, asparagus, eggplant, turnips, green beans, carrot and cauliflower. Food Chem 103:676–680
Karmakar P, Sagar V, Singh P (2022) Dynamics of anthocyanin and chlorophyll content in red fruited okra var. Kashi Lalima. Veg Sci 49(02):197–203
Khatun H, Rahman A, Biswas M, Islam AU (2011) Watersoluble fraction of Abelmoschus esculentus L. interacts with glucose and metformin hydrochloride and alters their absorption kinetics after Coadministration in rats. ISRN Pharm 2011(1):260537

Khomsug P, Thongjaroenbuangam W, Pakdeenarong N, Suttajit M, Chantiratikul P (2010) Antioxidative activities and phenolic content of extracts from okra (*Abelmoschus esculentus* L.). Res J Biol Sci 5:310–313

Konstantinos A, Dimitrios K, Evan D (2009) Flocculation behavior of mallow and okra mucilage in treating wastewater. Desalination 249(2):786–791

Kumar S, Roy A (2018) Industrial applications of okra polysaccharides. Carbohydr Polym 196:473–482

Kumar R, Patil MB, Patil SR, Paschapur MS (2009) Evaluation of *Abelmoschus esculentus* mucilage as paracetamol suspension. Int J Pharm Tech Res 1:658–665

Lazalde-Cruz R, Miranda-Romero LA, Tirado-González DN, Carrillo-Díaz MI, Medina-Cuéllar SE, Mendoza-Martínez GD, Lara-Bueno A, Tirado-Estrada G, Salem AZM (2021) Potential effects of Delphinidin-3-O-Sambubioside and Cyanidin-3-O-Sambubioside of *Hibiscus sabdariffa* L. on ruminant meat and milk quality. Animals (Basel) 11(10):2827

Lengsfeld C, Titgemeyer F, Faller G, Hensel A (2004) Glycosylated compounds from okra inhibit adhesion of *Helicobacter pylori* to human gastric mucosa. J Agric Food Chem 52:1495–1503

Liao H, Dong W, Shi X, Liu H, Yuan K (2012) Analysis and comparison of the active components and antioxidant activities of extracts from *Abelmoschus esculentus* L. Pharmacogn Mag 8(30):156–161

Liu J, Zhao Y, Wu Q, John A, Jiang Y, Yang J, Liu H, Yang B (2018) Structure characterisation of polysaccharides in vegetable "okra" and evaluation of hypoglycemic activity. Food Chem 242:211–216

Lobo V, Patil A, Phatak A, Chandra N (2010) Free radicals, antioxidants, and functional foods: impact on human health. Pharmacogn Rev 4(8):118–126

Lu Y, Demleitner MF, Song L, Rychlik M, Huang D (2016) Oligomeric proanthocyanidins are the active compounds in Abelmoschus esculentus Moench for its α-amylase and α-glucosidase inhibition activity. J Funct Foods 20:463–471

Macedo C, Costa PC, Rodrigues F (2023) Bioactive compounds from Actinidia arguta fruit as a new strategy to fight glioblastoma. Food Res Int 175:113770

Mairuae N, Connor JR, Lee SY, Cheepsunthorn P, Tongjaroenbuangam W (2015) The effects of okra (*Abelmoschus esculentus* Linn.) on the cellular events associated with Alzheimer's disease in a stably expressed HFE neuroblastoma SH-SY5Y cell line. Neurosci Lett 603:6–11

Marangoni F, Agostoni C, Borghi C, Catapano AL, Cena H, Ghiselli A, La Vecchia C, Lercker G, Manzato E, Pirillo A, Riccardi G (2020) Dietary linoleic acid and human health: focus on cardiovascular and cardiometabolic effects. Atherosclerosis 292:90–98

Maret W, Sandstead HH (2006) Zinc requirements and the risks and benefits of zinc supplementation. J Trace Elem Med Biol 20(1):3–18

Martins T, Barros AN, Rosa E, Antunes L (2023) Enhancing health benefits through chlorophylls and chlorophyll-rich agro-food: a comprehensive review. Molecules 28(14):5344

Marwat SK, Rehman FR, Khan MA, Ahmed A, Zafar M, Gulam S (2011) Medicinal folk recipes used as traditional phytotherapies in district Dera, Ismail Khan, KPK, Pakistan. J Bot 43:1453–1462

Maurya PK (2022) Health benefits of Quercetin in age-related diseases. Molecules 27(8):2498

Messing J, Thöle C, Niehues M, Shevtsova A, Glocker E, Borén T, Hensel A (2014) Antiadhesive properties of Abelmoschus esculentus (Okra) immature fruit extract against helicobacter pylori adhesion. PLoS One 9:e84836

Mishra A, Clark JH, Pal S (2008) Modification of Okra mucilage with acrylamide: Synthesis, characterization and swelling behavior. Carbohydr Polym 72(4):608–615

Mishra GP, Singh B, Seth T, Singh AK, Halder J, Krishnan N, Tiwari SK, Singh PM (2017) Biotechnological advancements and begomovirus management in okra (Abelmoschus esculentus L.): status and perspectives. Front Plant Sci 8:360

Mollick MMR, Bhowmick B, Mondal D, Maity D, Rana D, Dash SK, Chattopadhyay D (2014) Anticancer (in vitro) and antimicrobial effect of gold nanoparticles synthesized using *Abelmoschus esculentus* (L.) pulp extract via a green route. RSC Adv 4:37838–37848

Monika A, Rajani S, Anuradha M (2001) Study on flocculation efficiency of okra gum in sewage waste water. Macromol Mater Eng 286(9):560–563

Monte LG, Santi-Gadelha T, Reis LB, Braganhol E, Prietsch RF, Dellagostin OA, Lacerda ERR, Gadelha CAA, Conceição FR, Pinto LS (2013) Lectin of Abelmoschus esculentus (okra) promotes selective antitumor effects in human breast cancer cells. Biotechnol Lett 36:461–469

Nagy K, Tiuca ID (2017) Importance of fatty acids in physiopathology of human body. In: Catala A (ed) Fatty acids. IntechOpen, pp 3–22

Naveed A, Khan AA, Khan IA (2009) Generation mean analysis of water stress tolerance in okra (*Abelmoschus esculentus* L.). Pak J Bot 41:195–205

Ndjouenkeu R, Goycoolea FM, Morris ER, Akingbala JO (1996) Rheology of okra (*Hibiscus esculentus* L.) and dika nut (*Irvingia gabonensis*) polysaccharides. Carbohydr Polym 29:263–269

Ndunguru J, Rajabu A (2004) Effect of okra mosaic virus disease on the above-ground morphological yield components of okra in Tanzania. Sci Hortic 99:225–235

Ngoc T, Ngo N, Van T, Phung V (2008) Hypolipidemic effect of extracts from *Abelmoschus esculentus* L. (Malvaceae) on Tyloxapol-induced hyperlipidemia in mice. Warasan Phesatchasat 35:42–46

Nwachukwu EC, Nulit R, Go R (2014) Nutritional and biochemical properties of Malaysian okra variety. Adv Med Plant Res 2:16–19

Odedra, Nathabhai K (2009) Ethnobotany of Maher Tribe In Porbandar District, Gujarat, India. Thesis PhD, Saurashtra University

Ofoefule SI, Chukwu AN, Anayakoha A, Ebebe IM (2001) Application of *Abelmoschus esculentus* in solid dosage forms I: use as binder for poorly water-soluble drug. Indian J Pharm Sci 63:234–238

Okada Y, Okada M, Sagesaka Y (2010) Screening of dried plant seed extracts for adiponectin production activity and tumor necrosis factor-alpha inhibitory activity on 3T3-L1 adipocytes. Plant Foods Hum Nutr 65:225–232

Olano-Martin E, Rimbach GH, Gibson GR, Rastall RA (2003) Pectin and pectic-oligosaccharides induce apoptosis in in vitro human colonic adenocarcinoma cells. Anticancer Res 23:341–346

Ortaç D, Cemek M, Karaca T, Büyükokuroğlu ME, Özdemir ZÖ, Kocaman AT, Göneş S (2018) In vivo anti-ulcerogenic effect of okra (*Abelmoschus esculentus*) on ethanol-induced acute gastric mucosal lesions. Pharm Biol 56:165–175

Pawar SH, Upganlawar AB, Upasani CD (2023) Therapeutic benefits of phenolic acids in peripheral neuropathy. In: Treatments, nutraceuticals, supplements, and herbal medicine in neurological disorders. Academic Press, pp 753–771

Priya S, Bahadur V, Topno SE (2024) Study of plant growth regulators on growth, yield, and quality of Red Okra (*Abelmoschus esculentus* cv. Kashi Lalima) in Prayagraj, Uttar Pradesh, India. J Adv Biol Biotechnol 27(9):504–513

Rakesh K, Vishal P (2016) Phytochemical, nutritional and pharmacological evidences for *Abelmoschus esculentus* (L.). J Phytopharmacol 5(6):238–241

Reddy N, Yang Y (2015) Anti-inflammatory and anticancer properties of okra polysaccharides: an overview. J Food Sci Technol 52(5):2841–2850

Ribeiro KA, Chaves HV, Filho SM (2016) Alpha-2 adrenergic and opioids receptors participation in mice Gastroprotection of *Abelmoschus esculentus* lectin. Curr Pharm Desig 22:4736–4742

Sabitha V, Ramachandran S, Naveen KR, Panneerselvam K (2011) Antidiabetic and antihyperlipidemic potential of *Abelmoschus esculentus* (L.) Moench in streptozotocin-induced diabetic rats. J Pharm Bioallied Sci 3(3):397–402

Saha D, Jain B, Jain VK (2011) Phytochemical evaluation and characterization of hypoglycemic activity of various extracts of *Abelmoschus esculentus* Linn. Fruit. Int J Pharm Pharm Sci 3(2):183–185

Saravanan S, Pandikumar P, Pazhanivel N, Paulraj MG, Ignacimuthu S (2013) Hepatoprotective role of *Abelmoschus esculentus* (*Linn.*) Moench on carbon tetrachloride-induced liver injury. Toxicol Mech Methods 23(7):528–536

Savello PH, Martin FW, Hill JM (1980) Nutritional composition of okra seed meal. J Agric Food Chem 28:1163–1166

Semwal R, Joshi SK, Semwal RB, Semwal DK (2021) Health benefits and limitations of rutin-A natural flavonoid with high nutraceutical value. Phytochem Lett 46:119–128

Sengkhamparn N, Verhoef R, Schols HA, Sajjaanantakul T, Voragen AG (2009) Characterisation of cell wall polysaccharides from okra (*Abelmoschus esculentus* (L.) Moench). Carbohydr Res 344:1824–1832

Sengkhamparn N, Sagis LMC, de Vries R, Schols HA, Sajjaanantakul T, Voragen AGJ (2010) Physicochemical properties of pectins from okra (Abelmoschus esculentus (L.) Moench). Food Hydrocolloid 24:35–41

Shen DD, Li X, Qin YL, Li MT, Han QH, Zhou J, Lin S, Zhao L, Zhang Q, Qin W (2019) Physicochemical properties, phenolic profiles, antioxidant capacities, and inhibitory effects on digestive enzymes of okra (Abelmoschus esculentus) fruit at different maturation stages. J Food Sci Technol 56:1275–1286

Sheu SC, Lai MH (2012) Composition analysis and immuno-modulatory effect of okra (*Abelmoschus esculentus* L.) extract. Food Chem 134:1906–1911

Singh B, Karmakar P, Singh P, Maurya BK, Singh H, Sagar V, Sanwal SK (2023) Okra: breeding and genomics. Veg Sci 50(2):261–273

Soares GSF, Assreuy AMS, Gadelha CAA, Gomes VM, Delatorre P, Simões RC, Cavada BS, Leite JF, Nagano CS, Pinto NV, Pessoa HLF, Santi-Gadelha T (2012) Purification and biological activities of *Abelmoschus esculentus* seed lectin. Protein J 31:674–680

Sogo T, Terahara N, Hisanaga A, Kumamoto T, Yamashiro T, Wu S, Sakao K, Hou DX (2015) Anti-inflammatory activity and molecular mechanism of delphinidin 3-sambubioside, a hibiscus anthocyanin. Biofactors 41:58–65

Sreerama YN, Malleshi NG (2012) Composition and functional properties of okra (Abelmoschus esculentus) polysaccharides. Food Hydrocoll 28(1):62–67

Subrahmanyam GV, Sushma M, Alekya A, Neeraja CH, Harsha HS, Ravindra J (2011) Antidiabetic activity of Abelmoschus esculentus fruit extract. Int J Res Pharm Chem 1:17–20

Taha AY (2020) Linoleic acid—good or bad for the brain? NPJ Sci Food 4:1

Thanakosai W, Phuwapraisirisan P (2013) First identification of α-glucosidase inhibitors from okra (Abelmoschus esculentus) seeds. Nat Prod Commun 8(8):1085–1088

Thole C, Brandt S, Ahmed N, Hensel A (2015) Acetylated rhamnogalacturonans from immature fruits of *Abelmoschus esculentus* inhibit the adhesion of *Helicobacter pylori* to human gastric cells by interaction with outer membrane proteins. Molecules 20(9):16770–11687

Tomoda M, Shimizu N, Gonda R (1985) Isolation and characterisation of mucilage 'Okra Mucilage R' from the roots of *Abelmoschus esculentus*. Chem Pharm Bull 33(8):3330–3335

Tongjaroenbuangam W, Ruksee N, Chantiratikul P, Pakdeenarong N, Kongbuntad W, Govitrapong P (2011) Neuroprotective effects of quercetin, rutin and okra (*Abelmoschus esculentus* Linn.) in dexamethasone-treated mice. Neurochem Int 59(5):677–685

Tóth RJ, Csapó J (2018) The role of selenium in nutrition–a review. Acta Universitatis Sapientiae, Alimentaria 11(1):128–144

Vayssade M, Sengkhamparn N, Verhoef R, Delaigue C, Goundiam O, Vigneron P, Voragen AG, Schols HA, Nagel MD (2009) Antiproliferative and proapoptotic actions of okra pectin on B16F10 melanoma cells. Phy Ber Res 24:982–989

Wang K, Li M, Wen X, Chen X, He Z, Ni Y (2018) Optimization of ultrasound-assisted extraction of okra (Abelmoschus esculentus (L.) Moench) polysaccharides based on response surface methodology and antioxidant activity. Int J Biol Macromol 114:1056–1063

Weaver CM (2013) Potassium and health. Adv Nutr 4(3):368S–377S

WIKI (2023) Okra. From Wikipedia, the free encyclopedia. https://en.wikipedia.org/wiki/Okra

Williamson G, Kay CD, Crozier A (2018) The bioavailability, transport, and bioactivity of dietary flavonoids: a review from a historical perspective. Compr Rev Food Sci Food Saf 17:1054–1112

Woolfe ML, Martin FC, Otchere G (1977) Studies on the mucilages extracted from okra fruits (*Hibiscus esculentus* L.) and baobab leaves (*Adansonia digitata* L.). J Sci Food Agric 28:519–529

Wu DT, Nie XR, Shen DD, Li HY, Zhao L, Zhang Q, Lin DR, Qin W (2020) Phenolic compounds, antioxidant activities, and inhibitory effects on digestive enzymes of different cultivars of Okra (*Abelmoschus esculentus*). Molecules 25(6):1276

Xia F, Zhong Y, Li M, Chang Q, Liao Y, Liu X, Pan R (2015) Antioxidant and anti-fatigue constituents of okra. Nutrients 7(10):8846–8858

Yesilada E, Honda G, Sezik E, Tabata M, Fusita T, Takenda Y (1995) Traditional medicine in Turkey, folk medicine in the inner Taurus mountain. J Ethnopharmacol 46:133–152

Yuan K, Liao H, Dong W, Shi X, Liu H (2012) Analysis and comparison of the active components and antioxidant activities of extracts from Abelmoschus esculentus L. Pharmacogn Mag 8:156–161

Zaharuddin ND, Noordin MI, Ali K (2014) The use of *Hibiscus esculentus* (Okra) gum in sustaining the release of propranolol hydrochloride in a solid oral dosage form. Biomed Res Int 1:735891

Zeng H, Liu Q, Yu J, Wang M, Chen M, Wang R, He X, Gao M, Chen X (2015) Separation of α-amylase inhibitors from *Abelmoschus esculentus* (L.) Moench by on-line two-dimensional high-speed counter-current chromatography target-guided by ultrafiltration-HPLC. J Sep Sci 38:3897–3904

Zhang Z, Zheng Y (2014) Immunomodulatory effects of okra polysaccharides: a review. Int J Biol Macromol 64:218–224

Zhang T, Xiang J, Zheng G, Yan R, Min X (2018) Preliminary characterization and anti-hyperglycemic activity of a pectic polysaccharide from okra (*Abelmoschus esculentus* (L.) Moench). J Funct Food 41:19–24

Zhang Y, Zhang T, Zhao Q, Xie X, Li Y, Chen Q, Cheng F, Tian J, Gu H, Huang J (2021) Comparative transcriptome analysis of the accumulation of anthocyanins revealed the underlying metabolic and molecular mechanisms of purple pod coloration in okra (*Abelmoschus esculentus* L.). Food Secur 10:2180

Zheng W, Zhao T, Feng W, Wang W, Zou Y, Zheng D, Takase M, Li Q, Wu H, Yang L, Wu X (2014) Purification, characterization and immunomodulating activity of a polysaccharide from flowers of *Abelmoschus esculentus*. Carbohydr Polym 106:335–342

Part VI

Public-Private Partnership—Discussion on Possible Collaborative Projects

23 Potential for Breeding Consortium Projects on Okra—Private and Public Partnership

O. P. Dutta

23.1 Introduction

Okra *(Abelmoschus esculentus L. Moench)* is a multipurpose crop valued for its nutrient-rich, tender fruit mainly used in culinary preparations. Its mucilage is a viscous polysaccharide associated with proteins and minerals. It has enormous potential to be used as a functional food, having antitumor, antioxidant, antidiabetic, antimicrobial, and antiulcerogenic capacities as well as a plasma volume expander. Its ability to bind with cholesterol and bile acids helps remove toxins from the liver. In the food industry, it is mainly used as an emulsifying agent. In the pharmaceutical industry, it is used as a thickener, emulsifier, stabiliser, gelling agent, granulating agent, suspending agent, binding agent, encapsulating agent, and sustainable drug delivery agent. In the packaging industry, it is used as biodegradable and bioactive edible films with antimicrobial properties and increased shelf life. Its fibre is used in the paper and textile industry. Its seeds are rich in oil (14% lipids) and protein (21%). In water treatment plants, it provides an excellent opportunity to replace synthetic flocculants with plant-based flocculants, as the use of synthetic flocculants results in toxic by-products (Thamires et al. 2021; Sathish Kumar et al. 2013; Jain et al. 2012; Pal 2020; Gupta et al. 2021; Ali et al. 2021; Srinivasan 2022, Benjamin et al. 1951; Savello et al. 1980).

Okra is commercially cultivated in tropical, subtropical, and warm climates in 46 countries from Asia to Africa, Southern Europe, and America. The global production of okra was 10.8 million tonnes, led by India (>60%), Nigeria (18%), and Mali (6%) during the year 2021. These three countries contributed 86 per cent of okra production in the world. Though Nigeria had the largest area under okra cultivation, its productivity was very low (1.45 tonnes/ha) as compared to global productivity

O. P. Dutta (✉)
Division of Vegetable Crops, ICAR-Indian Institute of Horticultural Research, Bengaluru, Karnataka, India

S. K. Tikoo et al. (eds.), *Okra: Status, Challenges and Opportunities*,
https://doi.org/10.1007/978-981-97-9963-3_23

(4.36 tonnes/ha), India (12.17 tonnes/ha), and Mali (11.55 tonnes/ha) (FAOSTAT 2021). Presently, okra productivity needs to catch up to its potential with immediate attention.

23.2 Major Challenges to be Addressed in the Okra Improvement Programme are as Follows

1. Breaking yield barriers through plant ideotype breeding
2. Management of begomoviruses in okra
3. Developing okra varieties for industrial applications and export
4. Constraints in okra hybrid seed production.

23.2.1 Redesigning Okra Plant Ideotype

23.2.1.1 Some Possibilities for Transforming Okra Plant Architecture Are Summarised in Table 23.1

The yield per plant (<1 kg) is meagre. By increasing the productive branches, productive nodes per plant, nodal and internodal productivity per se, and the size and weight of tender fruit, one can increase the number of fruit per plant from 35 to 300, the fruit weight from 15 g to 50 g and the yield per plant from less than 1 to 15 kg.

23.2.1.2 Converting the Possibilities into Realities

Before exploiting the hybrid vigour in okra, it will be more useful to first explore its inbred vigour. The okra variety Heavy Hitter is one example of the much sought and inspired possibilities being converted into realities (Fig. 23.1). This variety was developed as a genetic variant of the okra variety Clemson Spineless after many years of selections by the Dry Creek Farm owner Ron Cook in Moodys-Oklahoma (2012). This variety has incredible branching potential, giving 65 productive branches per plant bearing 650 fruit (3–4 inches long) per plant. One can harvest 40 fruit per plant per day (Figs. 23.2 and 23.3).

Table 23.1 Transforming okra plant ideotype

Components	Possibilities	
	From	To
Yield/plant	<1 kg	15 kg
No. of productive branches/plants	2–3	40–50
No. of tender fruit/plant	25–30	300–350
No. of fruit/node	1	5
No. of fruit on main stem/branches (cauliflory/remiflory)	0	Many
Weight/tender fruit	15 g	50 g

Fig. 23.1 OkraVar. Heavy Hitter. (Courtesy: Baker Creek Heirloom Seeds Co./rareseeds.com)

Fig. 23.2 Normal okra plant

Fig. 23.3 Incredible branching habit. (Courtesy: Baker Creek Heirloom Seeds Co. (www.rareseeds.com)

Fig. 23.4 Variety Sky Rocket (5 fruit/cluster). (Courtesy: Baker Creek Heirloom Seeds Co. (www.rareseeds.com)

Okra Plant Ideotypes

Okra variety Sky Rocket is another variant of Heavy Hitter first observed by Ron Cook in 2018. This variety bears fruit in a cluster of 5 at a nodal interval of 3 inches—Fig. 23.4.

Such variants were also observed in an advanced generation of a cross between *Abelmoschus esculentus* and *Abelmoschus manihot* (Dutta 1991). The variants included plants bearing 2–5 fruit per node (Figs. 23.5 and 23.6).

Even the internodal space on the main stem and branches becomes productive, giving rise to cauliflory and ramiflory phenomena (Fig. 23.7; Dutta 1991). Krishna (2022) also observed okra plants bearing five fruit in a cluster at the nodal region in a cross between *Abelmoschus esculentus* and *A. moschatus*—Fig. 23.8.

Fig. 23.5 Two fruit/node

Fig. 23.6 Five fruit/cluster

Internodal Productivity

Fig. 23.7 Cauliflory and ramiflory

Nodal Productivity

Fig. 23.8 Five fruit/node). (Courtesy: Krishna, K. Namdhari seeds Pvt. Ltd., Bangalore)

Fig. 23.9 Tender fruit weight 50 g

A local landrace of okra from the west coast region of Karnataka, India, has excellent potential to improve upon the fruit size and weight of okra cultivars. The fruit are pale green with a length of 10 inches and girth of 2.5–3 inches, each weighing about 50 g at the tender stage (Fig. 23.9).

23.2.2 Managing *Begomoviruses* in Okra

Okra is susceptible to at least 27 Begomoviruses (Sanwal et al. 2014; Swanson and Harrison 1993), of which yellow vein mosaic disease was first reported in India (Kulkarni 1924), and okra enation leaf curl disease was also reported for the first time from India (Singh and Dutta 1986), causing significant yield losses ranging from 30–100 per cent, depending upon the different stages of the plant's growth (Singh 1996; Figs. 23.10 and 23.11).

In addition to India, significant yield loss due to begomovirus infection in okra has also been reported in Africa, South America, and Saudi Arabia (Ghanem 2003; Venkataravanappa et al. 2013, Vankataravanappa et al. 2014, Mishra et al. 2017; Ghanem 2003). Begomoviruses in okra are transmitted by the vector *Bemisia tabaci,* of which the polyphagous "B "biotype is highly virulent (Lisha et al. 2003).

Resistance to begomoviruses in okra is very unstable, with frequent breakdowns. The emergence of novel viral strains due to recombination and pseudo combination from mixed infection in susceptible okra plants poses serious challenges to both okra breeders and virologists (Figs. 23.12 and 23.13; Venkataravanappa and Reddy 2013).

Mixed virus infection in okra.

Identification of sources of resistance to yellow vein mosaic virus (YVMV) and enation leaf curl virus (ELCV) diseases in cultivated and wild species of okra, followed by gene pyramiding for combined disease resistance as well as the

Fig. 23.10 YVMV in okra

Fig. 23.11 ELCV in okra. (Healthy vs. infected leaf)

identification and mapping of major quantitative trait locus (QTL) for yellow vein mosaic virus and enation leaf curl virus, will help in developing durable resistance in okra varieties.

23.2.3 Developing Okra Varieties for Industrial Applications and Export

Industrial applications of okra polysaccharides are largely neglected globally. It has a huge potential in the medical sciences (functional food), the food industry (emulsifying agent), the pharmaceutical industry (stabilising, suspending, binding, encapsulating agent), the packaging industry (biodegradable and bioactive edible films), the paper and textile industry (bast fibre) (Haydar and Rahu 2022), the oil and biodiesel industry (seed and other plant parts) (Ali et al. 2021; Tanvi et al. 2021; Adejumo et al. 2020) and water treatment plants (natural polysaccharide flocculants). In big cities, several billion litres of water are treated annually. The required concentration of okra

Fig. 23.12 YVMV and petiole bending

Fig. 23.13 ELCV and YVMV

and tamarind polysaccharides is 1 g per litre of water (Rajani Srinivasan 2022). Large-scale production of necessary polysaccharides thus needs to be initiated. The widespread application of okra flocculants will spark a vast new market for okra. Once the okra is linked to industrial applications, the production of necessary polysaccharides, bast fibre, seed oil, and okra biomass has to be scaled up enormously. The okra production will shoot up dramatically, leading to an okra boom.

23.2.4 Okra Hybrid Seed Production

Increasing demand for organic food products and a rising inclination toward functional foods are promoting significant growth in the global seed market. The global okra seed market will reach USD 469.65 million during the forecast period 2023–2029 (Global Industry Analysis Inc.). In India, the okra seed market size was valued at USD 83.71 million in 2021; however, the Global Industry Analysis and Forecast Estimates for the 2023–2029 period indicated a value worth USD 225.37 million. Asia Pacific holds the largest market for okra seed production. Some of the major okra seed-producing countries in this region include India, Pakistan, Malaysia, and the Philippines.

Presently, hybrid seed production in okra is done by hand emasculation and pollination. It is a tedious and time-consuming job, causing hand injuries.

During the process of hand emasculation, some chemicals from the staminal column of the okra flower or okra petals cause wounds on the fingers of okra workers. Their fingers are swollen and injured, and their nails are dissolved partially (Figs. 23.14, 23.15, and 23.16).

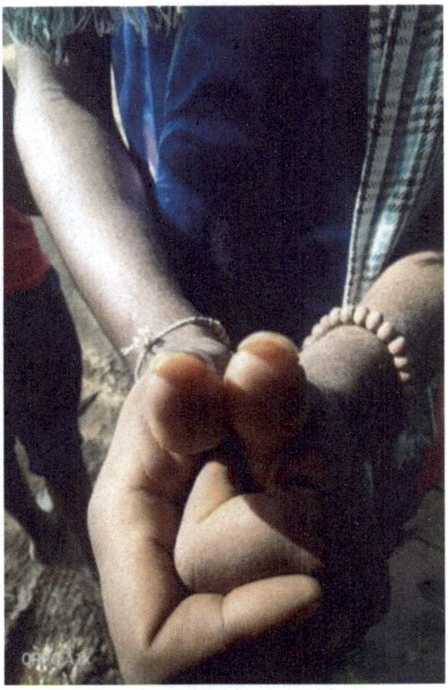

Fig. 23.14 Thumb swelling. (Courtesy: M. Pitchaimuthu, IIHR Bangalore)

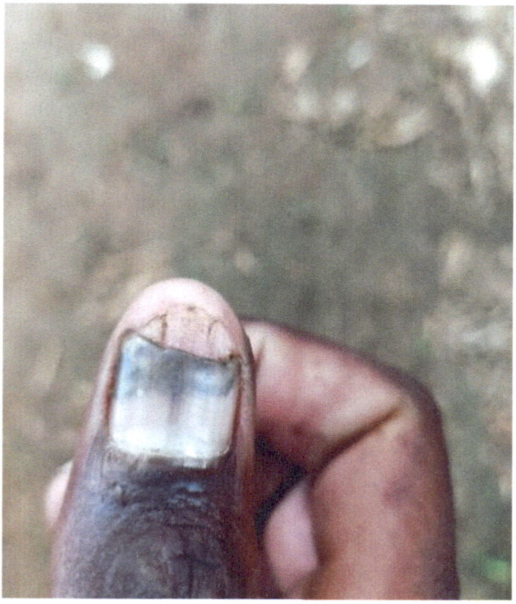

Fig. 23.15 Nail injury. (Courtesy: M. Pitchaimuthu, IIHR Bangalore)

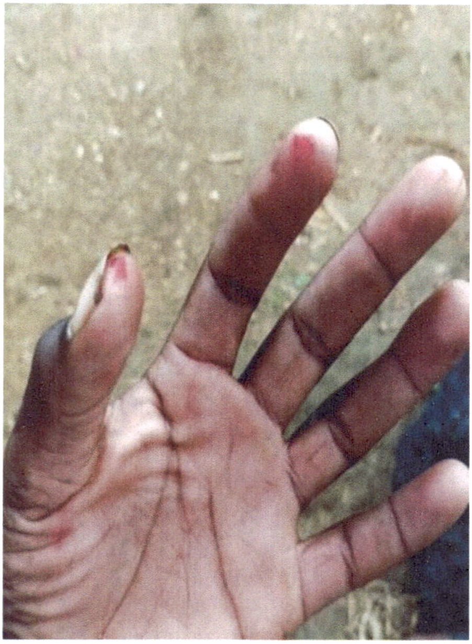

Fig. 23.16 Blood clots. (Courtesy: M. Pitchaimuthu, IIHR Bangalore)

Okra bud emasculation and hand injuries.

The entire process of emasculation is not only painful but harsh as well. One of the ways to avoid this human drudgery is to use male sterility in okra hybrid seed production. Induction of genetic male sterility in okra through gamma rays was first reported by the Indian Institute of Horticultural Research, Bangalore, India (Dutta 1971, 1980, 1991). Later, the genetic male sterility was incorporated into okra breeding lines for hybrid seed production commercially (Pitchaimuthu et al. 2012). Still, a better option would be the use of cytoplasmic-genic male sterility, presently unavailable in okra. It can be induced in okra with the help of chemical mutagens as well as by the possible interaction between the nucleus and the mitochondrial genome. The restorer of the fertility gene (Rf-gene) in the nuclear genome suppresses the expression of the CMS-causing gene, thereby restoring fertility. Thus, incorporating the cytoplasmic-genic male sterility in okra and locating the restorer genes will significantly facilitate the seed industry and the skilled workers engaged in okra hybrid seed production commercially.

23.3 Okra Networking Consortium—A Way Forward

A consortium approach to address a common set of goals in research and industry will be mutually beneficial. Such an approach will make the best use of the strengths of different organisations, bringing together diverse specialists in the right team. It facilitates unlocking the various sources of funding and support, thereby providing strategic continuity. In the beginning, one can start a consortium at the national level. The successful outcome of the consortium can then be used as a model for the international consortium. At the national level, the consortium should include members from both the public and private sectors. Keeping in view that okra is a multifunctional crop, one should identify the consortium members from the private seed industry, pharmaceutical industry, food industry, packaging industry, paper and textile industry, oil and biodiesel industry, and farmer associations. Public sectors may include premium research institutions and universities actively engaged in okra improvement programmes, national germplasm repositories, and organisations dealing with water treatment plants. Specific projects can be proposed in the following theme areas:

(a) *Improving okra productivity:*

   This can be achieved by redesigning the okra ideotype and incorporating resistance to multiple stresses with the help of marker-assisted and genome-assisted breeding programmes. Sharing and pooling of okra germplasm is imperative here for a successful outcome.

(b) *Developing okra varieties with specific traits for industrial applications, fresh consumption, and export:*

   Trait-specific okra varieties are presently unavailable at national and international levels. Such varieties are imperative for successful industrial applications and exports. Further, networking and interlinking the seed industry with the pharmaceutical, paper, and textile industries will be complimentary. Rejected seeds from the seed industry (after gravity separation) and rejected plant parts (okra stalks, roots, and dry okra capsules after fruit and seed harvest) can be gainfully utilised by the pharmaceutical industry for polysaccharide manufacturing, paper and textile industry, oil, and biodiesel industry.

(c) *Efficient and aggressive promotion of okra hybrid seed production:*

Incorporating cytoplasmic-genic male sterility in okra will provide a durable and economical solution for hybrid seed production.

A successful model of okra networking at the national level can be extended to an international networking consortium by involving participants from the Asia Pacific region and African countries.

References

Adejumo AOD, Ilori TA, Komolafe CA (2020) Biodiesel production from Okra(*Abelmoschus esculentus* (L.) Moench) plant. IOP Conf Ser Earth Environ Sci 445(1):012051. https://doi.org/10.1088/1755-1315/445/1/012051

Ali AA, Al-Shawi, Mustafa FH, Haneen KT (2021) Review on the biological applications of Okra polysaccharides and prospective research. Futur J Pharm Sci 7:102. https://doi.org/10.1186/s43094-021-00244-0

Benjamin HB, Ihrig HK, Roth DA (1951) The use of Okra as a plasma replacement. Rev Can Biol 10(3):215–21. PMID:14865338

Dutta OP (1971) Effect of gamma irradiation on seed germination, plant growth, floral biology, and fruit production in *Abelmoschus esculentus* (L.) Third international Symp. Trop. Hort. IIHR, Bangalore, India. pp 146–156

Dutta OP (1980) Male Sterility in Okra (*Abelmoschus esculentus* L.) Moench and Bottle gourd(*Lagenaria siceraria(*Mol.) Standl.) and its Utilization in Hybrid Seed Production. Ph.D. Thesis UAS.1980, Bangalore

Dutta OP (1991) Okra Germplasm Utilization at IIHR(Indian Institute of Horticultural Research) Bangalore. International Workshop on Genetic Resources enb.8–12 Oct. New Delhi India

FAOSTAT (2021) Food, and Agriculture Organization Statistics Division. Retrieved 19 Feb. 2023

Ghanem GAM (2003) Okra leaf Curl Virus: a monopartite Begomovirus infecting Okra crop in Saudi Arabia. Arab J Biotechnol 6:139–152

Gupta PK, Shamayita P, Kartick KS (2021) Potential of Okra for application in textile: a review. J Nat Fibers 18(11):1788–1800

Haydar UZ, Rahu AK (2022) A review on potential uses and characterization of Okra Fiber reinforced polymer composites. IJEC 8(2):2022

Jain N, Jain R, Jain V, Surendra J (2012) A review on *Abelmoschus esculentus*. Pharmacia 1(3):2012.ISSN 0976-9692

Krishna KA (2022) Cluster fruit bearing habit in okra: personal communications. Namdhari Seeds Pvt. Ltd., Bangalore

Kulkarni GS (1924) Mosaic and other related diseases of the crop in Bombay presidency. Poona Agric Coll Magaz 16:6–12

LishaVS, AntonyB, PalaniswamiMS, HenneberrTJ(2003) Bemisia tabaci (Homoptera:Aleyrodidae) biotypes in India. J Econ Entomol 96(2):322–327. https://doi.org/10.1093/jee/96.2.322

Mishra GP, Bijendra S, Tania S, Singh Jaydeep AK, Halder, Krishnan N, Shailesh KT, Prabhakar MS (2017) Biotechnological advancements and Begomovirus Management in Okra (*Abelmoschus esculentus* L.): status and perspectives. Front Plant Sci 8. https://doi.org/10.3389/fpls2017.0036

Pal K (2020) Synthesis, characterization of Okra Mucilage as a potential new age therapeutic intervention. MOL2NET 6:iSSN:2624-5078

Pitchaimuthu M, Dutta OP, Swami KRM (2012) Studies on inheritance of male Genic Male Sterility (GMS) and hybrid seed production in Okra. J Hortic Sci 7(2):199–202

Ron C (2012) Okra Variety Heavy Hitter. (DryCreekFarm). https://seedsavingnetwork.proboard.com/thread/15/heavy-hitter-okra

Sanwal SK, Singh M, Singh B, Naik PS (2014) Resistance to yellow vein mosaic virus and Okra Enation Leaf Curl Virus: challenges and future strategies. Curr Sci 11(106):1470–1471

Sathish Kumar D, Tony DE, Kumar AP, Kumar KA, Rao DBS, Nadendla RR (2013) A review on *Abelmoschus esculentus* (Okra). Int Res Pharm App Sci 3(4):129–132

Savello PA, Martin FW, Hill JM (1980) Nutrition composition of Okra seed meal. J Agric Food Chem 1980:281163–281166. https://doi.org/10.1021/jf60232a021

Singh SJ (1996) Assessment of losses in Okra due to enation Leaf Curl virus. Indian J Virol 12:51–53

Singh SJ, Dutta OP (1986) Enation leaf curl virus of Okra new virus disease. Indian J Virol 2:114–117

Srinivasan R (2022) Investigating the use of plant-derived polymers on the reduction of microplastics in freshwater samples. ACS Spring, March 2022, Poster-Final

Swanson MM, Harrison BD (1993) Serological relationships and epitope profiles of isolates of Okra Leaf Curl GeminiVirus from Africa and the Middle East. Biochimie 75(8):707–711. https://doi.org/10.1016/0300-9084(93)90101-w

Tanvi Hossein M, Islam T, MaheeZarin II, Reza T, Rahman M, Mahmud M (2021) A comprehensive study on Physico-mechanical characteristics of Okra (*Abelmoschus esculentus*) for textile applications. Indian J Sci Technol 14(9):765–775. https://doi.org/10.17485/IJST/v14i9.2268

Thamires LD, Flavia CB, Florentino ER (2021) Plan Theory 10(8):1683. https://doi.org/10.3390/plants10081683

Venkataravanappa V, Reddy MK (2013) Begomovirus characterization, and development of phenotypic and DNA-based diagnostics for screening of Okra genotype resistance against Bhendi yellow vein Mosaic Virus. 3Biotech 3:461–470. https://doi.org/10.1007/s13205-012-0107-z

Venkataravanappa V, Reddy CNL, Devaraju A, Jalali S, Rddy MK (2013) Association of a recombinant Cotton Leaf Curl Bangalore virus with yellow vein and Leaf Curl disease of Okra in India. J Virol 24:1–189. https://doi.org/10.1007/s13337-013-0141-4

Venkataravanappa V, Lakshminarayana Reddy CN, Sali J, Briddon RW, Reddy MK (2014) Molecular identification and biological characterization of a Begomovirus associated with Okra Enation Leaf Curl disease in India. Eur J Plant Pathol 141:217–235. https://doi.org/10.1007/s1065-014.0463.0

Potential Pre-breeding Projects in Okra for PPP Collaborations 24

Manish Bhatnagar

24.1 Introduction

Okra [*Abelmoschus esculentus (L.) Moench*] generally known as 'Bhindi', 'Gumbo', or 'Lady's finger', is a popular vegetable crop of the world. It is one of the economically important ancient crops cultivated throughout the subtropical and tropical zones for its fibrous but tender edible pods. This vegetable crop is a rich source of several minerals, vitamins, carbohydrates, calcium, proteins, and dietary fibre. A balanced proportion of amino acids like tryptophan and lysine comparable to that of other sources of proteins like legumes are present in the seed proteins of okra. Seeds of okra are a potential source for extracting edible oil (around 20%), with high linoleic acid (48%) content that has health benefits for patients suffering from cardiovascular problems (Gemede et al. 2014). The availability of high concentrations of mucilage in the plant is successfully being utilised for its medicinal value as a blood volume expander. It is also known as 'perfect villager's vegetable' in the developing countries due to the availability of complete nutrition contents. Even though rich in nutrients, there is a lack of suitable cultivars that provide higher sustainable yields under high disease pressures.

24.2 Major Challenges in Okra Cultivation

Being such an important crop, there are many challenges faced by growers in okra cultivation.

M. Bhatnagar (✉)
Growth Unit Head, Mahyco Private Limited, Mumbai, India

S. K. Tikoo et al. (eds.), *Okra: Status, Challenges and Opportunities*,
https://doi.org/10.1007/978-981-97-9963-3_24

1. *Diseases*
   (a) Yellow vein mosaic disease (YVMD) is caused by the *yellow vein mosaic virus (YVMV).*
   (b) Okra enation leaf curl disease (OELCD) caused by the *okra enation leaf curl virus (OELCV).*
   (c) *Cercospora* leaf spot caused by *Cercospora abelmoschi.*
   (d) Powdery Mildew (PM) caused by *Erysiphe cichoracearum.*
2. *Pests*
   (a) Jassids *(Amarasca biguttula biguttula).*
   (b) Mites *(Tetranychus* spp.*).*
   (c) Fruit and shoot borer *(Earias vitelli, E. insulana).*
3. *Low Yields*: Average yields in okra are 2.5–3 MT per acre, which are quite low in relation to the investments of the farmers. This is the area for improvement for the scientists working in this field.
4. *Frequent Resistance Breakdown*: In the past, many varieties have come up providing a good level of resistance towards YVMD and OELCD, but the breakdown of disease in these varieties was very fast and could not sustain the evolution in viral DNA. There are different strains causing different disease reactions in different regions of the same genotype/variety.
5. *Limited Molecular Work*: In okra there is a limited knowledge at molecular level.
6. *Amphidiploid and Large Redundant Genome*: The amphidiploid nature of cultivated types and large redundant genome present in okra DNA limit the technical advances, unlike in other crops.

24.3 Possible Solutions

24.3.1 Explore Wild Relatives for Durable Resistances

There are many diverse species across the world. Many wild relatives of okra express resistance towards major pests and diseases (Table 24.1). However, there are challenges due to the chromosomal differences and differences in ploidy levels. Interspecific crosses are successful with lower germination levels. But these can be acclimatised with thorough selections.

24.3.2 Exploring CRISPR/CAS and Other Permitted Technologies

CRISPR/Cas9 is the type of gene editing technology that edits genes by cutting DNA precisely and then letting the natural DNA repair process take over. This system consists of two parts: the Cas9 enzyme and a guide RNA. Cas9 is a bacterial RNA-guided endonuclease that uses base pairing to recognise and cleave the target DNAs with complementarity to the guide RNA.

This technology gives us immense possibilities and opportunities for crop improvement. In the absence of permission to using this technology, breeders and researchers are limited to use only the available genes or variability for developing

Table 24.1 Possible sources of resistance from the wild relatives of okra

S. No.	Major diseases	Wild species	Disease's reaction	References
1	YVMD	*A. caillei*	MR	Singh et al. (2006)
		A. tetraphyllus	HR	Prabu and Warade (2009)
		A. crinitus	HR	Kumar and Sreeparvathy (2010)
		A. angulous	HR	Shetty et al. (2013)
		A. pungens	HR	Sanwal et al. (2016)
		A. moschatus	MR	Gangopadhyay et al. (2017)
		A. manihot	HR	Manjua et al. (2018)
2	OELCD	*A. tetraphyllus*	R	Singh et al. (2006)
		A. ficulneus	HR	Dhaliwal and Sharma (2016)
		A. crinitus	HR	Mishra et al. (2017)
		A. angulous	HR	Yadav et al. (2018)
		A. pungens	HR	
		A. tuberculatus	R	
		A. moschatus	MR	
		A. manihot	MR	
3	Cercospora leaf spot	*A. caillei*	MR	Singh et al. (2006)
		A. crinitus	R	Kumar and Sreeparvathy (2010)
		A. angulous	R	Dhall (2015)
		A. pungens	R	
		A. moschatus	R	
		A. manihot	R	
4	Powdery mildew	*A. caillei (MR)*	MR	Singh et al. (2006)
		A. tetraphyllus	R	Dhall (2015)
		A. crinitus	R	
		A. angulous	MR	
		A. pungens	MR	
		A. moschatus	R	
		A. manihot	MR	

Source: Chowdhury and Kumar (2019) with modifications
R Resistant, *HR* Highly resistant, *MR* Moderately resistant

new varieties. But now we have the flexibility to create the desired variability and use for developments and innovations, provided the genes of interest are mapped.

24.3.3 Change in Plant Ideotype to Give Higher Yield

Medium-tall plants with a higher number of acute branches and good-quality fruits will help in increased yield. Usually, branches are observed from the base of the plant, but if we introduce branches from the middle or upper nodes, this will add further. There are genotypes that have this trait. Multiflory is another concept that can be explored.

24.3.4 More Focused and Aggressive Molecular Research

There is very limited work done on molecular aspects in okra. A collaborative approach towards this will help in advances in okra breeding.

Previously, okra was not focussed on molecular research for obvious reasons. Now it has become very popular and economically important for the private sector. Therefore, we may expect more attention and budgets being allocated to these less focussed aspects.

24.3.5 Awareness and Exploitation of Nutritive Value

Okra is rich in magnesium, folate, fibre, antioxidants, and vitamin C, K1, and A. It may help support healthy pregnancy, heart health, and blood sugar. It may even have anticancer properties. Creating awareness about its nutritive value among consumers will positively impact consumption and output prices for growers.

Moreover, improvement programmes need to emphasise for biochemical attributes in okra, which can further significantly contribute for nutritional, medicinal, as well as industrial applications.

24.4 Proposed Public-Private Partnership Projects

Considering the interest of private industry and the availability of genetic diversity and available technological advances, the following are the potential public-private partnership (PPP) projects that can be explored.

1. Transferring YVMD and ELCD resistance genes into cultivated backgrounds.
2. Okra genome mapping with trait association.
3. Discovery of novel traits; plant ideotype, branches, and multiflory.

References

Chowdhury S, Kumar S (2019) Okra breeding : recent approaches and constraints, Department of Genetics and Plant Breeding, School of Agriculture, Lovely Professional University, Phagwara-144 411 (Punjab), India

Dhaliwal MS, Sharma A (2016) Breeding for resistance to virus diseases in vegetable crops. Innovations in horticultural sciences. New India Publishing Agency, New Delhi, pp 303–327

Dhall RK (2015) Breeding for biotic stresses resistance in vegetable crops : a review. J Crop Sci Technol 4:13–27

Gangopadhyay KK, Singh A, Bag MK, Ranjan P, Prasad TV, Roy A, Dutta M (2017) Diversity analysis and evaluation of wild Abelmoschus species for agromorphological traits and major biotic stresses under the north-western agroclimatic conditions of India. Genet Resour Crop Evol 64:775–790

Gemede HF, Ratta N, Haki GD, Ashagrie Z, Beyene WF (2014) Global J Med Res K, Interdiscipl 14(5):Version 1.0

Kumar PS, Sreeparvathy S (2010) Researchar ticle studies on heterosis in okra [Abelmoschus esculentus (L.) Moench]. Elect J Plant Breed 1:1431–1433

Manjua KP, Lakshmia KV, Babub BS, Anithab K (2018) Evaluation of okra germplasm for their reaction to whitefly, Bemisia tabaci and okra yellow vein mosaic virus (OYVMV). J Entomol Zool Stud 6:2491–2496

Mishra GP, Singh B, Seth T, Singh AK, Halder J, Krishnan N, Tiwari SK, Singh PM (2017) Biotechnological advancements and begomovirus management in okra (Abelmoschus esculentus L.): status and perspectives. Front Plant Sci 8:360

Prabu T, Warade SD (2009) Biochemical basis of resistance to yellow vein mosaic virus in okra. Veg Sci 36:283–287

Sanwal SK, Venkataravanappa V, Singh B (2016) Resistance to bhendi yellow vein mosaic disease : a review. Indian J Agric Sci 86:835–843

Shetty AA, Singh JP, Singh D (2013) Resistance to yellow vein mosaic virus in okra : a review. Biol Agric Hortic 29:159–164

Singh B, Rai M, Kalloo G, Satpathy S, Pandey KK (2006) Wild taxa of okra (Abelmoschus species) : reservoir of genes for resistance to biotic stresses. In: International conference on indigenous vegetables and legumes. Prospects for fighting poverty, hunger and malnutrition 7, pp 323–328

Yadav Y, Maurya PK, Devi AP, Jamir I, Bhattacharjee T, Banerjee S, Dutta S, Debnath D, Mandal AK, Dutta S, Chattopadhyay A (2018) Enation leaf curl virus (ELCV) : a real threat in major okra production belts of India : a review. J Pharmacogn Phytochem 7:3795–3802

Possibilities of Genomic Research for Trait-Linked Marker Development in Okra

25

Dushant Singh Baghel, Sathish Kumar Peddamma, Ashish Kumar, A. Lalitha Hari Dharini, Sweta Yadav, Shraddha Mishra, Vidhya Singh, Suresh Iyer, and Baskaran Ramachandran

25.1 Introduction

Okra, valued for its tender pods, which are a rich source of vitamins, minerals, and dietary fiber, is cultivated in diverse agro-climatic regions. However, the improvement of important traits such as disease resistance, yield, quality, and tolerance to abiotic stresses using conventional breeding methods has been challenging in okra. Genomic research offers promising opportunities to revolutionize breeding programs in this crop. Most of the uncultivated (wild) species are a secondary gene pool expressing resistance toward major pests and diseases of cultivated okra, making them valuable resources for enhancing okra's genetic diversity.

25.2 Genomic Resources for Okra

Recent advances in DNA sequencing technologies have enabled the generation of high-quality genomic data for okra. Whole-genome sequencing, resequencing, and transcriptome analysis have provided researchers with a vast amount of genetic information. The allopolyploid nature of okra is characterized by the presence of multiple sub-genomes, resulting from the combination of divergent genomes. Recent research efforts have led to the first annotated genome assembly of okra, shedding light on its complex genomic architecture (Ronald Nieuwenhuis et al. 2021).

The genome analysis revealed intriguing features, including whole chromosome and chromosomal arm scaffolds, short interstitial telomeric repeats, and an S-type rRNA gene arrangement. The estimated genome size of approximately 1.45 Gb,

D. S. Baghel (✉) · S. K. Peddamma · A. Kumar · A. Lalitha Hari Dharini · S. Yadav
S. Mishra · V. Singh · S. Iyer · B. Ramachandran
Nucleome Informatics Pvt Ltd, Hyderabad, India
e-mail: dushyant@nucleomeinfo.com

S. K. Tikoo et al. (eds.), *Okra: Status, Challenges and Opportunities*,
https://doi.org/10.1007/978-981-97-9963-3_25

with 57% composed of repetitive sequences, provides insights into the genomic composition of okra. The nearly complete genome assembly, limited recombination among sub-genomes, and presence of at least 2 sub-genomes further support its allopolyploid nature.

Furthermore, over 130,000 putative genes were identified, with gene-rich regions enriched at the distal ends of scaffolds, while long terminal repeat (LTR) retrotransposons were more abundant in centrally located domains. Comparative analysis with related species unveiled putative enzyme-coding genes involved in the biosynthesis of dietary and therapeutic compounds in okra.

This genomic information opens up new possibilities for trait-linked marker development, enabling targeted improvement of important agronomic traits. By leveraging these insights, researchers and breeders can accelerate breeding programs and develop resilient and improved okra varieties to meet the challenges of a changing climate and increasing food demand. Here is a preview of a few techniques that can be used in the development of genomic resources and further development of trait-linked markers in Okra.

Various tools available are:

25.3 Genomic Sequencing on PacBio Sequel II HiFi Reads-Based Contig Assembly

PacBio Sequel II HiFi sequencing provides long and accurate reads, making it a powerful technology for de novo genome assembly in okra. The assembly process involves error correction and consensus calling, resulting in high-quality sequences. The long reads enable the resolution of complex genomic regions, repetitive elements, and structural variations, which are often challenging for short-read sequencing technologies.

25.4 Optical Mapping-Based Scaffolding and Super Scaffolding with Hi-C

After the initial assembly, optical mapping can be employed to further enhance the genome contiguity. Optical mapping technology uses single-molecule imaging to construct high-resolution physical maps of DNA molecules. By aligning these optical maps with the sequenced contigs, the genome assembly can be super scaffolded to connect contigs into larger chromosomal segments, improving the overall accuracy and contiguity of the genome.

25.5 ISO-Seq-Based Genome Annotation

Isoform sequencing (ISO-Seq) can be used to characterize the full-length transcripts of genes in okra. By combining RNA sequencing with long-read sequencing, comprehensive and accurate gene models can be generated. This ISO-Seq-based genome annotation facilitates the identification of gene structures, alternative splicing events, and non-coding RNAs, providing valuable insights into the functional elements of the okra genome.

25.6 Whole-Genome Resequencing for SNP and Structural Variant Discovery

Whole-genome resequencing of diverse okra accessions allows for the discovery of single nucleotide polymorphisms (SNPs) and structural variations across the population. By comparing the re-sequenced genomes with the reference genome, candidate SNPs associated with abiotic and biotic stress responses can be identified. Structural variants, such as deletions, insertions, and inversions, can also be detected, providing additional information about genetic diversity and adaptive traits.

25.7 GWAS and QTL Mapping

Genome-wide association studies (GWAS) and quantitative trait loci (QTL) mapping have emerged as powerful tools to uncover the genetic basis of complex traits in okra. By analyzing the genetic diversity within okra populations, researchers can identify marker-trait associations. These markers can be converted into robust trait-linked markers for use in breeding programs. Marker-assisted selection (MAS) can expedite the breeding process by enabling the identification of individuals carrying desirable traits early in the breeding cycle.

25.8 HiFi Long-Read Sequencing for Stress-Related Markers

HiFi long-read sequencing can be utilized to target specific genomic regions associated with biotic and abiotic stress-related traits. By focusing on candidate genes involved in stress responses, such as those related to drought tolerance, disease resistance, and temperature sensitivity, researchers can identify stress-related markers. These markers can be integrated into breeding programs to develop stress-tolerant okra varieties.

25.9 Trait Mapping and Marker Development

25.9.1 Candidate Gene Approach

Genomic research allows the identification of candidate genes involved in specific traits of interest. Once candidate genes are identified, they can be functionally validated to understand their role in trait expression. This knowledge can be used to develop functional markers linked to these genes, further enhancing the precision and efficiency of breeding efforts.

25.9.2 Genome Editing for Trait Improvement

The development of genome editing technologies, such as CRISPR-Cas9, offers exciting possibilities for trait improvement in okra. By precisely editing target genes, researchers can introduce desired traits directly into elite okra varieties. Genome editing can expedite the development of new cultivars with enhanced traits, reducing the time required for traditional breeding.

25.10 Creating a Mid-Density Chip for Genomic Selection and Low-Density Marker Test for Marker-Assisted Selection

A mid-density genotyping chip can be designed to include a set of informative markers distributed across the okra genome. These markers can be selected based on their association with key agronomic traits, stress responses, and other desirable characteristics like disease resistance. Genomic selection, leveraging the information from this chip, can enhance the accuracy and efficiency of breeding programs by enabling the selection of superior individuals based on their genomic profiles. For foreground and background selection, low-density markers targeting specific traits of interest can be used to screen and identify individuals carrying desirable traits, ensuring the desired genetic background from elite parental lines.

Finally, identified trait links can be incorporated into marker-assisted breeding programs for okra. By using these markers, breeders can efficiently select and cross parent plants with desired traits, accelerating the development of improved okra varieties. Additionally, genomic selection approaches can be applied to predict the performance of individuals in various environments, leading to the development of stress-resistant and high-yielding okra cultivars.

25.11 Conclusion

Leveraging genomic research and advanced sequencing technologies in okra breeding programs holds great promise for enhancing the efficiency and precision of trait improvement efforts. Integrating genome sequencing, optical mapping, Hi-C, and ISO-Seq-based annotation allows for the generation of high-quality reference genomes. SNP and structural variant discovery using HiFi long-read sequencing can aid in the identification of stress-related markers. The creation of mid-density chips and low-density marker tests facilitates genomic selection and marker-assisted breeding, ultimately contributing to the development of resilient and improved okra cultivars to meet the challenges of a changing climate and increasing food demand. Most of the uncultivated (wild) species are of secondary gene pool expressing resistance toward major pests and diseases of cultivated species, making them valuable resources for enhancing okra's genetic diversity.

Acknowledgments: We would like to acknowledge our collaborators in the Okra Genomics Program, Advanta Seeds, Namdhari Seeds, and Tierra Seeds, for their support and all the team members at Nucleome Informatics' NKC Centre for Genomics Research who worked in Okra SV discovery by HiFi sequencing and SNP discovery programs.

Reference

Nieuwenhuis R, Hesselink T, van den Broeck HC, Cordewener J, Schijlen E, Bakker L, Trivino SD, Struss D, de Hoop S-J, de Jong H, Peters SA (2021) Genome and transcriptome architecture of allopolyploid okra *(Abelmoschus esculentus)*. bioRxiv 20211118469076. https://doi.org/10.1101/2021.11.18.469076. Edited using ChatGPT, an AI tool by OpenAI

Conclusion

26

Surinder K. Tikoo, Sharan Angadi, Aparna Tiwari, Ramesh Kumar Yadav, B. S. Tomar, and Helen Niegiria Aderibigbe

26.1 Global Okra Round Table (GORT)—Summary

The first Global Okra Round Table (GORT 1) was held under the aegis of ATPBR with support of ICAR-IARI on October 10–12, 2022, at the NAAS complex, New Delhi. During the period 2010 to 2022, the hybridization of okra reached 85 percent coverage in India, and the market value of its seeds went from INR 90 crores to over INR 800 crores (USD 9.5 million). India grows 75% of the global acreage, followed by Nigeria at 15%. It displaced tomato as the highest-value vegetable crop for annual seed revenue. It was the first time that a full conference was organized for a single vegetable crop in India. The fact that over 250 participants from 54 organizations and 13 states, including five from abroad, attended the conference was proof that it was a decision in the right direction.

While the full papers of all presentations are compiled in this book and detailed summaries of the various sessions held over three days are also presented as summary of sessions, the key recommendations from the deliberations are presented below.

S. K. Tikoo (✉)
Plant Breeding, ATPBR- Foundation for Advanced Training in Plant Breeding, Aurangabad, India

S. Angadi
Genetics and Plant Breeding, ATPBR- Foundation for Advanced Training in Plant Breeding, Bangalore, Karnataka, India

A. Tiwari
ATPBR- Foundation for Advanced Training in Plant Breeding, Aurangabad, Maharashtra, India

R. K. Yadav · B. S. Tomar
Division of Vegetable Science, Indian Agriculture Research Institute, New Delhi, Delhi, India

H. N. Aderibigbe
Farming System, National Horticultural Research Institute, Ibadan, Nigeria

S. K. Tikoo et al. (eds.), *Okra: Status, Challenges and Opportunities*,
https://doi.org/10.1007/978-981-97-9963-3_26

26.1.1 Crop Improvement

1. Many new sources of resistance to key viruses, yellow vein mosaic virus (YVMV) and okra leaf curl virus (OLCV) in okra, especially from wild species, were reported during the conference. It is now necessary to study the inheritance and search for trait-linked molecular markers to overcome the frequent breakdown of current virus resistance in several areas.
2. Considering the wide variability of pathogenicity of these viruses across locations, a study on identifying hot spots by not only phenotypic screening but also studying viral genome variability will help work on developing broad-spectrum resistances.
3. Genome mapping of the okra pan genome and identifying trait-linked associations need heavy investments. Public-private consortiums need to be developed.
4. Investments in studies of nutritional quality in okra germplasm need to be strengthened to help identify nutrient-dense materials and initiate breeding projects on developing high-quality okra in combination with the agronomic qualities and disease resistance. The efforts in this direction made in South Africa in a paper presented during the conference indicated promise. Health benefits of purple-colored okra need to be carried out and popularized to improve consumer acceptance.
5. Discovery of novel traits in the national germplasm repository, for example,
   (a) Branching behavior.
   (b) Shorter internodes.
   (c) Multiflora and sympodial branching.
   (d) Nutritional quality diversity.
   (e) Breed okra hybrids with low volatile expression to minimize whitefly infection.
   (f) Emphasize breeding new ecotypes with iodine content, vitamins, minerals, and antioxidants.
   (g) Focus on ideal ecotypes suitable for mechanical harvesting and protection.
   (h) For North Africa, there is an urgent need for developing hybrids/improved varieties for tolerance to drought, salinity, and disease resistance.
   (i) Work on root architecture for improving nutrient and water use efficiency.
6. Focus on value added trait breeding. Industry has started processing okra into chips, and traits that will enhance chip production need to be studied. New ways of cooking okra, as presented by a speaker from Nigeria, are another way of diversifying okra segmentation and value. A few value-added traits discussed were baby okra, okra as a leafy vegetable, easy picking, thick flesh, sympodial branching, shelf life, sucking pest resistance, etc.
7. While the genetic male sterility (GMS) trait presented by Indian Institute of Horticultural Research (IIHR) shows promise to reduce hybrid seed production costs, studies on utilizing this trait more cost-efficiently are still lacking as hand pollination is still required. Studies on the use of bees to get crossed seeds naturally need to be intensified.

8. Searching for a CGMS and restorer system in the germplasm is another area that needs attention.
9. Explore opportunities for international collaboration to exchange knowledge and germplasm for okra improvement.
10. Encourage collaboration between public and private institutions to work on areas at pre-compete levels, e.g., create pre-breeding consortiums like those created by WVC, Taiwan; ICRISAT, Hyderabad; IRRI, Los Banos; CIMMYT, Mexico.
11. Promote the documentation and sharing of successful case studies, like the story of the development of path-breaking YVMV varieties like 'Parbhani Kranti; Arka Anamika,' which has not only benefited Indian farmers but has also been used as a donor of YVMV resistance by breeders from public and private institutions. Sharing such success stories can inspire and guide future initiatives.

26.1.2 Agronomy

1. Identify factors contributing to poor productivity in regions like Nigeria by focusing on capacity building for smallholder farmers. Provide education and resources to improve agronomic knowledge, ensure access to quality inputs, and establish better marketing linkages.
2. Leverage promotional schemes and initiatives, as demonstrated by Agricultural and Processed Food Products Export Development Authority (APEDA), to incentivize the adoption of best practices, quality standards, and export-oriented approaches among okra farmers and exporters.
3. Fertilizer use efficiency should be improved through recent methods of plant delivery systems. For example, modern methods of automated irrigation based on sensors should be researched for higher nutrient and water use efficiencies.
4. Solid matrix priming on-farm technique for rapid uniform seed emergence should be promoted among okra growers.
5. The role of plant growth regulators should be explored intensively in vegetable and seed crops of okra.

26.1.3 Marketing

1. Invest in making okra a global crop by increasing awareness among farmers about the benefits of growing okra, helping them choose the best hybrids/varieties, helping them identify the problems faced by them, and giving them solutions. Explore opportunities for expanding the okra market in Southeast Asia, Europe, Australia, and the Americas.
2. Capitalize on the export potential of okra by collaborating with organizations like APEDA to ensure standardized packaging, post-harvest management, and Good Agricultural Practices (GAP) to enhance the quality of okra produce for

international markets. Encourage farmers to diversify okra usage beyond household consumption, exploring opportunities in the Hotel, Restaurant, and Cafe (HORECA) sectors.
3. Product development protocols as presented by Advanta need to be replicated to help improve the selection of new products for specific areas/zones.

26.1.4 Panel Discussion

1. A special committee needs to be constituted with members from the public sector and private sector to list concrete projects and the road map to implement under Public-Private Partnership (PPP) mode using the consortium approach.
2. Encouraged with discussion, Dr. B. S. Tomar, Head Division of Vegetable Science, came forward with a proposal that he will organize a meeting of all concerned under the Chairmanship of DG-ICAR to discuss the issues and policy matters related to the sharing of IARI technologies among the seed companies under PPP mode.
3. Steps need to be taken to import elite lines of okra/selected germplasm of okra primarily of African origin from World Veg Centre following due procedures as available on the website of World Veg Centre. Once such seeds are available, the importing seed company may multiply the same and distribute among the interested seed companies as per the terms and conditions of the Material Transfer Agreement signed between World Veg Centre and the germplasm importing seed company.
4. Extending research facilities by private seed companies for postgraduate students thesis work within the same city on mutually agreed terms and conditions.
5. Listing of some of the private seed company's lead breeders as adjunct faculty members in the universities located in the same city for lectures on selected topics of few courses based on mutually agreed terms and conditions. The typical example is that of Dr. Roland Schafleitner of the World Vegetable Center, listed as an adjunct professor at the Institute of Tropical Plant Sciences and Microbiology at National Cheng Kung University, Tainan.
6. Pooling together all types of wild okra from various sources, their seed multiplication, and karyotype analysis for accurate identification and then distribution to the users.

26.1.5 General

1. Bringing out joint publication/manual between public sector scientists and private seed company scientists on hybrid seed production of okra and then other vegetable crops as practiced on farmers' fields for the overall benefit of students and faculty members.
2. Extending research facilities by private seed companies for postgraduate students thesis work within the same city on mutually agreed terms and conditions.

3. Make GORT a biennial platform for knowledge exchange and discussions among okra stakeholders around the globe. It was clear that okra is on way to become a USD 300 million crop in less than a decade.
4. The felicitation of scientists for their major achievements in okra was liked by all, and it is recommended to continue this on the GORT platform in the future too.
5. Create a global network of experts in the okra industry for easy access to all sectors involved in improving the crop's profile and value.

26.2 Global Okra Round Table (GORT)—Way Forward

GORT 1 established a platform that brought together all stakeholders in the okra value chain to discuss the latest advancements in okra crop improvement, seed production technology, and marketing across various countries. These discussions fostered a shared understanding of the crop and outlined the challenges ahead. The sessions concluded with a panel discussion aimed at forming a consortium of public and private sector partners to fund projects on complex traits, including nutritional quality, at a pre-competitive level. The goal is for stakeholders to use the results of these projects to deliver value-added traits to farmers and consumers.

As a first follow-up step, an all-India crop trial is being conducted at multiple locations, with entries from both the public and private sectors. Few private sector companies and ICAR-IIVR have come together to conduct these trials. The results will help both sectors to know the adaptability of their products across hot spots in India, and after this mini step of PPP collaboration, it is expected to lead to joint trait development through prebreeding long-term projects that will help both sectors to provide long-term solutions to farmers and consumers besides allowing academic excellence to the researchers.

The success of GORT 1 also brings to notice that crop-specific meetings and resulting consortia can yield focused outcomes, aiding the agricultural community in developing better varieties for sustainable agricultural development.

The manufacturer's authorised representative in the EU is Springer Nature Customer Service Centre GmbH, Europaplatz 3, 69115 Heidelberg, Germany. If you have any concerns regarding our products, please contact ProductSafety@springernature.com

Printed and bound by CPI Group (UK) Ltd, Croydon, CR0 4YY
01/10/2026
02225932-0007